Advanced X-ray Detector Technologies

Krzysztof (Kris) Iniewski
Editor

Advanced X-ray Detector Technologies

Design and Applications

 Springer

Editor
Krzysztof (Kris) Iniewski
Emerging Technologies CMOS Inc.
Port Moody, BC, Canada

ISBN 978-3-030-64281-5   ISBN 978-3-030-64279-2   (eBook)
https://doi.org/10.1007/978-3-030-64279-2

This Springer imprint is published by the registered company Springer Nature Switzerland AG
The registered company address is: Gewerbestrasse 11, 6330 Cham, Switzerland

Preface

Currently, radiation detectors for medical and industrial applications are based on effectively integrating the X-ray photons emitted from the X-ray tube. This standard technique is vulnerable to noise due to variations in the magnitude of the electric charge generated per X-ray photon. Higher energy photons deposit more charge in the detector than lower energy photons so that in a quantum integrating detector, the higher energy photons receive greater weight. This effect is undesirable in many detection applications because the higher part of the energy spectrum provides lower differential attenuation between materials, and hence these energies yield images of low contrast.

Direct conversion X-ray quantum counting detectors solve the noise problem associated with photon weighting by providing better weighting of information from X-ray quanta with different energies. In an X-ray quantum counting system, all photons detected with energies above a certain predetermined threshold are assigned the same weight. Adding the energy windowing capability to the system theoretically eliminates the noise associated with photon weighting and significantly decreases the required X-ray dosage compared to energy integrating systems.

High-Z materials like CdTe, CdZnTe (CZT) and GaAs offer the best implementation possibility of direct conversion X-ray detectors. In addition, emerging class of materials, perovskite crystals are starting to be used for gamma and X-ray detection. Advanced X-ray detection technologies based on these materials are subject of this book. We discuss material challenges, detector operation physics and technology, and readout integrated circuits required to detect signals processes by high-Z sensors. These advanced X-ray detection technologies are expected to revolutionize X-ray detection for medical imaging and industrial applications for years to come. There is a large number of technological developments in the field of computed tomography (CT) scanners in hospitals, X-ray scanners at airports for baggage scanning, dirt bomb detection in security applications, and imaging in non-destructive testing applications like food inspection. Advanced materials and detectors are needed to pursue worldwide developments of these technologies.

The book is divided into two parts. The first deals with X-ray sensors and materials, the second with system applications. First two chapters are devoted to emerging perovskite materials and highlight major recent advances in the field. Perovskite materials are calcium titanium oxide mineral composed of calcium titanate (CaTiO3) and the term is also applied to the class of compounds which have the same type of crystal structure as CaTiO3 (XIIA2+VIB4+X2−3), known as the perovskite structure. Many different cations can be embedded in this structure, allowing the development of diverse engineered materials. Perovskite crystals are considered to be the future material for low-cost and high-performance radiation sensors. Wolszczak et al from Wake Forest University, USA, discuss material challenges and physical concepts behind material developments in Chap. 1, while Nie et al from Los Alamos National Laboratory, USA, focus on Lead-Halide perovskites in Chap. 2.

The remaining chapters in the first part deal with more developed high-Z solid state detectors. Researchers from Leicester University, UK, outline spectroscopic performance of these pixelated detectors using sophisticated simulation tool in Chap. 3. Tsigaridas and Ponchut from ESRF, France, describe detector performance for synchrotron applications in Chap. 4. Finally, Grimm from ETH, Switzerland, guides us through high-Z sensors used for space X-ray detection in Chap. 5.

The second application part of the book starts with Skrzynski, Curie National Research Institute of Oncology, Poland, providing overview of advanced X-ray detector for medical imaging in Chap. 6. In the following chapter, Kang, Chonnam National University, Korea, describes DEXA, dual energy X-ray absorptiometry. This technique sends two X-ray beams at different peak energy frequencies to the target bones and soft tissue. Chapter 8 deals with a very important emerging technology, spectral computed tomography, and is written by the leading research group on that topic from Siemens. The CdTe-based scanner from Siemens is currently undergoing clinical trials at Mayo Clinic in the USA. Chapter 9 by Chi from Beijing Normal University, China, provides an interesting perspective on X-ray fluorescence computed tomography followed by Chap. 10 by Lee from Korea University, Korea, on XRF and Compton imaging. The book closes with Chap. 11 by Kaissas, Greek Atomic Energy Commission, Greece, describing coded aperture approaches in identification of radioactive spots, followed by Chap. 12 by Yin, Lanzhou University, China, and Komarov, Washington University in St. Louis, USA, exploring possibility of building positron emission tomography (PET) instrumentation using CZT detectors.

This book will give the reader a good overview of some of most recent advances in the field of advanced materials used for gamma and X-ray imaging. We have tried to strike a balance between general chapters in both high-Z materials (CdTe and CdZnTe) and perovskite crystals. The book provides an in-depth review of the research topics from leading world specialists in the field and should be an excellent reference for people already working in the field as well as for those wishing to enter it.

Port Moody, BC, Canada   Krzysztof (Kris) Iniewski
February 2021

Contents

Contributors

Sarah L. Bugby Leicester University, Leicester, UK

David L. Carroll Center for Nanotechnology and Molecular Materials, Wake Forest University, Winston-Salem, NC, USA

Zhijun Chi Beijing Normal University, Beijing, China

Thomas Flohr Siemens Healthcare GmbH, Computed Tomography, Forchheim, Germany

Oliver Grimm Institute for Particle Physics and Astrophysics, ETH, Zürich, Switzerland

Andre Henning Siemens Healthcare GmbH, Computed Tomography, Forchheim, Germany

Krzysztof (Kris) Iniewski Emerging Technologies CMOS Inc., Port Moody, BC, Canada

Ioannis Kaissas Greek Atomic Energy Commission, Patriarchou Grigoriou & Neapoleos, Agia Paraskevi, Attiki, Greece

Jihoon Kang Chonnam National University, Yeosu, Korea

Younghak Kim Korea University, Seoul, Korea

Kjell A. L. Koch-Mehrin Leicester University, Leicester, UK

Sergey Komarov Department of Radiology, Washington University, St. Louis, MO, USA

Wonho Lee Korea University, Seoul, Korea

John E. Lees Leicester University, Leicester, UK

Fangze Liu LANL (Los Alamos National Lab), Los Alamos, NM, USA

Wanyi Nie LANL (Los Alamos National Lab), Los Alamos, NM, USA

Martin Petersilka Siemens Healthcare GmbH, Computed Tomography, Forch-heim, Germany

Cyril Ponchut European Synchrotron Radiation Facility (ESRF), Grenoble, France

Bernhard Schmidt Siemens Healthcare GmbH, Computed Tomography, Forch-heim, Germany

Shreetu Shrestha LANL (Los Alamos National Lab), Los Alamos, NM, USA

Witold Skrzynski Medical Physics Department, Maria Sklodowska-Curie National Research Institute of Oncology, Warsaw, Poland

Jeremy Tisdale LANL (Los Alamos National Lab), Los Alamos, NM, USA

Hsinhan Tsai LANL (Los Alamos National Lab), Los Alamos, NM, USA

Stergios Tsigaridas European Synchrotron Radiation Facility (ESRF), Grenoble, France

Stefan Ulzheimer Siemens Healthcare GmbH, Computed Tomography, Forchheim, Germany

Richard T. Williams Center for Nanotechnology and Molecular Materials, Wake Forest University, Winston-Salem, NC, USA

Weronika W. Wolszczak Center for Nanotechnology and Molecular Materials, Wake Forest University, Winston-Salem, NC, USA
Faculty of Applied Sciences, Department of Radiation Science and Technology, Section Luminescence Materials, Delft University of Technology, Delft, Nether-lands

Yongzhi Yin School of Nuclear Science and Technology, Lanzhou University, Lanzhou, China

Changyeon Yoon Korea Hydro & Nuclear Power Co., Seoul, Korea

Toward Perovskite-Related Scintillators with Necessary Stokes Shift and Thickness for Hard X-Ray Radiography and Gamma Spectroscopy

Weronika W. Wolszczak, David L. Carroll, and Richard T. Williams

1 Attractiveness of Pb-Halide Perovskites for Scintillation Detectors of Ionizing Radiation

It often happens in the advance of technology that intense research and development on a given material class driven by obvious promise for new consumer products have carried on its back other technology that would probably not have been developed in isolation, at least not on the same timescale. The last decade of intense research activity on lead-halide perovskites, already producing advances in solar cells and light emitters, seems such a moment. Leveraging the tremendous activity and investment in these materials may allow fundamental new advances in allied areas, e.g., x-ray radiography in medicine, security inspections, and industry.

In addition to the ongoing material research activity that may benefit development of new ionizing radiation detectors, it is attractive that the Pb-halide perovskites can be grown and processed in solution at relatively low temperature and that the basic raw materials are earth abundant. Furthermore, the $APbX_3$ perovskites have reasonably high-density and high-Z elements (Pb, possibly A = Cs, and possibly X = I), which are helpful in a radiation detector to stop and convert the incident x-ray or gamma photon to electron-hole pairs.

W. W. Wolszczak
Center for Nanotechnology and Molecular Materials, Wake Forest University, Winston-Salem, NC, USA

Faculty of Applied Sciences, Department of Radiation Science and Technology, Section Luminescence Materials, Delft University of Technology, Delft, Netherlands
e-mail: wolszcw@wfu.edu

D. L. Carroll · R. T. Williams (✉)
Center for Nanotechnology and Molecular Materials, Wake Forest University, Winston-Salem, NC, USA
e-mail: carrolld@wfu.edu; williams@wfu.edu

© Springer Nature Switzerland AG 2022
K. Iniewski (ed.), *Advanced X-ray Detector Technologies*,
https://doi.org/10.1007/978-3-030-64279-2_1

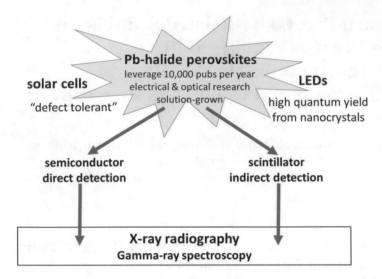

Fig. 1 Extraordinary activity in photovoltaic and lighting research on perovskites may be leveraged for two corresponding routes to radiation detection, particularly radiography

The lead-halide perovskites exhibit the so-called defect tolerance, which is partly credited for transport properties beneficial in solar cells and potentially in direct semiconductor detection of ionizing radiation. Defect tolerance is taken to mean that defect trap levels lie more within the bands than in the gap or are shallow if within the gap. Reviews and explanations have been offered [3, 6, 26, 32]. The attractive properties of metal halides are accompanied by the fact that such salts generally are soft insulators prone to point and line defects. Many of the electron trap levels that would appear within the bandgaps of most other halide insulators become enveloped within the Pb-derived conduction band that is mainly responsible for narrowing the gap from typical metal halide insulator to $APbX_3$ semiconductor.

As suggested by the branching in the lower part of Fig. 1, it appears that lead-halide perovskites may be poised to furnish radiographic detectors by two different routes, one depending on defect-tolerance-associated carrier lifetime in semiconductor mode for direct radiographic detection and the other depending on high-quantum-yield light emission from excitons in semiconducting quantum dots (QDs) and possibly single crystals. The chapter in this book by Wanyi Nie et al. on *Emerging Lead-Halide Perovskite Semiconductors for Solid-State Detectors* examines progress along the left branch using charge collection. We focus in this chapter on the right branch seeking to exploit scintillating light emission for indirect radiography.

Quantum confinement in $CsPbX_3$ nanoparticles renders the nearly non-emissive (at room temperature) bulk semiconductor into a QD bright emitter with photoluminescence quantum yield reported as high as 90% to 95% for interband excitation [8, 9, 13, 48]. The luminescence is from carrier recombination and exciton decay (free and/or bound depending on temperature), so one may reasonably

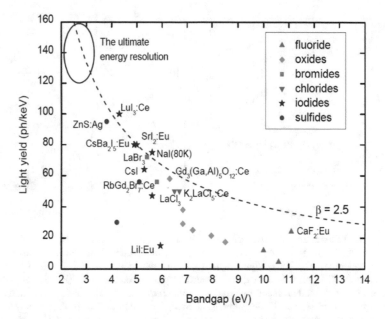

Fig. 2 Light yield of different scintillators versus bandgap energy E_g, with the bandgap-based theoretical limit represented by the dashed curve. From [70]

infer total electron-hole radiative recombination with 90% quantum efficiency in the perovskite nanocrystal. Another promising property of Pb-halide perovskites that should make them good scintillators is their small bandgap compared to the insulators employed for most conventional scintillators. A well-known plot of scintillator light yield versus bandgap [15] is shown in Fig. 2 with a superimposed curve of "maximum theoretical light yield" plotting the quantity $1 \text{ keV}/\beta E_{gap}$, where $\beta \approx 2.5$ and βE_{gap} approximates the average energy deposited per electron-hole pair produced in an x-ray stopping event. Then the dashed curve expresses maximum light yield in units of photons/keV to be compared to the experimental points.

2 Temperature Dependence of Photoluminescence and Radioluminescence from Single-Crystal Pb-Halide Perovskites

Given a 90% quantum efficiency, Fig. 2 suggests that with 2.3 eV bandgap, inorganic $CsPbBr_3$ or hybrid $(CH_3NH_3)PbBr_3(=MAPbBr_3)$ might attain a scintillation light yield as high as 156 photons/keV, which would be revolutionary among scintillators if it could be realized in practice at room temperature. So far, it is not realized even approximately at room temperature. However, low-temperature data are encouraging with regard to the prediction of Fig. 2 for two materials: $MAPbBr_3$ with 2.3 eV bandgap and $MAPbI_3$ with 1.66 eV bandgap [4]. From the

4   W. W. Wolszczak et al.

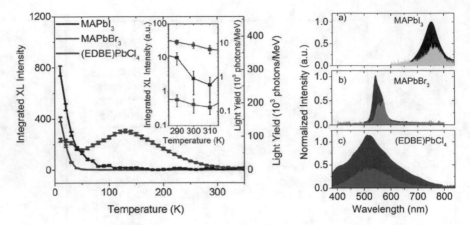

Fig. 3 (Left) Temperature dependence (10–350 K) of light yields of MAPbI$_3$, MAPbBr$_3$, and (EDBE)PbCl$_4$ single crystals determined from x-ray-excited luminescence (left scale) by comparing to the (EDBE)PbCl$_4$ signal which was independently calibrated at 300 K by pulse height spectrum. The inset is an expanded vertical plot around 300 K. (Right) X-ray-excited luminescence (light shaded spectrum) and photoluminescence (dark shaded spectrum) of (a) MAPbI$_3$, (b) MAPbBr$_3$, and (EDBD)PbCl$_4$ at room temperature. The peak-normalized x-ray-excited spectra were reduced by ½ relative to the photoluminescence spectra for display. All plots from [4]

radioluminescence data reproduced in Fig. 3, [4] estimated that under x-rays from a 45 kV Cu-anode tube, single-crystal MAPbBr$_3$ at 10 K has a light yield (LY) of 152 photons/keV, while MAPbI$_3$ with 1.66 eV bandgap has an estimated LY of 296 photons/keV at 10 K. As shown in Fig. 4a, Mykhaylyk et al. report that an all-inorganic Bridgman-grown single crystal of CsPbBr$_3$ at 7 K has a LY of 50 photons/keV excited by 12 keV x-rays and 109 photons/keV under 5.5 MeV alpha particle radiation from a ^{241}Am source [43]. Taken together, these values indicate that the fantastic bandgap-limited light yield is in principle attainable in semiconducting Pb-halide perovskite materials under the right circumstances, where right circumstances apparently mean temperature of 10 K or lower. Unfortunately, at room temperature, the light yield is far smaller. In the experiments of Birowosuto et al., the radioluminescence light yields of single-crystal MAPbBr$_3$ and MAPbI$_3$ fell to ∼ 0.2 photon/keV and ∼ 1 photon/keV, respectively, at room temperature, a value barely usable if at all [4]. (See Fig. 3 including inset.) What causes that temperature dependence is an important issue to consider if attempting to use extended three-dimensional Pb-halide perovskite semiconductors as room-temperature scintillators.

The right side of Fig. 4 presents temperature dependence of the free exciton, donor-bound exciton, and donor-acceptor pair contributions to PL of single-crystal Bridgman-grown CsPbBr$_3$ as measured and assigned by Wang et al. [67]. Combining this information with the temperature dependence of the amplitudes of radioluminescence (left side Fig. 4), the following inferences about Bridgman-grown single-crystal CsPbBr$_3$ are suggested: (a) the room-temperature luminescence comes from free-exciton radiative decay (on the basis of the Wang et al. spectrum

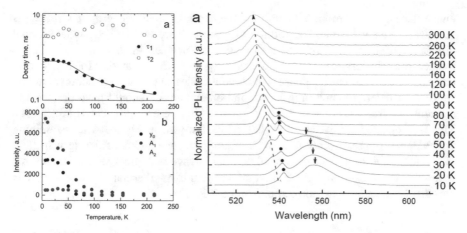

Fig. 4 (Left) Temperature dependence (7–230 K) of single-crystal CsPbBr$_3$ luminescence decay times (a) and intensity amplitude components (b), from fitting 12 keV x-ray pulse-excited luminescence to $y = A_1 \exp(-t/\tau_1) + A_2 \exp(-t/\tau_2) + y_0$, from [43]; (right) temperature dependence (10–300 K) of photoluminescence spectra from a single crystal of CsPbBr$_3$. The dashed line indicates the peak evolution of the free-exciton emission; solid circles mark the peaks of donor-bound exciton emission; and arrows denote the donor-acceptor pair emission. From [67]

identification), but it is very weak relative to the low-temperature luminescence (based on Mykhaylyk et al. intensity data); (b) spectrally, the low-temperature luminescence (e.g., at 10 K) comes from defect-bound excitons and donor-acceptor pairs, but hardly at all from free excitons [67], and it is very bright relative to room temperature luminescence [43]. These observations suggest the following mechanism for at least part of the observed thermal quenching of luminescence in single-crystal CsPbBr$_3$: defects classed as donors and acceptors may each on the one hand bind an exciton, and on the other hand the same or different donors and acceptors may trap electrons and holes separately and mediate radiative decay by the donor-acceptor pair mechanism. In the case of exciton binding at a defect site that allows radiative decay, there are two advantages that the trap-mediated radiative decay may have over free-exciton radiative decay: (1) the defect facilitates momentum conservation in the radiative transition, whereas free-exciton radiative decay is restricted to the population of a small range of exciton band states around the photon momentum connecting to the ground state at zero momentum; and (2) binding at a defect that ultimately promotes radiative decay prevents the exciton for a period of time from moving to find defects that promote nonradiative decay. Electrons and holes trapped on donors and acceptors close enough to undergo donor-acceptor radiative decay benefit from the same two mechanisms promoting their radiative decay relative to free-exciton radiative decay. Starting from the luminescence of defect-bound excitons and donor-acceptor pairs at low temperature in Figs. 3 and 4, warming the crystal leads to de-trapping of the bound excitons and carriers so that the luminescence defaults to radiative recombination of free excitons competing with transport of the mobile excitons to point and line defects and surfaces. Furthermore, although the free-exciton binding energy in CsPbBr$_3$ is somewhat greater than $k_B T$ at room temperature, there will be substantial thermal ionization of free excitons and reformation of new excitons from some of the ionized free carriers. They are in thermal equilibrium and at risk for loss due to trapping in the free-carrier phase, which contributes to thermal quenching. But the surviving carriers recombining by second-order kinetics should have the spectral signature of free excitons, as long as exciton binding energy exceeds $k_B(295 \text{ K})$. This is supported by the spectral data of [67] shown in Fig. 4. After examining the evidence for reabsorption of exciton luminescence and considering it in light of the temperature dependence in Figs. 3 and 4, we will propose in addition that reabsorption of exciton light is part of the empirical thermal quenching that is measured particularly in radioluminescence (Fig. 3).

3 Thermal Quenching Avoided in Nanocrystalline Pb-Halide Perovskites: Success of LEDs

Prospects for making room-temperature perovskite scintillators benefited from research and development on perovskite light-emitting diodes (LEDs), where it had been found that perovskite nanocrystals as light emitters avoid the room-temperature quenching of luminescence that occurs in extended 3D bulk crystals. The retention of light emission at room temperature in nanocrystals is generally attributed to quantum confinement of electrons and holes together [48, 73], but understanding of how the thermal quenching in microscopic and macroscopic crystals occurs and how quantum confinement defeats the room-temperature quenching would be helpful to know better. In light of high-luminescence quantum yield persisting to room temperature in nanocrystal Pb-halide perovskites, it is fair to inquire: "Has the use of nanocrystal 3D Pb-halide emitters with ≥90% PLQY permitted the promise of bandgap-limited light yield around 150,000 photons/MeV to be attained at room temperature in detector blocks or thick films?" No. Still not even close. Except for some success in imaging with soft x-rays, nanocrystalline 3D Pb-halide scintillators at useful thickness and room temperature have only poor to moderate light yield, not yet sufficient for energy resolution or competitive hard-x-ray imaging.

4 What Is Hindering Transfer of LED Success to Scintillation Success Using 3D Pb-Halide Perovskites?

What is the other issue that has prevented realizing a good scintillator, granting the following three outstanding Pb-halide perovskite attributes: (1) high quantum yield of interband recombination luminescence in perovskite nanocrystals at room temperature, (2) small bandgap assuring many carriers deposited upon x-ray stopping, and (3) reasonably high-density and effective Z of Pb-halide perovskites assuring x-ray stopping in a reasonable thickness? Why have the perovskite QD emitters that work so well in LEDs not been able to convert *x*-rays or gamma rays to a large number of *detected* luminescence photons?

One possible problem is suggested by noting that the measurements of high PLQY of halide perovskites at room temperature have been made in thin films or dilute suspensions of nanoparticles where it does not much matter that the exciton radiative recombination has small Stokes shift (typical of excitons in semiconductors), because the medium is optically thin. Unlike LEDs, gamma scintillation and hard-x-ray radiography require a substantial thickness of dense scintillator to stop ionizing radiation, so a small Stokes shift relative to a strong absorption edge means the scintillation light may be substantially reabsorbed while exiting the material. Such considerations led us to point out that one should find or introduce a more substantial Stokes shift in halide perovskite materials or systems in order to use them as a scintillator detector of hard radiation [69]. Other groups have noted the reabsorption problem in Pb-halide perovskites as well, such as [2, 4, 18, 33, 43, 78], and some have started investigations of how one might mitigate the problem, for example, by a wavelength shifting dye [18], introduction of Mn^{2+} as an activator and wavelength shifter in a perovskite β-particle detector [78], or utilizing the large Stokes shift of self-trapped excitons in low-dimensional perovskites [2].

5 Absorption and Photoluminescence in Single-Crystal CsPbBr₃

It is worth a look at the magnitude of reabsorption of excitonic luminescence in representative Pb-halide perovskite materials based on published data. We will first examine emission and reabsorption of exciton luminescence in single crystals of CsPbBr₃ and then in nanocrystals of CsPbBr₃ and related perovskites. The groups of M. Kanatzidis and D. Y. Chung at Northwestern University and Argonne National Laboratory have grown purified CsPbBr₃ crystals from the melt principally for use in semiconducting direct detection of radiation [23]. He et al. [22] measured transmission, diffuse-reflectance determination of relative absorbance, and excitonic photoluminescence in a Bridgman-grown CsPbBr₃ single crystal, reproduced below in Fig. 5a [23]. The excitonic PL is Stokes-shifted from the peak of room-temperature band-edge absorption by about 100 meV in these data.

The plotted quantity $(1-R)^2/2R$, where R is diffuse reflectance, is a statement of the Kubelka-Munk transform equaling the quantity α/s where α is the optical absorption coefficient (cm^{-1}) and s is the semiempirical scattering coefficient to account for internal scattering. The parameter s is dominated by particle size, packing, and refractive index. It is not a strong function of wavelength and is commonly considered roughly constant over the spectrum for a given sample. Thus the measured curve of $(1-R)^2/2R$ is approximately the absorption coefficient

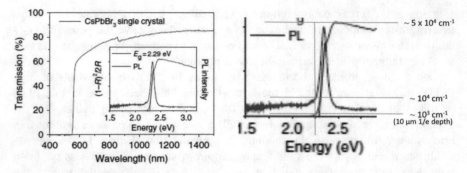

Fig. 5 (Left) Measurements [23] of the transmission spectrum of $5 \times 5 \times 3$ mm^3 single-crystal CsPbBr$_3$ and (inset) diffuse-reflectance results plotted as a quantity approximately proportional to absorption coefficient, overlaid by the photoluminescence of this Bridgman-grown single crystal of CsPbBr$_3$ excited at 440 nm. (Right) We expanded the inset and assigned the peak of absorption to be the maximum absorption edge coefficient $\sim 5 \times 10^4$ cm^{-1} measured on solvent-suspended CsPbBr$_3$ nanocrystals by [11]

spectrum with arbitrary scale factor s^{-1}, and we must estimate s from additional information. In this case we estimated the maximum absorption $\sim 5 \times 10^4$ cm^{-1} from measurements on suspended CsPbBr$_3$ nanocrystals measured by [11] for assigning an absorption coefficient scale in Fig. 5b. Relative to that maximum, horizontal lines are drawn at $\alpha = 10^4$ and 10^3 cm^{-1}. The latter value corresponds to a 10 μm optical attenuation depth for the peak of photoluminescence. A vertical line is drawn at the intersection of the absorbance curve and the 10^3 cm^{-1} line, and its intersection with the PL spectrum suggests that most of the PL except its reddest tail should be strongly attenuated in ≤10 μm of CsPbBr$_3$ at room temperature.

We have made measurements to test the self-absorption attenuation length of exciton luminescence in single-crystal CsPbBr$_3$ using a two-photon confocal microprobe to scan depth of the excitation spot below the surface of the sample. Results below were measured on Bridgman-grown CsPbBr$_3$ provided by M. Kanatzidis and D. Y. Chung, prepared by the same method as the sample characterized in [23] in Fig. 5a. The two-photon depth-scanned photoluminescence experiment is useful to study reabsorption of exciton luminescence in melt-grown single crystals because of the difficulty to cut and polish slices of CsPbBr$_3$ crystal in the few μm to 100 μm range that would be needed for usual transmittance measurements of single-crystal material near the exciton edge. Photoluminescence from a specified depth in melt-grown single-crystal CsPbBr$_3$ is extracted in our experiment using a 0.65 NA microscope objective in a two-photon photoluminescence experiment [45] represented in the schematic at the left side of Fig. 6. The 850 nm, 130 fs laser pulses are transmitted by CsPbBr$_3$ at ordinary (unfocused) intensities but are efficiently two-photon absorbed in a small focal volume at the beam waist. For the 0.65 NA objective, the diffraction-limited focal volume of the infrared light at the 1/e contour is 1.58 μm in lateral diameter $\times$ 2.19 μm Rayleigh length (z). The two-photon absorption contour is smaller, corresponding to 1.12 μm diameter $\times$ 1.55 μm

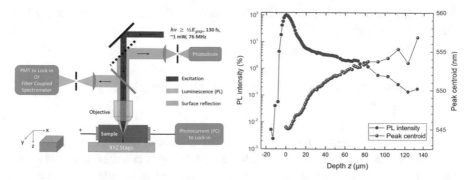

Fig. 6 (Left) Two-photon photoluminescence microprobe setup for measuring luminescence spectra from melt-grown single-crystal CsPbBr$_3$ detected through a 350–700 nm band-pass filter with 0.65 NA objective collection optics as a function of depth z of the excitation/collection focal spot below the sample surface. (Right) PL intensity (red points) and centroid of the spectral peak (blue points) as a function of depth z of the excited focal volume

length. As is typical in two-photon fluorescence microscopy, the luminescence is collected back through the objective using a dichroic beamsplitter. At the PL wavelength of 545 nm, the collected focal volume is similar in size to the excitation volume.

The right side of Fig. 6 displays data for photoluminescence intensity normalized to 100% at the maximum versus z, and PL redshift plotted in nm shift of the centroid of a fitted Gaussian spectrum versus depth z of the excited focal volume. The redshift reaches 31 meV relative to the surface-excited spectrum at a depth (escape length) of 60 μm. The PL intensity is attenuated to 5% of its maximum when extracted from a depth of 20 μm, implying an effective absorption coefficient over the entire PL band of approximately 1500 cm^{-1} assuming the excitation efficiency is constant in the top 20 μm of the sample. The slope of the PL attenuation plot becomes more gradual when the excited spot is deeper than 20 μm. The attenuation from 20 μm to 60 μm indicates an effective absorption coefficient of about 240 cm^{-1}. A gradual lessening of the effective absorption coefficient can be expected as the surviving PL spectrum redshifts farther down the sloping tail of Urbach-broadened exciton absorption, but the sudden change of slope at 20 μm is not reproduced in our model of absorption and redshifting based on an Urbach absorption edge. The two-slope phenomenon might involve altered properties of the top 20 μm of this sample because of cutting and polishing damage and/or reaction with atmosphere, the latter of which has been shown to increase PL from the near-surface region of CsPbBr$_3$ single crystals [67]. In any case, the PL attenuation data vs extraction depth measured in Fig. 6 is reasonably consistent with expectation based on our simple analysis of the He et al. spectral measurements in Fig. 5 [23], namely, that the reabsorption of excitonic PL in CsPbBr$_3$ single crystal is quite severe already at about 20 μm path length. The expected redshift as well as the approximate absorption coefficient over the PL band is confirmed by the measurements in Fig. 6.

6 Absorption and Photoluminescence in Single-Crystal MAPbBr₃

Staub et al. have conducted a study of the photoluminescence intensity and spectrum from single-crystal MAPbBr₃ comparing the cases of front-side and backside excitation, where front-side denotes collection of luminescence from the laser-pulse-irradiated surface and backside means collection from the crystal face opposite the irradiated side. The outline of their experiment is summarized on the left side of Fig. 7, where the blue and red wavy arrows represent the blue side and red side of the photoluminescence spectral band [58]. The case of front-side illumination with the 404 nm laser excitation pulse is on the left side of the schematic illustration representing early time at top and later time at bottom. Likewise, backside illumination is depicted on the right side of the schematic. The peak-normalized spectra plotted on the right side of Fig. 7 reveal a substantial redshift in the backside luminescence because it transits the sample thickness subject to the sloping absorption spectrum of MAPbBr₃ under the PL band. The peak of the PL band (measured front-side) experiences a transmittance of only about 10^{-3} on transiting 1.7 mm of crystal thickness (backside). Though highly attenuated, this is larger transmittance than warranted by the absorption coefficient

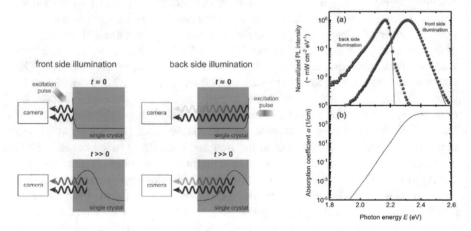

Fig. 7 (Left) Illustrating the setups for front-side excitation (relative to detection) and backside excitation. (Right) Comparison in (a) of peak-normalized spectra of MAPbBr₃ single-crystal PL under front- and backside illumination at 3.07 eV with pulse photon density 2.9×10^{14} cm^{-2}. Modeled spectra are shown by solid lines. (b) Absorption coefficient determined by fitting the model to the spectra in (a). From [58]

of CsPbBr$_3$ deduced in Figs. 5 and 6. Reasons for the difference could be that MAPbBr$_3$ has lower absorption coefficient than CsPbBr$_3$ at the PL band peak, that carrier diffusion plays more of a role in energy transport in MAPbBr$_3$, and/or that details of photon energy downshift during photon diffusion. They analyzed the time- and spectrally resolved data in a model that included carrier diffusion, recombination luminescence, optical reabsorption/reemission (photon recycling), and surface recombination [58].

The lower right side of Fig. 7 displays the absorption coefficient spectrum deduced from model fitting of the front- and backside photoluminescence data. Note that the absorption spectrum is a straight line on the semilogarithmic plot, indicating an Urbach rule shape spanning many decades of absorption coefficient. A direct measurement of the Urbach edge spanning 4.5 decades in MAPbI$_3$ measured by De Wolf et al. [12] will be shown later. The results of Staub et al. and De Wolf et al. provide strong evidence that there is Urbach broadened edge absorption underlying the whole of the excitonic photoluminescence spectrum in MAPbBr$_3$ and MAPbI$_3$, respectively. The utility of identifying Urbach absorption edge broadening in Pb-halides of interest for scintillation is that it then becomes a reasonably reliable basis for extrapolating the absorption coefficient across the entire PL or scintillation emission band, including the red-tail range of the surviving luminescence band where absorption coefficient cannot be reliably measured by transmittance due to dominance by scattering losses. From analysis of their data, Staub et al. concluded that it is carrier diffusion rather than photon recycling which accounts for most of the spatial energy redistribution. By switching photon recycling on and off in their model, they determined that it does not much affect spatial energy distribution but does have an important effect in determining the carrier concentration in the sample versus time [58]. The significant contribution of carrier diffusion to extracting luminescence from a given depth may be different in a single crystal as measured by Staub et al. versus a nanocomposite of perovskite quantum dots in an insulating host as discussed below.

In solar cells, reabsorption of excitonic luminescence is desirable and also known as photon recycling [35]. Excitonic recombination luminescence is lost energy in a solar cell if it escapes as a photon, but reabsorption back into an exciton or free-carrier state gives a second bite at the apple for charge separation and collection. There are multiple bites with multiple emission and reabsorption cycles, which occur more frequently, the smaller is the Stokes shift. This is termed photon recycling in works on solar cells. Small excitonic Stokes shifts are sought after in that context. GaAs offers one of the smallest exciton Stokes shifts, 4.1 meV at room temperature [65]. Thus in the two most investigated applications of Pb-halide perovskites – LEDs and solar cells – reabsorption of excitonic luminescence is, respectively, nearly irrelevant and highly sought after.

7   Absorption and Photoluminescence in Nanocrystals of Three-Dimensional CsPbBr₃

When nanocrystals of Pb-halide perovskites are the emitters, photoluminescence quantum efficiency does not fall off at room temperature in the way it does in bulk single crystals, the latter case being exemplified in Figs. 3 and 4. PLQY of nanocrystal $CsPbBr_3$ achieves values reported earlier in our introduction as 90% to 95% at room temperature [8, 9, 13, 48]. Let's examine the magnitude of reabsorption of excitonic luminescence by samples composed of nanocrystals. The absorption coefficient at the peak of the interband/excitonic edge in $CsPbBr_3$ nanocrystals has been reported by [11] as $\alpha \approx 5 \times 10^4$ cm^{-1} and by [40] as $\alpha \approx 3 \times 10^4$ cm^{-1}. The excitonic luminescence Stokes shift in $CsPbBr_3$ depends on NC size. Taking 9 nm as a rough midpoint of the measured sizes studied by ([5, 47], the corresponding Stokes shift for this size is ~ 40 meV. The luminescence peak in their data for this size overlies an absorption coefficient of $\alpha \approx 1.5 \times 10^4$ cm^{-1}, roughly 40% below the maximum absorption at the interband edge, but still very substantial. For a conservative round number of absorption coefficient taken as $\alpha \approx 1 \times 10^4$ cm^{-1} at the PL band peak in medium-sized nanocrystals, the peak of the excitonic luminescence band should be attenuated 63% (to 1/e) by reabsorption upon traversal of roughly 1 μm in nanocrystalline $CsPbBr_3$. This is somewhat more severe reabsorption attenuation than deduced for single-crystal $CsPbBr_3$ in Fig. 5 and measurements reported in Fig. 6. Part of this difference is that the exciton luminescence Stokes shift is smaller in nanocrystal $CsPbBr_3$ than in the single-crystal $CsPbBr_3$ measured by He et al. [He et al], Fig. 5a. Another is that we are addressing attenuation of the PL band peak in the present discussion of nanocrystals, but the data of Fig. 6 measured extraction of the whole PL band including the surviving red wing. A third possibility discussed in [58] is the greater role of carrier diffusion for energy transport in semiconducting macrocrystal $CsPbBr_3$. In any case, the measured and/or deduced optical attenuation lengths of exciton luminescence in nanocrystal and macrocrystal $CsPbBr_3$ lie in the range of a few μm to tens of μm. In comparison, the typical light-emitting layer thickness in an LED of about 100 nm can be seen to pose a negligible problem of reabsorption for the escaping light. On the other hand, for 100 μm, 1 mm, or 1 cm thicknesses of scintillators needed to stop x-rays of, respectively, 30 keV, 100 keV, and 500 keV photon energy in $CsPbBr_3$, it can readily be seen that optical attenuation of the escaping exciton luminescence light is a serious problem.

8   X-Ray Attenuation Length Compared to Luminescence Attenuation Length

An ionizing radiation detector needs to stop (i.e., convert primary photon energy to secondary electron-hole pairs) most of the flux of incident x-rays to make a good imaging detector. It should stop all of the energy of most of the counted

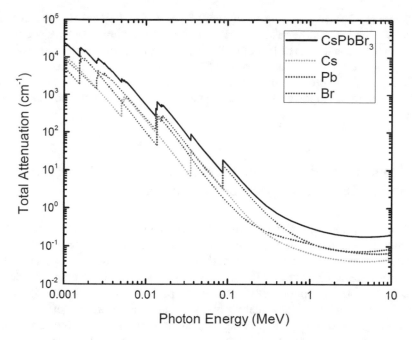

Fig. 8 X-ray attenuation coefficient of CsPbBr₃ along with those of elemental Cs, Pb, and Br weighted according to their mass fraction in CsPbBr₃. These results are plotted using data from the NIST Standard Reference Database 8 (XCOM Photon Cross Sections)

gamma photons to make a good energy-resolving detector. Figure 8 shows the x-ray attenuation plot versus photon energy for CsPbBr₃ and its elemental constituents individually. The x-ray attenuation coefficient μ (cm^{-1}) expresses the x-ray flux $I(x)$ remaining after traversal of thickness x by an incident flux I_0, according to:

$$I(x) = I_0 \exp(-\mu x). \tag{1}$$

The total attenuation coefficient includes contributions from the photoelectric effect, Compton scattering, and pair production. The photoelectric effect is largest at low energies, Compton scattering dominates at intermediate energies around 1 MeV and for lighter elements, and pair production dominates at high energies, above roughly 5 MeV.

For x-ray photon energy of 30 keV, used at the softer end of the spectrum for much of medical x-ray imaging, stopping 63% ($= 1 - e^{-1}$) of the photon flux requires a thickness of CsPbBr₃ equal to about 100 μm. For security scanning of shipping containers, industrial inspection of metal structures, and some medical procedures such as positron-emission tomography (PET), x-ray and gamma photon energies within the range of 0.511 MeV up to 10 MeV should be detected and sometimes energy-resolved. Figure 8 shows that attenuation to 1/e requires a thickness of CsPbBr$_3$ ranging from about 2 cm at 511 keV up to about 3 cm at 10 MeV. Using our earlier estimate of exciton band peak luminescence attenuation to 1/e in only one to tens of μm of CsPbBr$_3$, you can see that scintillation traversal of the whole thickness required to stop 511 keV and 10 MeV gamma or x-rays in CsPbBr$_3$ is largely prohibited when luminescence attenuation length is compared to x-ray attenuation length, and there is even a penalty of light extraction for 30 keV x-rays.

9   Optical Absorption Coefficient of Pb-Halide Perovskites in the Range of Exciton Luminescence

Optical absorption spectra in the edge and interband range of Pb-halide perovskites are often presented in the literature with a normalized or arbitrary scale of absorption coefficient. Quantitative absorption coefficient spectra for CsPbBr$_3$ nanocrystals in suspension were reported by [11] and by [40], which can be averaged to deduce an approximate absorption coefficient at the peak of the interband edge onset, $\alpha \approx 4 \times 10^4$ cm^{-1}. Spectra that report quantitative absorption coefficient based on transmittance of Pb-halide perovskite films are often displayed on a linear scale of about one decade of absorption, trailing off to a baseline that includes Fresnel interface and Rayleigh scattering losses, hindering assessment of absorption in the red wing of the exciton PL band. Photothermal deflection spectroscopy (PDS) and Fourier transform photocurrent spectroscopy (FTPS) measure temperature rise and carrier generation by the probe beam, respectively, and so can measure the spectrum of energy deposition continuing below the floor dictated by scattering losses in transmission measurements. However, such measurements give the accurate spectral decrease of probe energy deposition relative to some spectral maximum of deposition, but not directly the absorption coefficient at the maximum. PDS and FTPS were measured over multiple decades of band-edge absorption in MAPbI$_3$ by [12], shown in Fig. 9. FTPS is able to measure the spectrum of absorption into free-carrier states over a span of at least 4.5 decades. Although absorption coefficient is not expressed directly in the plot of [12], one may use a transmission measurement of quantitative absorption coefficient at the maximum absorption of the band edge in MAPbI$_3$, $\sim 2 \times 10^4$ cm^{-1} [30], to attach an approximate absorption coefficient scale to Fig. 9.

10   Urbach Edge Broadening

The straightness of the FTPS absorption edge spectrum over four decades in the semilog plot of Fig. 9 indicates an exponential falloff of absorption coefficient below the interband/exciton edge in this room-temperature measurement according

Fig. 9 Transmittance, PDS, and FTPS measurements of absorbance relative to a maximum normalized (for PDS) at 3.0 eV in CsPbI$_3$. From [12]

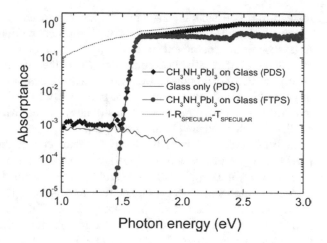

to the Urbach rule. Urbach edge broadening purely by electron-phonon interaction, as exhibited particularly in AgBr [66] and alkali halides [21, 34, 61], displays characteristic broadening of the excitonic absorption edge versus photon energy and temperature according to:

$$\alpha\,(h\nu, T) = A\,\exp\left[-\sigma\,(E_0 - h\nu)\,/k_{\mathrm{B}}T\right] \tag{2}$$

where $\alpha(h\nu, T)$ is the absorption coefficient, E_0 is the photon energy corresponding to the common intersection of the extrapolated data for different temperatures T, A is the absorption coefficient at the intersection point (not physical), and the constant σ is the Urbach slope parameter. In less ionic and/or disordered materials, the Urbach edge broadening is often not fully due to electron-phonon interactions leading to the form of Eq. (2) but can have a temperature-independent part of exponential broadening due to disorder or defects, as first shown by [10] and quantified by theories such as the electric microfield effect [17]. To include broadening with temperature dependence (or independence) more general than Eq. (2), Dow and Redfield among others defined a "spectral Urbach rule" stated as Eq. (3) below.

Temperature dependence data over sufficient absorption decades to support or test Eq. (2) have not been measured for Pb-halide perovskites to our knowledge. Instead, it is common practice in works on Pb-halide perovskites at room temperature to use a restatement of the Urbach rule valid specifically at room temperature in the following form:

$$\alpha\,(h\nu) = A\,\exp\left[-\,(E_0 - h\nu)\,/E_{\mathrm{U}}\right] \tag{3}$$

E_{U} is called the Urbach energy in this form of the rule. Comparison to the original Urbach rule in Eq. (2) identifies the Urbach energy as $E_{\mathrm{U}} = k_{\mathrm{B}}(295\ \mathrm{K})/\sigma$. The constants A and E_0 are defined differently in Eqs. (3) and (2). In Eq. (3), there is no concept of an intersection point of extrapolated edges at different temperatures.

Instead, the constants A, E_0, and E_U are parameters in an empirical fit of the measured straight-line semilog plot of absorption coefficient at room temperature. A is absorption coefficient at the highest point of the straight-line plot, E_0 is the energy of that reference point, and E_U is the energy interval over which the absorption coefficient falls to $1/e$. The Urbach edge broadening characterized by E_U in perovskites has gained importance particularly for solar cell research. Disorder-induced Urbach edge broadening in the alloy system $MAPb(I_{1-x}Br_x)_3$ has been studied by sub-bandgap photocurrent [60].

Galvani et al. calculated the electron-phonon contribution to edge broadening in MAPbI₃, finding $E_U = 9.5$ meV [19], compared to the experimental measurement of $E_U = 15$ meV [12, 60] in MAPbI₃. They attributed the difference to the disorder component of edge broadening in MAPbI₃. The flip side of this is that more than half of the Urbach broadening in MAPbI₃, and by extension other Pb-halide perovskites, is attributable to electron-phonon broadening with temperature dependence characterized by Eq. (2). The near-edge spectral range in T-dependent Urbach broadening becomes exponentially more transparent as temperature is lowered. In alkali halides, the near-edge absorption can increase as much as five decades of absorption coefficient from 10 K up to room temperature, e.g., measured 0.3 eV below the exciton peak in KCl [34, 61].

11 Temperature-Dependent Urbach Broadening as a Contributor to the Experimentally Observed Thermal Quenching of Excitonic Radioluminescence in Single-Crystal 3D Pb-Halide Perovskites

The Urbach edge is not as broad in Pb-halide perovskites as in alkali halides, but it is worth considering that if the classic temperature dependence of the electron-phonon part of Urbach broadening is partially represented in Pb-halide perovskites, as the theory of Galvani et al. on MAPbI₃ [19] suggests, then a substantial part of the reabsorption of excitonic luminescence could be strongly temperature dependent. Figure 3b compared the room temperature photoluminescence and radioluminescence (45 kV, Cu target) spectra of MAPbBr₃, MAPbI₃, and (EDBE)PbCl₄ single crystals measured by [4]. The radioluminescence spectra of MAPbBr₃ and MAPbI₃, coming from the roughly 200 μm penetration depth of the polychromatic x-rays, are redshifted compared to the photoluminescence spectra coming from the surface (roughly the top 200 nm). This is clear evidence of reabsorption of the exciton luminescence while exiting from the x-ray penetration depth at room temperature, confirming again much of the discussion in preceding parts of this chapter. The (EDBE)PbCl₄ sample did not display this redshift because it is a 2D perovskite with significant Stokes shift of its self-trapped exciton emission away from the absorption edge (see discussion later). While there were no spectral data reported at low temperature, recall from our earlier discussion of Fig. 3 that the light yields of MAPbBr$_3$ and MAPbI$_3$ rose to very near the theoretical maximum bandgap-limited value when temperature was lowered to 10 K. If the light yield at 10 K is at the theoretical maximum, then there cannot be much reabsorption of the luminescence into the exciton edge at 10 K during escape of light from the x-ray penetration depth. The scintillation reabsorption problem in these perovskites has been somehow solved at low temperature!

We suggest that the temperature-dependent Urbach edge in Eq. (2), credited by [19]) for about 63% of the exciton edge broadening in MAPbI$_3$, could be part of the explanation. That part of the absorption becomes exponentially more transparent to exciton luminescence as the temperature approaches 10 K. Also contributing to this trend is the fact that the bound exciton luminescence and especially the donor-acceptor pair recombination luminescence centers which become dominant at low temperature are more highly Stokes-shifted than is the free exciton, which is dominant at room temperature [67]. Considering this, we suggest that the rise of light yield toward low temperature (conversely, the quenching of light yield toward room temperature) observed by [4] in Fig. 3 for hybrid perovskites and [43] in Fig. 4 for inorganic perovskites may be caused by at least two different mechanisms acting together. One is the thermal ionization of radiative and momentum-conserving bound exciton and donor-acceptor pair centers discussed earlier in regard to the right side of Fig. 4 from [67]. This ionization occurs at temperatures above about 80 K, leaving free excitons and free carriers existing in thermal equilibrium as room temperature is approached. Another "quenching" mechanism is the substantial phonon-assisted temperature-dependent component of the Urbach broadening, which according to Eq. (2) predicts exponentially increasing reabsorption as temperature is raised.

12 The Stokes Shift Challenge

Many conventional scintillators employ ionic insulator hosts with doped activator ions, and others rely on self-activation by self-trapped excitons formed spontaneously on excitation of certain hosts. In both cases, the Stokes shifts are large and/or reckoned from the activator-excited state which was already chosen to lie well below the conduction band edge. As a result, reabsorption of the light in conventional scintillators is not usually a serious barrier to performance. To be sure, some of the rare-earth dopants like Eu^{2+} in SrI$_2$ have a small enough Stokes shift that reabsorption becomes a problem. The difficulty in that case is pulse lengthening during photon recycling, which can express itself as loss of detected light within a specified gate, rather than actual loss of light to nonradiative channels [1]. In contrast, semiconductor luminescent materials including the three-dimensional Pb-halide perovskites have typically small luminescence Stokes shift, and it is often reckoned directly from a strong excitonic absorption edge rather than an isolated sharp dopant state as in ion-activated scintillators. Even in a semiconductor with 90% or 95% PLQY, nonradiative channels can drain excited states during photon recycling as light exits a scintillator thick enough to stop x-rays and gamma rays of medium to high energy. To be sure, there have been several publications demonstrating scintillation from Pb-halide perovskites suitable for imaging applications. Most were demonstrated at medium to soft x-ray energies ([4, 7, 8, 24]), but some also showed response and broad photopeaks for high-energy photons up to 662 keV gamma rays [31, 81]. Our point is not to deny that scintillation can be observed in 3D Pb-halide perovskites – because it can be. The point is to specify what seems to be a problem with this material class, i.e., insufficient Stokes shift relative to excitonic absorption, in order to try to overcome it and benefit fully from the other attractive properties already pointed out. Returning to Fig. 2, imagine getting rid of the loss to reabsorption at room temperature and thereby moving closer to realizing the maximum light yield suggested from the combination of PLQY and βE_{gap}.

The problem then is how to get a bigger Stokes shift in Pb-halide perovskites or related materials that retain some or all of the other properties that make them attractive for ionizing radiation detectors. Three possible strategies will be discussed:

1. Perovskite-related Pb-halides and other metal halides of dimensionality lower than 3 are often found to exhibit self-trapped excitons (STE) emitting highly Stokes-shifted broadband luminescence. This is a route to get a bigger Stokes shift in perovskite emitters themselves and thus suppress reabsorption.
2. Wavelength shifters, particularly dyes, have been in use for some time in organic scintillators either to avoid reabsorption or to match the response of photodetectors. Wavelength-shifting ions are employed as co-dopants in inorganic crystal scintillators, e.g., [71, 72]. Wavelength-shifting dyes have been combined with QDs in polymer hosts at high QD loading for scintillator use [29, 37] and recently with $CsPbBr_3$ QDs in PMMA to suppress reabsorption [18]. However the initial report in [18] was only for 2% QD loading of $CsPbBr_3$ in PMMA.
3. Dope a perovskite or perovskite-related host with an activator ion, e.g., a rare-earth ion or thallium. This provides a light-emitting state that is already separated below the exciton/interband edge and that furthermore exhibits a configuration-coordinate-described substantial Stokes shift commensurate with a localized excited state. The hope is to also benefit from attractive properties of a halide perovskite host which could include narrow gap for light yield and (defect tolerant?) semiconductor transport for efficient transfer of charges and/or excitons from ionizable host to emitting activator.

13   Strategy #1: Self-Trapped Excitons in Lower-Dimensional Perovskite-Related Materials

Excitons are found to self-localize by lattice relaxation in a number of materials including ionic halides, oxides, and condensed rare gases as reviewed, e.g., in [57], and perovskite-related metal halides of dimensionality lower than 3 as studied more recently by, e.g., [2, 14, 25, 36, 39, 41, 54, 55, 77, 79, 81–84]. Self-trapped excitons (STE) typically emit highly Stokes-shifted broadband luminescence. The phenomenon has become a subject of considerable interest for developing white-light LEDs [14, 25, 36, 39, 41, 54, 55, 77, 79, 84]. It is also attracting growing attention in the search for halide perovskite scintillators with sufficient Stokes shift to avoid the reabsorption problem [2, 81, 84].

Besides the role of Stokes shift in granting freedom from reabsorption of luminescence, other benefits of a self-activated scintillator based on self-trapped excitons are that there is in principle no energy transfer step from host to dilute activators. Every lattice site of the pure crystal is a potential self-trapping site and therefore a light-emitting site. One is not limited by solubility concerns for raising the concentration of activators, since the concentration of such recombination sites is 100%. There are already commercial self-activated scintillators using STE luminescence instead of doped-in activators for the electron-hole recombination luminescence counted as scintillation – e.g., cadmium tungstate ($CdWO_4$), cesium hafnium chloride (Cs_2HfCl_6), and undoped CsI. The Stokes shift of the scintillation light is reckoned from the strong exciton/interband absorption edge, just as it is in $CsPbBr_3$. The Stokes shift must be large to get clear of the exciton absorption edge, which has Urbach broadening extending as much as 0.6 eV below the exciton peak at room temperature in alkali halides and not much less in the other ionic crystals listed above. The Stokes shift of STE luminescence is indeed large in the examples of self-activated scintillators: 0.5 eV in $CdWO_4$, 2.9 eV in Cs_2HfCl_6, and 1.8 eV in CsI. Large Stokes shift and accompanying broad emission bandwidth are characteristic attributes of STE luminescence. The STE's very existence is owing to electron-phonon interaction, so it is not surprising that phonon broadening and Stokes shifting can be dramatic in the luminescence [68].

Li et al. [36] have presented a recent perspective on self-trapped excitons in halide perovskites. Their illustration of general processes of electron-hole radiative recombination is reproduced below in Fig. 10a. The configuration-coordinate representation of STE ground and excited electronic states are illustrated in Fig. 10b, c. The concepts represented in such diagrams were worked out initially from experimental and theoretical studies of STEs in alkali halides and other wide-gap ionic crystals with strong exciton-phonon interaction and have now been found to properly describe broadband intrinsic luminescence found in Pb-halide as well as Pb-free halide perovskite-related materials of lower dimensionality than 3, by multiple research works ([36] and references therein).

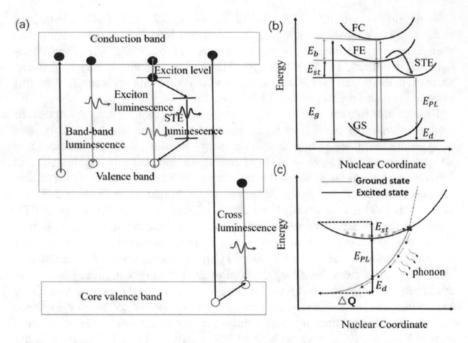

Fig. 10 (**a**) Schematic representation of various intrinsic luminescence phenomena including band-to-band luminescence, exciton luminescence, STE luminescence, and cross luminescence. (**b**) Configuration coordinate schematic of the energy-level structure of STE (*GS* ground state, *FE* free-exciton state, *FC* free-carrier state, *STE* self-trapped exciton state, E_g bandgap energy, E_{st} self-trapping energy, E_d lattice deformation energy, E_{PL} emission energy). (**c**) Schematic of the nonradiative recombination process for STE when the excited-state relaxation is large. From [36]

14 Ion-Size Control of STE Relaxation: Parallels Between Alkali Halides and Halide Perovskites

In alkali halides, ion size is found to exert control over how much lattice relaxation occurs in the excited state, thus how much energy relaxation occurs on the way to the emission coordinate in the configuration coordinate curves of excited and ground states, and from that how much Stokes shift. The more spacious and "loose" is the ionic lattice to allow relaxational ion motion in the electronic excited state, the larger is the STE Stokes shift in general. In alkali halides, where large "off-center relaxation" of the halogen is mainly responsible for the largest Stokes shifts, the important ion-size parameter is the halogen to alkali ratio [56]. To illustrate, the Stokes shifts of the triplet STE luminescence in potassium halides are 7.4 eV in KF, 5.5 eV in KCl, 4.4 eV in KBr, and 2.6 eV in KI [57], differing only in the halide ion. Besides the big Stokes shifts, an additional consequence of the large lattice relaxation is that the STE luminescence of every crystal listed above is thermally quenched far below room temperature. As Li et al. pointed out in their discussion of STEs in perovskites, overly large excited-state relaxation (leading to very large Stokes shift) risks a crossing of the ground and excited-state adiabatic potential curves, resulting in dumping of the excited-state population depicted in Fig. 10c [36]. In the present context of STEs in lower-dimensional perovskite-related materials, the issue of openness of the lattice, such as around the B-site affecting distortion of the BX_6 octahedra, seems reminiscent of the ion-size effects on STE Stokes shift in alkali halides.

CsPbBr$_3$ and other APbX$_3$ perovskites with Cs or organic groups as the A-site ion have not been found to exhibit self-trapped excitons, with one exception in the low-temperature ($T < 176$ K) phase of CsPbCl$_3$ [22]. The reason is basically that the APbX$_3$ perovskites are semiconductors, more so (smaller gaps) for the heavier halogens on X-sites. As semiconductors, they have relatively large band curvature implying rapid site-to-site electron and exciton transfer. Toyozawa stated a simple rule governing when spontaneous self-trapping may occur: the energy lowering from polarization, bond formation, or other relaxation around the excited state (once it is localized) should exceed half the bandwidth of the energy dispersion curve if the self-trapping relaxation is to be stable [62]. Half the bandwidth is the energy cost of localizing from a band state, where the full band width 2B is the range of translational kinetic energies due to site-to-site transfer. Their range of ±B about the energy of the localized state can be illustrated, e.g., by constructing the band from anti-bonding and bonding combinations of local atomic states. The energy payback from local lattice relaxation (including chemical bonding and polarization) has to be sufficient to justify the cost of localization from the band. Generally, semiconductors with bandgaps roughly 3 eV and below do not exhibit STEs, mainly because the exciton band half-width B is substantial in such materials. Toyozawa et al. defined $g = E_{LR}/B$ as the exciton-lattice coupling constant, where E_{LR} is the energy lowering of the exciton by lattice relaxation upon localization [52, 64]. In terms of g, the conceptual criterion discussed above for self-trapping becomes roughly $g > 1$. Schreiber and Toyozawa elaborated on this in more detail, showing a connection (inverse proportionality) between g and the Urbach rule slope constant σ in Eq. (2):

$$\sigma = s/g \tag{4}$$

where s is a dimensionless constant that depends on the dimensionality and geometry of the crystal lattice. For example, $s = 1.5$ for simple cubic lattices in 3 dimensions, and $s = 1.24$ for simple square lattices in 2 dimensions [52]. The broader the Urbach edge (smaller σ), the more likely to find self-trapped excitons, in general.

Toyozawa and Shinozuka showed that there is always an activation barrier between free and self-trapped excitons in three-dimensional materials, may or may not be a barrier in 2D materials depending on the value of g, and that there is no barrier in 1D materials and 0D materials [63]. The barrier present in 3D and 2D materials may be surmounted with thermal energy, and in such cases, free excitons and self-trapped excitons may coexist. Materials with 1D and 0D structure should be unstable to spontaneous lattice distortion in the excited state, although we note in this context that Yuan et al. have reported self-trapped excitons and free excitons coexisting in 1D $C_4N_2H_{14}PbBr_4$ [80].

The evidence for self-trapping in perovskite materials of reduced dimensionality (as well as coexistence of free and self-trapped excitons in 2D but not 0D) is becoming quite clear by now in the literature. Two-dimensional sheet-structured perovskite materials, generally with organic ligand spacers between sheets, have exhibited Stokes-shifted broadband emission attributed to STEs starting from work on (EDBE)PbBr$_4$ [14], $(C_6H_{11}NH_3)_2PbBr_4$ [77], (EDBE)PbCl$_4$ [4], (DMEN)PbBr$_4$, and (DMAPA)PbBr$_4$ [41], and multiple other 2D perovskite-related materials reviewed by [36, 54], for example. Some of the 2D perovskite materials exhibit both the large Stokes-shifted, broadband emission of STEs and a sharp peak or shoulder attributable to free excitons in coexistence [25, 54, 77], which implies a thermal activation barrier for self-trapping, illustrated in Fig. 10b. Material parameter dependence of the free/self-trapped luminescence intensity ratio has been studied [55]. Stokes-shifted broadband emission has also been reported in one-dimensional perovskites including $(C_4N_2H_{14})PbBr_4$ [80] and $(C_4N_2H_{14})SnBr_4$ [84]. The so-called zero-dimensional perovskite-related materials such as Cs$_4$SnBr$_6$ [2], $(C_4N_2H_{14}Br)_4SnBr_6$ [84], and $(C_9NH_{20})_2SbCl_5$ [83] also exhibit broad, Stokes-shifted STE luminescence and achieve higher PLQY, up to near unity. In a survey of the literature, Li et al. observe that the 0D perovskite materials display STE luminescence but not in coexistence with a free-exciton luminescence peak, concluding that there is no barrier to self-trapping in the 0D perovskites [36].

Li et al. further noted that the PLQY of many of the 2D perovskite-related broadband STE emitters is relatively low. They suggested that one cause of the low PLQY may be a consequence of overly large excited-state relaxation [36]. The 2D perovskites allow both inter-octahedral tilting as well as distortion of the octahedra themselves as relaxation modes determining the STE energy. Their meaning of "overly large" is that it may lead to a crossing of the excited- and ground-state adiabatic potential energy curves as represented in Fig. 10c, with consequent dumping of the excited state being one mechanism of thermal quenching. Again noting parallels with STE phenomena in alkali halides (AX), the STE relaxation in those materials can involve large "off-center" relaxation of the X_2^- core of the STE, by almost a lattice constant, in addition to the "on-center" diatomic bonding relaxation that forms the X_2^- halogen core of the STE itself [57]. The ion-size ratio of halide to alkali in NaI restricts the large halide displacement needed for the STE to undergo off-center relaxation of its I_2^- core relative to the whole lattice. In most other alkali halides, it is the off-center X_2^- relaxation that accounts for the main part of very large Stokes shifts noted four paragraphs earlier, as well as consequent thermal quenching by crossing to the ground state at relatively low temperature [56, 57]. In contrast, the tight lattice dictated by ion-size ratio in NaI restricts the STE to stay on center (i.e., on the center point of the two iodide ions that bond in the excited state). Because of this, NaI has a modest Stokes shift compared to most other alkali halides and does not thermally quench by crossing directly to the ground state at room temperature. The partial quenching of NaI STEs at room temperature and above is instead attributed to thermally activated diffusion of the STE, given its relatively shallow trapping depth in NaI compared to other alkali halides [44]. On the basis of one- and two-photon excitation studies, Nagata et al. proposed that the thermal quenching in NaI occurs when the thermally mobilized STEs encounter defects, including point, line, and especially interface defects [44]. This relative thermal stability of the STE against ground-state crossing in the "tight lattice" of NaI and thus its survival to transport excited-state energy may be considered an important contributor to why NaI is one of only three activated alkali halide hosts that find use as scintillators. CsI, which allows only a small amount of off-center relaxation (the so-called type II STE), is a second one, useful both as the activated scintillator CsI:Tl and the self-activated STE scintillator CsI. LiI, which is even more restricted by ion-size ratio than NaI, is the third. It is worth noting that they are all alkali iodides, and despite the inconvenience of alkali iodides being hygroscopic, two of them have continued to find very wide use as scintillators for roughly the last 70 years.

In the 0D perovskite-related materials Cs_4PbBr_6 and $(C_4N_2N_{14}Br)_4SnBr_6$, theoretical calculations of STE mobility by resonance transfer performed by [20] indicate that greater thermal quenching of the STE luminescence in Cs_4PbBr_6 compared to $(C_4N_2N_{14}Br)_4SnBr_6$ can be attributed to a relatively tighter lattice (small interstitial space) in Cs_4PbBr_6, restricting excited-state relaxation and associated Stokes shift. Because of the smaller Stokes shift, resonant overlap of tails of the STE emission and absorption bands is not entirely removed in Cs_4PbBr_6, and the STE is proposed to migrate in that material by resonant transfer to neighboring octahedra. In the compared case of $(C_4N_2N_{14}Br)_4SnBr_6$, the larger interstitial space of the lattice permitted larger STE relaxation, and the consequent larger Stokes shift removed all overlap of exciton absorption and STE emission, immobilizing the STE against resonant transfer. The more shallow-trapped STEs in Cs_4PbBr_6 were proposed to diffuse readily at room temperature and find nonradiative defects [20], in partial analogy to the conclusion of [44] on STE quenching in NaI discussed above.

The relative looseness of the $(C_4N_2N_{14}Br)_4SnBr_6$ lattice compared to Cs_4PbBr_6 may be partly attributed to the large organic ligand in the A ion position allowing more tilting of the octahedra during relaxation. However, Han et al. made it clear that substitution of the smaller Sn^{2+} for Pb^{2+} in the B ion position has the main effect on STE relaxation, opening space for larger tetrahedral distortion of the $(B)Br_6$ octahedron when the B-site ion is Sn^{2+} compared to Pb^{2+}. Once again we draw analogy/contrast to the alkali halide experience, where the ion-size ratio in NaI restricted STE relaxation, preventing its crossing to the ground state but enabling thermal hopping mobility of the STE. Because of the thermally activated mobility, quenching can occur by finding nonradiative defects. Furthermore energy transport to Tl^+ activator ions, if present, can occur helping make the widely used activated scintillator NaI:Tl. In the perovskite comparison, ion size of the B and X ions composing octahedra can restrict or allow deformation of the octahedra and thus control the degree of STE relaxation and consequent Stokes shift. The Stokes shift (relaxation) is not so big as to threaten quenching to the ground state (Fig.

10c), but Han et al. showed that it is big enough with Sn replacing Pb to shut down resonant transfer mobility of the STE and thus suppress quenching on defects at room temperature.

We also note the experiment of [82] in which hydrostatic pressure was applied to $(C_6H_5C_2H_4NH_3)_2PbBr_4$ [$(PEA)_2PbBr_4$], a layered perovskite of inorganic 2D sheets separated by large organic ligands. Increased pressure caused broadband Stokes-shifted STE emission and reduced the near-band-edge-bound exciton emission. Forces on the $PbBr_6$ corner-sharing octahedra exerted in the plane of a sheet are transmitted by the relatively stiff inorganic layer, whereas forces perpendicular to the sheet are transmitted by relatively softer deformable ligands. Zhang et al. studied deformation of the $PbBr_6$ octahedra under pressure by x-ray diffraction. We speculate that because distortions by in-plane compression of the octahedra partially coincide with local distortion modes that occur during STE relaxation in $(PEA)_2PbBr_4$, there may result a cooperative achievement of a stable STE energy minimum under pressure. There are parallels in the earlier literature on applied lattice strain and self-trapped excitons in alkali halides. Itoh et al. melted and resolidified films of NaI and NaBr inside the narrow gaps of a succession of fused quartz cells [27, 28]. Because of the difference in expansion coefficients of cell and contents, the resulting films were under dilatational strain. As commented earlier, NaI and NaBr are the two alkali halide examples where halide and alkali relative ion sizes make a tight lattice that restricts translational motion of halide ions even in the excited state such that these two alkali halides exhibit only the on-center or Type I STE configuration characterized by small Stokes shift. Itoh et al. observed that in proportion to decreasing cell thickness (increasing dilatational strain), a more highly Stokes-shifted STE luminescence band, by 1 eV in NaI and 1.5 eV in NaBr), appeared and grew in intensity at the expense of the on-center STE band. They identified the new luminescence band as due to moderately off-center type II STEs, made possible when the lattice was "opened up" to the possibility of halogen excited-state translation in the lattice. Lattice expansion promotes additional self-trapping relaxation in NaI and NaBr because it was the tightness of the lattice that inhibited the larger ion and atom displacements, whereas in $(PEA)_2PbBr_4$ discussed above, hydrostatic pressure may compress the Br ions in the inorganic sheet, which is in the direction of relaxation of the midplanes of octahedra shown to be involved in self-trapping of excitons in this perovskite.

The lead-free double perovskite $Cs_2AgInCl_6$ emits Stokes-shifted broadband STE luminescence, but only weakly so because the luminescence is parity forbidden [38, 39]. By doping it with Na, the inversion symmetry was broken with the result that the optimal alloy $Cs_2(Ag_{0.6}Na_{0.4})InCl_6$ achieved a PLQY three orders of magnitude higher than $Cs_2AgInCl_6$, at 86% [Luo]. Other studies of STE emission in alloyed double perovskites $Cs_2Na_xAg_{1-x}InCl_6$ [75] and $Cs_2AgSb_xBi_{1-x}Br_6$ [74] have been conducted.

Shinozuka and Toyozawa showed that the extrinsic exciton localization potential of a defect, plus the intrinsic exciton-lattice coupling constant g, can lead to a stable localized exciton where neither coupling alone could achieve it [53, 64]. They termed this "impurity-assisted self-trapping." Yu et al. have shown that doping

PEA$_2$PbI$_4$ with Sn initiates self-trapping of excitons with broadband emission, whereas the STE did not occur in the pure material [79].

B. Yang et al. studied a lead-free metal halide, Rb$_2$CuBr$_3$, with crystal structure describable as one-dimensional aligned ribbons of distorted CuBr$_4$ tetrahedra spaced by Rb ions [76]. While not a perovskite, it has some similarity as a low-dimensional self-activated solution-grown metal halide whose STE emission is Stokes-shifted enough to avoid reabsorption. With high exciton binding energy in the 1D structure, the STE in Rb$_2$CuBr$_3$ is found to maintain high quantum efficiency at room temperature, so that together with the absence of reabsorption, an impressively high radioluminescence light yield of 91,000 photons/MeV has been measured. Other reported properties of the material make it appear promising for evaluation as a scintillation x-ray detector [76].

15 Sacrifice of Stopping Power to Get Stokes Shift?

From the above discussion, one can surmise that tailoring perovskite materials to support self-trapped excitons that are stable against quenching at room temperature is a potentially useful way to gain Stokes shift needed to avoid reabsorption. However, there seems to be a performance price to pay in most instances of achieving the stable STE and its advantages. One adds bulky organic spacer ligands to achieve 2D layering and/or substitutes the large Pb^{2+} ion at the B-site of the inorganic octahedra by a smaller ion to gain easier deformability of the octahedron during excited-state relaxation for a more Stokes-shifted and thermally stable STE. Both of those actions lower the density and the effective Z of the material, therefore lowering the stopping power for ionizing radiation. The attenuation coefficient μ of an element of atomic number Z and density ρ is proportional to ρZ^4. Rodnyi defined an effective atomic number of a compound scintillator, Z_{eff}, which allows rough estimation of stopping power given the high-order dependence on Z:

$$Z_{\text{eff}} = \sqrt[4]{\sum_i w_i Z_i^4} \tag{5}$$

where w_i is the ith element mass fraction and Z_i is the corresponding atomic number [50]. Table 1 lists several perovskite and perovskite-related materials of interest for scintillation along with their mass density ρ, effective atomic number Z_{eff}, and attenuation lengths for 30 keV, 100 keV, and 1 MeV x-rays. Several conventional scintillators are included for comparison.

Whereas the density of CsPbBr$_3$ is 4.87 g/cm^3, examples of broadband-emitting 2D Pb-bromide perovskites have densities of 2.39 g/cm^3 in (C$_6$H$_{11}$NH$_3$)$_2$PbBr$_4$ [77] and 2.860 g/cm^3 in (DMABA)PbBr$_4$ [41], a reduction to about half the density of CsPbBr$_3$ and substantially less than half that of some conventional inorganic scintillators, e.g., in Table 1. The good news is that for the Stokes-shifted 2D and 0D perovskite scintillators, the x-ray attenuation coefficient need not be compared

Table 1 Density, effective atomic number, and gamma/x-ray photon attenuation coefficient μ of perovskite materials and conventional scintillators

	Compound	Density (g/cm³) (References)	Z_{eff}	μ (cm⁻¹) (30 keV)	μ (cm⁻¹) (100 keV)	μ (cm⁻¹) (1 MeV)
3D perovskites	$CsPbBr_3$	4.87 (orthorhombic, 300 K) [59]	57	98.7	13.4	0.3
	$(CH_3NH_3)PbBr_3$	3.58 [4]	67	78.4	9.9	0.23
	$(CH_3NH_3)PbI_3$	3.95 [4]	67	60.8	12.1	0.17
2D perovskites	$(EDBE)PbCl_4$	2.19 [4]	66	29.3	5.3	0.15
	$(C_6H_{11}NH_3)_2PbBr_4$	2.39 [77]	61	47.0	5.5	0.19
	$(DMABA)PbBr_4$	2.86 [41]	63	52.9	6.2	0.18
	$\alpha\text{-}(DMEN)PbBr_4$	3.11 [41]	63	60.0	7.0	0.20
Conventional scintillators	NaI(Tl)	3.67 [51]	51	27.0	6.1	0.26
	CsI(Tl)	4.51 [51]	54	40.8	9.2	0.26
	$CdWO_4$	7.90 [51]	64	184.9	21.9	0.50
	$Lu_2SiO_5{:}Ce^{3+}$	7.40 [51]	66	115.5	23.1	0.48
	$Bi_4Ge_3O_{12}$	7.13 [51]	75	168.5	28.3	0.49
	$(Y,Gd)_2O_3{:}Eu^{3+}$	6.90 [51]	56	103.7	13.5	0.42
	$Gd_2O_2S{:}Tb^{3+}$	7.34 [51]	61	92.0	19.2	0.45

to the optical reabsorption coefficient, which would become prohibitive for the extra stopping thickness required by halving the x-ray attenuation coefficient. That issue is resolved by the STE Stokes shift in some 2D and 0D perovskites. In summary, the x-ray attenuation coefficients of the lower-dimensional and/or lead-free perovskites are less than x-ray attenuation coefficients of conventional scintillators in Table 1, but still serviceable for some radiography applications.

16   Strategy #2: Wavelength Shifter in the Perovskite Emitter or Its Supporting Matrix

Rather than seeking a bigger Stokes shift in the luminescence of the halide perovskite itself, one can utilize energy transfer to a secondary emitter which is designed to shift the emission wavelength enough to avoid reabsorption. Organic wavelength-shifting dyes have been used extensively in plastic scintillators to achieve better spectral match with photodetectors and/or to suppress reabsorption. Co-doping inorganic scintillators with ions chosen to receive transferred energy from an activator ion and reemit at a shifted frequency have been employed for similar purposes of matching a photodetector's spectral response or suppressing reabsorption, e.g., [71, 72].

A recent paper by [18] used a custom-synthesized perylene dye with HOMO-LUMO levels well matched to $CsPbBr_3$ excitonic emission and found nearly 100% efficient energy transfer to the Stokes-shifting dye. However, their initial work reports loadings of the perovskite nanoparticles only up to 2% by weight in the PMMA+dye host polymer. Much higher loading of the high-Z perovskite nanocrystals relative to the polymer support and wavelength shifter will ultimately be needed for hard x-ray radiography. The Pei group at UCLA has achieved 60% loading of $Cd_xZn_{1-x}S/ZnS$ core-shell semiconducting QDs in PVT + dye [37] and 80% loading of YbF_3 QDs in PVT + dye [29]. Their method utilizes partial ligand exchange, substituting part of the usual oleic acid short ligands by longer double-headed ligands where one end binds to the QD and the tail gets incorporated in polymerization of the host to anchor the QD against migration and aggregation that lead to dot quenching and light scattering.

The issue of diluting the dense, high-Z, radiation-stopping material with an organic low-Z binder and wavelength shifter to get a net Stokes shift brings up the same consequences of reduced stopping power as discussed in the previous section.

17   Strategy #3: Ion Activator in Perovskite Host

A third approach to employing a perovskite in a scintillator with suitable Stokes shift to avoid reabsorption is to use the perovskite as the host, with an activator ion as the light emitter. If the activator is an isolated ion (atomic size) rather than a semiconducting nanoparticle, its more localized excited-state wave function will cause greater local lattice relaxation in the excited state for greater Stokes shift and better extraction of scintillation light. This is nothing more than a conventional ion-activated scintillator where the host happens to be a perovskite. What could be the advantage of a perovskite in that role? We have previously noted in the discussion of Fig. 2 that the small bandgap of a Pb-halide perovskite host or other semiconductor can in principle permit very high scintillation light yield. In addition, a semiconducting host in place of the more typical insulating host may be more effective in transport of carriers from points of generation in the dense host to the activator ions, and if defect tolerance of a Pb-halide perovskite host aids in the energy transport to activator ions, it would be a further bonus. Against these possible advantages, the challenge is to find a suitable activator with levels that fit energetically between the band edges of the perovskite semiconductor host, or if f-f transitions between states overlying a band are used, that there be a means of efficiently coupling host excitations to the ion excited state. There are experiments to suggest that these challenges are not prohibitive at least when $CsPbCl_3$ is the host.

Pan et al. measured luminescence and excitation spectra of Ce, Sm, Eu, Tb, Dy, Er, and Yb lanthanide dopants in hot-injection solution-grown nanocrystals of $CsPbCl_3$ [46]. They determined that the lanthanide ion most likely substitutes for Pb^{2+} on the perovskite B-site. Doping concentrations up to about 10 mole% were achieved. The luminescence spectra they measured in the undoped and lanthanide-doped $CsPbCl_3$ are shown in Fig. 11a. Notably, the Ce^{3+}-doped sample exhibits a broadband of Ce 5d-4f luminescence with peak at about 430 nm. Its decay time is 9.7 ns, significantly faster than Ce luminescence in macrocrystals of almost all other hosts. The quantum efficiency of Ce luminescence was found to be about 25%. All of the other lanthanide dopants in $CsPbCl_3$ studied by Pan et al. exhibited only f-f transitions of the lanthanide, with decay times ranging from 600 to 900 µs. The quantum efficiency was in the range of 20–30% in all lanthanide dopants studied except Yb, which exhibited surprisingly high quantum efficiency of 143%. The authors attributed this to a quantum cutting process involving a defect near mid-gap which could participate in two-step transitions from the conduction band ultimately to the valence band, each step able to couple to Yb^{3+} $^2F_{7/2} \rightarrow {}^2F_{5/2}$ excitation. Milstein et al. also reported on this emission in Yb^{3+}-doped $CsPbCl_3$ [42].

Vacuum referred binding energy (VRBE) diagrams of lanthanides are a very useful tool for studying lanthanide luminescence properties, e.g., [16]. We have constructed a VRBE diagram for lanthanide dopants relative to the $CsPbCl_3$ bands, shown in Fig. 11b. The conduction and valence band energies were obtained from cyclic voltammetry [49]. Exploiting the fact that Pan et al. observed the Ce^{3+} luminescence transition $4f^05d^1 \rightarrow 4f^1$ (5d-4f for short) just below the ~410 nm free-exciton emission of $CsPbCl_3$ [46], we fitted the 5d excited state and Ce^{3+} ground state within the bandgap in Fig. 11a. This resulted in the estimated model parameters of $U = 7.2$ eV and Ce^{3+} redshift $D_3 = 3.3$ eV. The divalent and trivalent ground and ionized states of all other lanthanides follow from the VRBE construction.

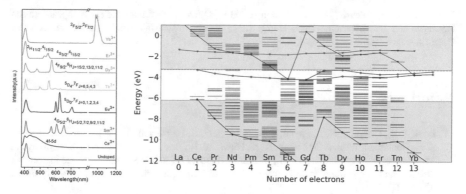

Fig. 11 (Left) Luminescence spectra of lanthanide-doped $CsPbCl_3$ excited at 365 nm, from [46]; (right) VRBE diagram of lanthanide ions in $CsPbCl_3$. Energy is referred to the vacuum level (0 eV). Blue and red colors correspond to 3+ and 2+ charge states of the lanthanides. The continuous lines with filled circle markers connect the ground states of 3+ lanthanides, while the open circles indicate the energy of the lowest 4f5d excited state. Blue shading shows the valence band, while the light orange is the conduction band

Except for Ce^{3+}, the trivalent ground states of all other lanthanides lie within or below the valence band of $CsPbCl_3$ rather than in the gap. The only possible 5d to 4f luminescence transition which is within the bandgap is that of Ce^{3+}. Due to the low-lying conduction band of $CsPbCl_3$, almost all states of the divalent lanthanides lie in the conduction band. The only exceptions are Eu^{2+} and Yb^{2+} ground states, which makes Eu^{3+} and Yb^{3+} dopants capable of acting as an electron trap. Despite the fact that the 4f5d excited states (marked by open circles) lie below the conduction band minimum, the energy to excite them from the trivalent ground state is larger than the bandgap energy in all except Ce^{3+}, which will result in full reabsorption of the resulting emission. However, Pan et al. observed 4f-4f emission from all the studied lanthanides (Ce, Sm, Eu, Tb, Dy, Er, Yb). This surprising fact can be best illustrated with Yb^{3+} dopant, for which a single sharp transition was observed around 980 nm and identified as the $^2F_{5/2} \rightarrow {}^2F_{7/2}$ transition. The Stokes shift of the Yb^{3+} f-f transition excited at the $CsPbCl_3$ band edge is enormous, ∼ 1.9 eV. Pan et al. observed this emission under 365 nm excitation. Since Yb^{3+} does not have any electronic transitions in that energy range to be able to directly absorb excitation light, this observation highlights the importance of the $CsPbCl_3$ host in the energy transfer process. If we consider free excitons in $CsPbCl_3$ as the primary products of electron-hole creation, there is no overlap of the exciton emission with absorption to the Yb^{3+} $^2F_{5/2}$ state. This indicates that some intermediate state is crucial for obtaining lanthanide luminescence in $CsPbCl_3$, such as the defect near mid-gap suggested by Pan et al. for its role in quantum cutting, while enabling resonant energy transfer from the $CsPbCl_3$ exciton to the Yb^{3+} $^2F_{7/2} \rightarrow {}^2F_{5/2}$ excitation.

Acknowledgments WWW acknowledges support from the Netherlands Organization for Scientific Research (NWO) through a Rubicon grant. RTW acknowledges support from the

National Nuclear Security Administration (NNSA), Office of Defense Nuclear Nonproliferation Research and Development (DNN R&D) LB15-V-GammaDetMater-PD2Jf, under subcontract from Lawrence Berkeley National Laboratory.

References

1. Beck, P. R., Cherepy, N. J., Payne, S. A., Swanberg, E. L., Nelson, K. E., Thelin, P. A., Fisher, S. E., Hunter, S., Wihl, B. M., Shah, K. S., Hawrami, R., Burger, A., Boatner, L. A., Momayezi, M., Stevens, K. T., Randles, M. H., & Solodovnikov, D. (2014). Strontium iodide instrument development for gamma spectroscopy and radioisotope identification. *Proceedings of SPIE, 9213*, 92130N.
2. Benin, B. M., Dirin, D. N., Morad, V. M., Wörle, M., Yakunin, S., Rainò, G., Nazarenko, O., Fischer, M., Infante, I., & Kovalenko, M. V. (2018). Highly emissive self-trapped excitons in fully inorganic zero-dimensional tin halides. *Angewandte Chemie, International Edition, 57*, 11329–11333.
3. Berry, J., et al. (2015). Hybrid Organic-Inorganic Perovskites (HOIPs): Opportunities and challenges. *Advanced Materials, 27*, 5102–5112.
4. Birowosuto, M., Cortecchia, D., Drozdowski, W., Brylew, K., Lachmanski, W., Bruno, A., & Soci, C. (2016). X-ray scintillation in lead halide perovskite crystals. *Scientific Reports, 6*, 37254.
5. Brennan, M. C., Zinna, J., & Kuno, M. (2017). Existence of a size-dependent stokes shift in $CsPbBr_3$ perovskite nanocrystals. *ACS Energy Letters, 2*, 1487–1488.
6. Brandt, R. E., Poindexter, J. R., Gorai, P., Kurchin, R. C., Hoye, R. L. Z., Nienhaus, L., Wilson, M. W. B., Polizzotti, J. A., Sereika, R., Zaltauskas, R., Lee, l. C., MacManus-Driscoll, J. L., Bawendi, M., Stevanovic, V., & Buonassisi, T. (2017). Searching for "defect-tolerant" photovoltaic materials: Combined theoretical and experimental screening. *Chemistry of Materials, 29*, 4667–4674.
7. Cao, F., Yu, D., Ma, W., Xu, X., Cai, B., Yang, Y. M., Liu, S., He, L., Ke, Y., Lan, S., Choy, K.-L., & Zeng, H. (2020). Shining emitter in a stable host: Design of halide perovskite scintillators for x-ray imaging from commercial concept. *ACS Nano, 14*, 5183–5193.
8. Chen, Q., Wu, J., Ou, X., Huang, B., Almutlaq, J., Zhumekenov, A. A., Guan, X., Han, S., Liang, L., Yi, Z., Li, J., Xie, X., Wang, Y., Li, Y., Fan, D., The, D. B. L., All, A. H., Mohammed, O. F., Bakr, O. M., Wu, T., Bettinelli, M., Yang, H., Huang, W., & Liu, X. (2018a). All-inorganic perovskite nanocrystal scintillators. *Nature, 561*, 88–93.
9. Chen, X., Zhang, F., Ge, Y., Shi, L., Huang, S., Tang, J., Lv, Z., Zhang, L., Zou, B., & Zhong, H. (2018b). Centimeter-sized Cs_4PbBr_6 crystals with embedded $CsPbBr_3$ nanocrystals showing superior photoluminescence: Nonstoichiometry induced transformation and light-emitting applications. *Advanced Functional Materials, 28*, 1706567. (1–7).
10. Chopra, K. L., & Bahl, S. K. (1970). Structural, electrical, and optical properties of amorphous germanium films. *Physical Review B, 1*, 2545–2556.
11. De Roo, J., Ibáñez, M., Geiregat, P., Nedelcu, G., Walravens, W., Maes, J., Martins, J. C., Van Driessche, I., Kovalenko, M. V., & Hens, Z. (2016). Highly dynamic ligand binding and light absorption coefficient of cesium lead bromide perovskite nanocrystals. *ACS Nano, 10*, 2071–2081.
12. De Wolf, S., Holovsky, J., Moon, S.-J., Löper, P., Niesen, B., Ledinsky, M., Haug, F.-J., Yum, J.-H., & Ballif, C. (2014). Organometallic perovskites: Sharp optical absorption edge and its relation to photovoltaic performance. *Journal of Physical Chemistry Letters, 5*, 1035–1039.
13. DiStasio, F., Christodoulou, S., Huo, N., & Konstantatos, G. (2017). Near-unity photoluminescence quantum yield in $CsPbBr_3$ nanocrystal solid films via postsynthesis treatment with lead bromide. *Chemistry of Materials, 29*, 7663–7667.

14. Dohner, E. R., Jaffe, A., Bradshaw, L. R., & Karunadasa, H. I. (2014). Intrinsic white-light emission from layered hybrid perovskites. *Journal of the American Chemical Society, 136,* 13154.
15. Dorenbos, P. (2010). Fundamental limitations in the performance of $Ce^{3+}-$, $Pr^{3+}-$, and Eu^{2+} – activated scintillators. *IEEE Transactions on Nuclear Science, 57,* 1162–1167.
16. Dorenbos, P. (2013). Lanthanide 4f-electron binding energies and the nephelauxetic effect in wide band gap compounds. *Journal of Luminescence, 136,* 122–129.
17. Dow, J. D., & Redfield, D. (1972). Toward a unified theory of Urbach's rule and exponential absorption edges. *Physical Review B, 5,* 594–610.
18. Gandini, M., Villa, I., Beretta, M., Gotti, C., Imran, M., Carulli, F., Fantuzzi, E., Sassi, M., Zaffalon, M., Brofferio, C., Manna, L., Beverina, L., Vedda, A., Fasoli, M., Fironi, L., & Brovelli, S. (2020). Efficient, fast and reabsorption-free perovskite nanocrystal-based sensitized plastic scintillators. *Nature Nanotechnology.* https://doi.org/10.1038/s41565-020-0683-8.
19. Galvani, B., Suchet, D., Delamarre, A., Bescond, M., Michelini, F., Lannoo, M., Guillemoles, J.-F., & Cavassilas, N. (2019). Impact of electron-phonon scattering on optical properties of $CH_3NH_3PbI_3$ hybrid perovskite material. *ACS Omega, 4,* 21487–21493.
20. Han, D., Shi, H., Ming, W., Zhou, C., Ma, B., Saparov, B., Ma, Y.-Z., Chen, S., & Du, M. H. (2018). Unraveling luminescence mechanisms in zero-dimensional halide perovskites. *Journal of Materials Chemistry C, 6,* 6398–6405.
21. Haupt, U. (1959). Über temperatureabhängigheit und form der langwelligsten excitonenbande in KJ-kristallen. *Zeitschrift für Physik, 157,* 232–246.
22. Hayashi, T., Kobayashi, T., Iwagana, M., & Watanabe, M. (2001). Exciton dynamics related with phase transitions in $CsPbCl_3$ single crystals. *Journal of Luminescence, 94,* 255–259.
23. He, Y., Matei, L., Jung, H. J., McCall, K. M., Chen, M., Stoumpos, C. C., Peters, J. A., Chung, D. Y., Wessels, B. W., Wasielewski, M. R., Dravid, V. P., Burger, A., & Kanatzidis, M. G. (2018). High spectral resolution of gamma-rays at room temperature by perovskite $CsPbBr_3$ single crystals. *Nature Communications, 9,* 1609.
24. Heo, J. H., Shin, D. H., Park, J. K., Kim, D. H., Lee, S. J., & Im, S. H. (2018). High-performance next-generation perovskite nanocrystal scintillator for nondestructive x-ray imaging. *Advanced Materials, 30,* 1801743.
25. Hu, T., Smith, M. D., Dohner, E. R., Sher, M.-J., Wu, X., Trinh, M. T., Fisher, A., Corbett, J., Zhu, X.-Y., Karunadasa, H. I., & Lindenberg, A. M. (2016). Mechanism for broadband white-light emission from two-dimensional (110) hybrid perovskites. *Journal of Physical Chemistry Letters, 7,* 2258–2263.
26. Huang, H., Bodnarchuk, M. I., Kershaw, S. V., Kovalenko, M. V., & Rogach, A. L. (2017). Lead halide perovskite nanocrystals in the research spotlight: Stability and defect tolerance. *ACS Energy Letters, 2,* 2071–2083.
27. Itoh, M., Hashimoto, S., & Ohno, N. (1990). Effect of dilatational strain on the self-trapped exciton luminescence of alkali halides. *Journal of the Physical Society of Japan, 59,* 1881–1889.
28. Itoh, M., Hashimoto, S., & Ohno, N. (1991). Self-trapped exciton luminescence in dilated NaI crystals – Relaxation process of excitons in alkali halides. *Journal of the Physical Society of Japan, 60,* 4357–4365.
29. Jin, Y., Kishpaugh, D., Liu, C., Hajagos, T. J., Chen, Q., Li, L., Chen, Y., & Pei, Q. (2016). Partial ligand exchange as a critical approach to the synthesis of transparent ytterbium fluoride polymer nanocomposite monoliths for gamma ray scintillation. *Journal of Materials Chemistry C, 4,* 3654.
30. Kanemitsu, Y. (2017). Luminescence spectroscopy of lead-halide perovskites: Materials properties and application as photovoltaic devices. *Journal of Materials Chemistry C, 5,* 3427.
31. Kawano, N., Koshimizu, M., Okada, G., Fujimoto, Y., Kawaguchi, N., Yanagida, T., & Asai, K. (2017). Scintillating organic-inorganic layered perovskite-type compounds and the gamma-ray detection capabilities. *Scientific Reports, 7,* 14754. (1–8).

32. Kang, J., & Wang, L.-W. (2017). High defect tolerance in Lead halide perovskite CsPbBr$_3$. *Journal of Physical Chemistry Letters, 8*, 489–493.
33. Kobayashi, M., Omata, K., Sugimoto, S., Tamagawa, Y., Kuroiwa, T., Asada, H., Takeuchi, H., & Kondo, S. (2008). Scintillation characteristics of CsPbCl$_3$ single crystals. *Nuclear Instruments and Methods in Physics Research A, 592*, 369–373.
34. Kurik, M. V. (1971). Urbach Rule. *Physica Status Solidi, 8*, 9–45.
35. Lee, S., Choi, K., Min, C. H., Woo, M. Y., & Noh, J. H. (2020). Photon recycling in halide perovskite solar cells for higher efficiencies. *MRS Bulletin, 46*. https://doi.org/10.1557/mrs.2020.145.
36. Li, S., Luo, J., Liu, J., & Tang, J. (2019). Self-trapped excitons in all-inorganic halide perovskites: Fundamentals, status, and potential applications. *Journal of Physical Chemistry Letters, 10*, 1999–2007.
37. Liu, L., Zhou, T. J., Hajagos, D., Kishpaugh, D. Y., & Chen, Q. P. (2017). Transparent ultra-high-loading quantum dot/polymer nanocomposite monolith for gamma scintillation. *ACS Nano, 11*, 6422–6430.
38. Locardi, F., Cirignano, M., Baranov, D., Dang, Z., Prato, M., Drago, F., Ferretti, M., Pinchetti, V., Fanciulli, M., Brovelli, S., DeTrizio, L., & Manna, L. (2018). Colloidal synthesis of double perovskite Cs$_2$AgInCl$_6$ and Mn-doped Cs$_2$AgInCl$_6$ nanocrystals. *Journal of the American Chemical Society, 140*, 12989–12995.
39. Luo, J., Wang, X., Li, S., Liu, J., Guo, Y., Niu, G., Yao, L., Fu, Y., Gao, L., Dong, Q., Zhao, C., Leng, M., Ma, F., Liang, W., Wang, L., Jin, S., Han, J., Zhang, L., Etheridge, J., Wang, J., Yan, Y., Sargent, E. H., & Tang, J. (2018). Efficient and stable emission of warm-white light from Lead-free halide double perovskites. *Nature, 563*, 541–545.
40. Maes, J., Balcaen, L., Drijvers, E., Zhao, Q., De Roo, J., Vantomme, A., Vanhaecke, F., Geiregat, P., & Hens, Z. (2018). Light absorption coefficient of CsPbBr$_3$ perovskite nanocrystals. *Journal of Physical Chemistry Letters, 9*, 3093–3097.
41. Mao, L., Wu, Y., Stoumpos, C. C., Wasielewski, M. R., & Kanatzidis, M. G. (2017). White-light emission and structural distortion in new corrugated two-dimensional lead bromide perovskites. *Journal of the American Chemical Society, 139*, 5210–5215.
42. Milstein, T. J., Kroupa, D. M., & Gamelin, D. R. (2018). Picosecond quantum cutting generates photoluminescence quantum yields over 100% in ytterbium-doped CsPbCl$_3$ nanocrystals. *Nano Letters, 18*, 3792–3799.
43. Mykhaylyk, V. B., Kraus, H., Kapustianyk, V., Kim, H. J., Mercere, P., Rudko, M., DaSilva, P., Antonyak, O., & Dendebera, M. (2020). Bright and fast scintillations of an inorganic halide perovskite CsPbBr$_3$ crystal at cryogenic temperatures. *Scientific Reports*. https://doi.org/10.1038/s41598-020-65672-z.
44. Nagata, S., Fujiwara, K., & Nishimura, H. (1991). Dynamical aspects of excitons in NaI. *Journal of Luminescence, 47*, 147–157.
45. Onken, D. R. (2018). *Applying novel material characterization techniques using ultrafast laser excitation and neutron diffraction in radiation detector crystals*. Ph.D. Dissertation, Wake Forest University, August 2018.
46. Pan, G., Bai, X., Yang, D., Chen, X., Jing, P., Qu, S., Zhang, L., Zhou, D., Zhu, J., Xu, W., Dong, B., & Song, H. (2017). Doping lanthanide into perovskite nanocrystals: Highly improved and expanded optical properties. *Nano Letters, 17*, 8005–8011.
47. Protosescu, L., Yakunin, S., Bodnarchuk, M. I., Krieg, F., Caputo, R., Hendu, C. H., Yang, R. X., Walsh, A., & Kovalenko, M. V. (2015). Nanocrystals of cesium lead halide perovskites (CsPbX3, X=Cl, Br, and I): Novel optoelectronic materials showing bright emission with wide color gamut. *Nano Letters, 15*, 3692–3696.
48. Quan, L. N., Quintero-Bermudez, R., Voznyy, O., Walters, G., Jain, A., Fan, J. Z., Zheng, X., Yang, Z., & Sargent, E. H. (2017). Highly emissive green perovskite nanocrystals in a solid state crystalline matrix. *Advanced Materials, 1605945*(1–6), 29.
49. Ravi, V. K., Markad, G. B., & Nag, A. (2016). Band edge energies and excitonic transition probabilities of colloidal CsPbX3 (X=Cl, Br, I) perovskite nanocrystals. *ACS Energy Letters, 1*, 665–671.

50. Rodnyi, P. A. (1997). *Physical processes in inorganic scintillators*. Boca Raton: CRC Press.
51. Ronda, C. R., & Srivastava, A. M. (2008). In C. R. Ronda (Ed.), *Chapt. 5 in Luminescence: from theory to applications*. New York: Wiley.
52. Schreiber, M., & Toyozawa, Y. (1982). Numerical experiments on the absorption lineshape of the exciton under lattice vibrations. III. The Urbach rule. *Journal of the Physical Society of Japan, 51*, 1544–1550.
53. Shinozuka, Y., & Toyozawa, Y. (1979). Self-trapping in mixed crystal – Clustering, dimensionality, percolation. *Journal of the Physical Society of Japan, 46*, 505–514.
54. Smith, M. D., & Karunadasa, H. I. (2018). White-light emission from layered halide perovskites. *Accounts of Chemical Research, 51*, 619–627.
55. Smith, M. D., Jaffe, A., Dohner, E. R., Lindenberg, A. M., & Karunadasa, H. I. (2017). Structural origins of broadband emission from layered Pb-Br hybrid perovskites. *Chemical Science, 8*, 4497.
56. Song, K. S., Leung, C. H., & Williams, R. T. (1989). A theoretical basis for the Rabin-Klick criterion in terms of off-centre self-trapped-exciton relaxation. *Journal of Physics. Condensed Matter, 1*, 683–687.
57. Song, K. S., & Williams, R. T. (1993). In M. Cardona (Ed.), *Self-trapped excitons* (Springer series in solid-state sciences) (Vol. 105). Berlin/Heidelberg/New York: Springer-Verlag.
58. Staub, F., Anusca, I., Lupascu, D. C., Rau, U., & Kirchartz, T. (2020). Effect of reabsorption and photon recycling on photoluminescence spectra and transients in lead-halide perovskite crystals. *Journal of Physics Materials, 3*, 025003.
59. Stoumpos, C. C., Malliakas, C. D., Peters, J. A., Liu, S., Sebastian, M., Im, J., Chasapis, T. C., Wibowo, A. C., Chung, D. Y., Freeman, A. J., Wessels, B. W., & Kanatzidis, M. G. (2013). Crystal growth of the perovskite semiconductor CsPbBr$_3$: A new material for high-energy radiation detection. *Crystal Growth & Design, 13*, 2722–2727.
60. Sutter-Fella, C. M., Miller, D. W., Ngo, Q. P., Roe, E. T., Toma, F. M., Sharp, I. D., Loneran, M. C., & Javey, A. (2017). Band tailing and deep defect states in CH3NH3Pb(I1-xBrx)3 perovskites as revealed by sub-bandgap photocurrent. *ACS Energy Letters, 2*, 709–715.
61. Tomiki, T. (1967). Optical constants and exciton states in KCl single crystals. II. The spectra of reflectivity and absorption constant. *Journal of the Physical Society of Japan, 23*, 1280–1296.
62. Toyozawa, Y. (1980). Electrons, holes, and excitons in deformable lattice. In R. Kubo & E. Hanamura (Eds.), *Relaxation of elementary excitations* (Springer Series in Solid-State Sci) (Vol. 18). Berlin/Heidelberg: Springer.
63. Toyozawa, Y., & Shinozuka, Y. (1980). Stability of an electron in deformable lattice – Force range, dimensionality, and potential barrier. *Journal of the Physical Society of Japan, 48*, 472–478.
64. Ueta, M., Kanzaki, H., Kobayashi, K., Toyozawa, Y., & Hanamura, E. (1986). *Excitonic Processes in Solids* (Springer Ser. Solid-State Sci) (Vol. 60). Berlin/Heidelberg: Springer. Chap. 4.
65. Ullrich, B., Singh, A. K., Barik, P., Xi, H., & Bhowmick, M. (2015). Inherent photoluminescence stokes shift in GaAs. *Optics Letters, 40*, 2580–2583.
66. Urbach, F. (1953). The long-wavelength edge of photographic sensitivity and of the electronic absorption of solids. *Physics Review, 92*, 1324.
67. Wang, Y., Ren, Y., Zhang, S., Wu, J., Song, J., Li, X., Xu, J., Sow, C.-H., Zeng, H., & Sun, H. (2018). Switching excitonic recombination and carrier trapping in cesium lead halide perovskites by air. *Communications on Physics, 1*, 96. & Supplementary Fig. 9(a).
68. Williams, R. T., & Song, K. S. (1990). The self-trapped exciton. *Journal of Physics and Chemistry of Solids, 51*, 670–716.
69. Williams, R. T., Wolszczak, W., Yan, X., & Carroll, D. (2020). Perovskite quantum-dot-in-host for detection of ionizing radiation. *ACS Nano, 14*, 5161–5169.
70. Wolszczak, W. (2019, September). *Into darkness: from high density quenching to near-infrared scintillators*. Ph. D. Dissertation, Delft University of Technology.

71. Wolszczak, W., Krämer, K. W., & Dorenbos, P. (2019). CsBa$_2$I$_5$:Eu^{2+},Sm^{2+}—The first high-energy resolution black scintillator for γ -ray spectroscopy. *Physica Status Solidi RRL: Rapid Research Letters, 1900158*, 13.

72. Wolszczak, W., Krämer, K. W., & Dorenbos, P. (2020). Engineering near-infrared emitting scintillators with efficient Eu^{2+} $\rightarrow$ Sm^{2+} energy transfer. *Journal of Luminescence, 222*, 117101.

73. Xu, J., Huang, W., Li, P., Onken, D. R., Dunn, C., Guo, Y., Ucer, K. B., Lu, C., Wang, H., Geyer, S. M., Williams, R. T., & Carroll, D. L. (2017). Imbedded nanocrystals of CsPbBr$_3$ in Cs$_4$PbBr$_6$: Kinetics, enhanced oscillator strength, and application in light-emitting diodes. *Advanced Materials, 1703703*(1–10), 29.

74. Yang, B., Hong, F., Chen, J., Tang, Y., Yang, L., Sang, Y., Xia, X., Guo, J., He, H., Yang, S., Deng, W., & Han, K. (2019a). Colloidal synthesis and charge-carrier dynamics of Cs2AgSb1-yBiyX6 (X: B, Cl; 0≤y≤1) double perovskite nanocrystals. *Angewandte Chemie, International Edition, 58*, 2278.

75. Yang, B., Mao, X., Hong, F., Meng, W., Tang, Y., Xia, X., Yang, S., Deng, W., & Han, K. (2018). Lead-free direct bandgap double perovskite nanocrystals with bright dual-color emission. *Journal of the American Chemical Society, 140*, 17001–17006.

76. Yang, B., Yin, L., Niu, G., Yuan, J.-H., Xue, K.-H., Tan, Z., Miao, X.-S., Niui, M., Du, X., Song, H., Lifshitz, E., & Tang, J. (2019b). Lead-free halide Rb$_2$CuBr$_3$ as sensitive x-ray scintillator. *Advanced Materials, 31*, 1904711.

77. Yangui, A., Garrot, D., Lauret, J. S., Lusson, A., Bouchez, G., Delaporte, E., Pillet, S., Bendeif, E. E., Castro, M., Triki, S., Abid, Y., & Boukheddaden, K. (2015). Optical investigation of broadband white-light emission in self-assembled organic-inorganic perovskite (C6H11NH3)2PbBr4. *Journal of Physical Chemistry C, 119*, 23638–23647.

78. Yu, D., Wang, P., Cao, F., Gu, Y., Liu, J., Han, Z., Huang, B., Zou, Y., Xu, X., & Zen, H. (2020). Two-dimensional halide perovskites as β-ray scintillator for nuclear radiation monitoring. *Nature Communications, 11*, 3395. https://doi.org/10.1038/s41467-020-17114-7.

79. Yu, J., Kong, J., Hao, W., Guo, X., He, H., Leow, W. R., Liu, Z., Cai, P., Qian, G., Li, S., & Chen, X. (2019). Broadband extrinsic self-trapped Exciton emission in Sn-doped 2D Lead-halide perovskites. *Advanced Materials, 1806385*(1–9), 31.

80. Yuan, Z., Zhou, C., Tian, Y., Shu, Y., Messier, J., Wang, J. C., van de Burgt, L. J., Kountouriotis, K., Xin, Y., Holt, E., Schanze, K., Clark, R., Siegrist, T., & Ma, B. (2017). One-dimensional organic lead halide perovskites with efficient bluish white-light emission. *Nature Communications, 8*, 14051. https://doi.org/10.1038/ncomms14051

81. Zhang, Y., Sun, R., Ou, X., Fu, K., Chen, Q., Ding, Y., Xu, L.-J., Liu, L., Han, Y., Malko, A. V., Liu, X., Yang, H., Bakr, O. M., Liu, H., & Mohammed, O. F. (2019a). Metal halide perovskite nanosheet for x-ray high-resolution scintillation imaging screens. *ACS Nano, 13*, 2520–2525.

82. Zhang, L., Wu, L., Wang, K., & Zou, B. (2019b). Pressure-induced broadband emission of 2D organic-inorganic hybrid perovskite (C$_6$H$_5$C$_2$H$_4$NH$_3$)$_2$PbBr$_4$. *Advancement of Science, 6*, 1801628.

83. Zhou, C., Lin, H., Tian, Y., Yuan, Z., Clark, R., Chen, B., van de Burgt, L. J., Wang, J. C., Zhou, Y., Hanson, K., Meisner, Q. J., Neu, J., Besara, T., Siegrist, T., Lambers, E., Djurovich, P. I., & Ma, B. (2018). Luminescent zero-dimensional organic metal halide hybrids with near-unity quantum efficiency. *Chemical Science, 9*, 586–593.

84. Zhou, C., Tian, Y., Wang, M., Rose, A., Besara, T., Doyle, N. K., Yuan, Z., Wang, J. C., Clark, R., Hu, Y., Siegrist, T., Lin, S., & Ma, B. (2017). Low-dimensional organic tin bromide perovskites and their photoinduced structural transformation. *Angewandte Chemie, International Edition, 56*, 9018–9022.

Emerging Lead-Halide Perovskite Semiconductor for Solid-State Detectors

Hsinhan Tsai, Jeremy Tisdale, Shreetu Shrestha, Fangze Liu, and Wanyi Nie

1 Introduction

Lead-halide perovskites are emerging semiconducting materials which have high crystalline, defect tolerance, excellent long carrier lifetime for the next-generation solution-processed optoelectronic devices [1–3]. Notably, the progress on photovoltaics made from perovskite thin films has been unprecedented and is still driving extensive research and development efforts in science and industry [4–11]. In this chapter, we will introduce lead-halide perovskites as a new class of materials for next-generation SSRDs. The three-dimensional perovskite structures follow the chemical formula ABX_3, where A is a cation, B is a metallic cation, and X is a halide anion. The structures can be tuned to adapt desired functionalities; in particular, heavy metals such as Pb, Sn, and Bi can be selected at the B-site. Along with halides at the X-site (Br, Cl, I), the average atomic number for perovskite compounds can be quite high, comparable to, or higher than the popular zinc-doped cadmium telluride (CZT), to efficiently stop high-energy radiation photons. The various perovskite structures available and expected properties are discussed in Sect. 2. The perovskites are unique because large, high-quality single crystals can be synthesized via low-cost methods, which facilitate large-scale fabrication (Sect. 3). Furthermore, the semiconducting properties are essential to consider for radiation detectors, where the product of the charge mobility and carrier lifetime should be maximized and bulk/surface defect density should be minimized. A detailed summary on the electronic properties of perovskite materials is shown in Sect. 4. Finally, pioneer works on the perovskite-based semiconducting radiation detectors have been reported in the literature and are reviewed in Sect. 5. This

H. Tsai · J. Tisdale · S. Shrestha · F. Liu · W. Nie (✉)
LANL (Los Alamos National Lab), Los Alamos, NM, USA
e-mail: jtisdale@lanl.gov; shreetu@lanl.gov; fliu@lanl.gov; wanyi@lanl.gov

© Springer Nature Switzerland AG 2022
K. Iniewski (ed.), *Advanced X-ray Detector Technologies*,
https://doi.org/10.1007/978-3-030-64279-2_2

section includes the reported progress on detectors made with both single-crystal and polycrystalline films. We also point out the potential challenges for detector performance optimizations.

2 Material Structure

Perovskite is a class of common mineral materials found on earth, with widely known perovskite oxides like $CaTiO_3$ and $SrTiO_3$. The structure was first discovered by a German mineralogist Gustav Rose in the Ural Mountains of Russia in 1839 and named after the Russian mineralogist Lev Perovski [12]. Perovskite materials have a general formula of ABX_3 with a three-dimensional structure, (Fig. 1a) and the lead-halide perovskites are a new class of materials in the perovskite family which are composed of an organic (or inorganic) cation (A-site), metal cores, (B-site), and halides (X-site). Recently, hybrid perovskites have demonstrated breakthroughs in photovoltaics, light-emitting diodes, and radiation detectors, which have driven intense research interests to push these materials toward next-generation optoelec-tronics. One of the most famous hybrid organic-inorganic perovskite materials is methylammonium lead triiodide ($CH_3NH_3PbI_3$ or $MAPbI_3$) that is widely used for photovoltaic devices [13].

To understand the perovskite structure, the Goldschmidt's tolerance factor (TF) [14–16] is often employed to determine the structural stability:

$$TF = \left(r_{A,\text{eff.}} + r_X\right)/\sqrt{2\left(r_B + r_X\right)}$$

where r_A and r_B are the radii of the A-site and B-site cations and r_X is the radius of the halides in the X-site. For hybrid lead-halide perovskites, a tolerance factor between 0.8 and 1.0 will lead to a stable 3D perovskite phase [15, 16]. An ideal

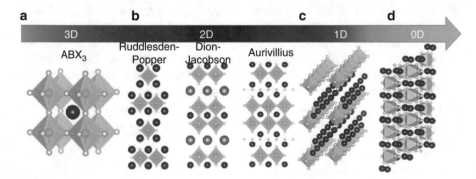

Fig. 1 Perovskite crystal structure. (**a**) Three-dimensional (ABX_3); (**b**) Two-dimensional (Ruddlesden-Popper, Dion-Jacobson, and Aurivillius phase); (**c**) one-dimensional structure; and (**d**) zero-dimensional structure

perovskite structure stabilizes in a symmetric cubic structure with a corner-sharing octahedron as the backbone. If the size of A is smaller than B, the tetragonal or orthorhombic structure will be favored. In addition, if A is larger, the crystal will adopt layered two-dimensional structures (Fig. 1b). For instance, Ruddlesden-Popper phases [17–21], Dion-Jacobson phases, [22, 23], and Aurivillius phases [24, 25] have been synthesized and reported. Furthermore, tuning the perovskite structure by chemical approach can further reduce the dimension down to zero-dimensional nanocrystals [26–30] (Fig. 1d), which provide more opportunities for desired functionalities and applications. For example, the low-dimensional perovskites are of particular interest for light-emitting applications and radiation scintillators.

The hybrid perovskite materials have many excellent properties such as direct band gaps, good absorption coefficients, low defect densities, long carrier lifetimes, and long diffusion lengths, which will be described in detail in the Electronic Properties section. More importantly, high-quality hybrid perovskite materials can be produced with solution processes. However, these properties highly depend on the crystal structures, material qualities, and processing conditions. Therefore, previous reports have developed many approaches for high-quality perovskite material growth for optoelectronic applications. For example, Bakr et al. reported using an inverse temperature method to grow single crystals for $MAPbI_3$, $MAPbBr_3$, and $MAPbCl_3$ from solution [31–33] that have low defect density. Liu et al. used a continuous flow method in a confined space to laterally grow large-area single-crystal wafers with high sensitivity photodetector demonstration [34]. The single-crystal growth kinetics are described in Sect. 3. Nie et al. pre-heated the solution and substrate to coat perovskite thin films via a "hot-casting" process, which can produce high-quality, mm-scale large-grain perovskite thin films for reproducible perovskite solar cells [35–39]. Apart from the 3D perovskites, significant efforts are devoted in 2D perovskite-based optoelectronics. Tsai et al. reported nearly single-crystal thin film growth with controllable orientation for 2D lead-halide perovskite materials [19, 40–42]. This is essential to facilitate the charge transport, to suppress the carrier recombination in the semiconductor, as well as to improve the device operation stability under light and humidity conditions [19, 41, 43]. The powerful synchrotron grazing incidence wide-angle X-ray scattering (GIWAXS) tool has also been applied to characterize the crystal orientation and crystallinity in 2D perovskite materials [19, 40, 42, 44, 45]. More recently, Chen et al. utilized strain engineering for epitaxial stabilization of halide perovskite thin films, which exhibit enhanced semiconductor device performance [6]. Furthermore, these studies have shown that high-quality perovskite growth greatly impacts the device stability and performance in both optoelectronics and radiation detection (Fig. 2).

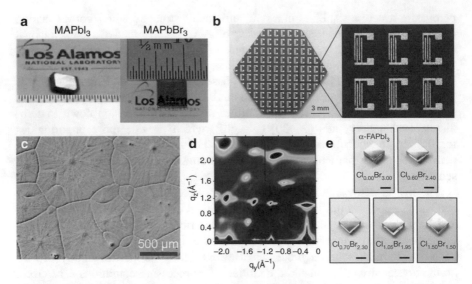

Fig. 2 Solution-processed high-quality perovskite materials. (**a**) Single-crystal MAPbI₃ and MAPbBr₃ and (**b**) perovskite wafer. (**c**) Hot-casting approach for large-grain perovskite thin film. (**d**) Nearly single-crystalline 2D perovskite thin film. (**e**) Strain engineering for epitaxial stabilization of halide perovskite thin film. Panel b is reproduced with permission from Ref. [34]. Panel d is reproduced with permission from Ref. [42]. Panel is reproduced with permission from Ref. [6]

3 Crystal Growth and Characterizations

3.1 Solution-Based Single-Crystal Growth

In 2015, high-quality, hybrid perovskite single-crystal growth was first published using a low-cost, rapid growth technique, now known as inverse temperature crystallization (ITC) [31–33]. Soon thereafter, large, ongoing research efforts have been placed in a variety of areas in solution-based hybrid perovskite single-crystal growth. This growth method was first used with the discovery of the inverse relationship between temperature and solubility with hybrid perovskite materials, i.e., as temperature of the solution increases, its solubility decreases, yielding crystalline precipitation (as shown in Fig. 3a). In this method, a solution is typically prepared using two to three precursors in a specific solvent, depending on the composition. Once the high concentration hybrid perovskite precursor solution is prepared (1 M) and complete dissolution, it is then placed in an oil bath on top of a hot plate. Over the course of minutes, hours, or days, depending on the type of growth desired, the temperature of the oil bath is slowly heated to begin nucleation. By controlling the ramp rate of the temperature, the nucleation and crystallization can be accurately controlled to ensure that one large, high-quality crystal is grown in a single vial [46, 47]. This process is displayed in Fig. 3b. Once crystals are

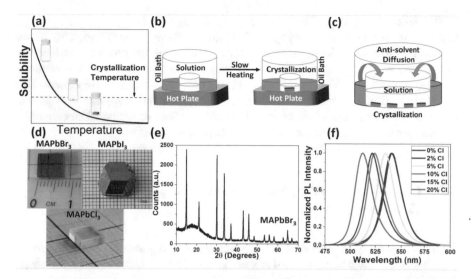

Fig. 3 (**a**) Graphical representation of common solubility versus temperature for hybrid perovskite materials for where crystallization begins. (**b**) Schematic of inverse temperature crystallization growth method. (**c**) Schematic of anti-solvent diffusion single-crystal growth method. (**d**) Pictures of different hybrid perovskite composition single crystals as grown via solution growth methods (MAPbBr$_3$, orange/MAPbI$_3$, black/MAPbCl$_3$, clear). (**e**) Example cubic phase characterization of MAPbBr$_3$ via powder X-ray diffraction. (**f**) Photoluminescence characterization data of MAPbBr$_{3-x}$Cl$_x$ to display accuracy in anionic doping

grown, they are carefully removed from solution and dried to remove excess solvent from the surface. There are other important single-crystal growth techniques used for hybrid perovskites that will be discussed in the next section.

There are many unique benefits to using solution-based growth techniques for single-crystalline hybrid perovskites. The first being simplicity of the growth cycle from beginning to end without the need of complex instrumentations (i.e., vacuum chamber, high-temperature oven). Most solution-based growth methods only require equipment such as hot plates, crystallization dishes, silicone oil, scintillation vials, etc. Because of the inexpensive equipment, along with inexpensive precursor materials and solvents, this type of single-crystal growth is orders of magnitude cheaper than other types of single crystals grown for similar applications. Most estimates for growth costs have been <$10/cm^3, as it has been shown that ultrahigh-purity precursors are not needed to grow high-quality single crystals capable of accurate semiconductor applications [48–53]. Also, an important note is that all of these solution-based single-crystal growth methods occur at low temperatures (less than 150 °C) as compared to other traditional high-temperature single-crystalline semiconductor growths (such as CdZnTe, which is grown around 800 °C) [54]. Not only does low-temperature processing reduce the cost of crystal growth, it also reduces the possibility of issues such as thermal stress cracking in the single crystals. Also, because the solution-based growth techniques depend on solubility manipulation, many have shown the ease of growing very large, high-quality single crystals by continuously flowing fresh solution during growth [46, 55]. Shape control has also been clearly demonstrated by modifying the growth container size and shape, which yields possibilities in growing specific dimensions needed for specific applications [32, 34]. It is important to note that because these growths take place in solution, it is possible to have issues with remaining solvent after growth on the surface or inside of the crystal. However, this can easily be taken care of by low-temperature baking of the single crystals post-growth.

3.2 Commonly Used Hybrid Perovskite Single-Crystal Growth Techniques

There are a variety of other efficient solution-based and non-solution-based growth techniques that have been successfully demonstrated to produce high-quality single crystals of hybrid and non-hybrid perovskite single crystals. These include anti-solvent vapor diffusion, temperature-lowering methods, slow solvent evaporation, and melt growths [33, 46, 56–59]. For solution-based methods, anti-solvent diffusion crystal growth is among one of the most common techniques used for hybrid perovskite single-crystal growth [33]. In this growth technique, the crystallization solution is prepared similarly to the process explained in the previous section for the ITC method. The difference between these two methods is the driving force used for nucleation/crystallization. Briefly, the solution container is placed inside of a second outside container. The outside container is filled with a known anti-solvent to the crystallization solution. The anti-solvent evaporates to form a vapor, which then diffuses into the crystallization solution to decrease the solubility. The solubility then slowly decreases to the point where nucleation and crystallization can occur. This process is depicted in Fig. 3c. At first, although high-quality single crystals were grown using this technique, it was proven to be difficult to grow large single crystals. However, over time this technique has been properly modified, by factors such as lowering the temperature to reduce the vaporization rate of anti-solvents or covering the crystallization solution with a mesh or foil with controlled pore size to reduce the diffusion rate [33]. With these modifications, the anti-solvent diffusion growth method has become viable in producing large, high-quality perovskite single crystals, while maintaining low cost of simplicity of growth.

Other less commonly used solution-based techniques have been demonstrated to produce high-quality perovskite single crystals. A top seeded solution growth (TSSG) technique has been shown to produce high-quality $MAPbI_3$ single crystals [56]. In this method, a previously grown seed crystal is placed on an air-cooled silicon substrate inside of a heated supersaturated solution to grow the seed into a much larger crystal. Although this method has produced high-quality crystals for semiconductor applications, the method requires more solution components such as acids, which increase the cost of this process, without significantly increasing the quality of product. Another method is solution temperature lowering where seed crystals are typically grown via rapid nucleation. Then seeds are chosen for further crystal growth, and the concentrations are gradually saturating as the temperature is slowly lowered. Both MAPbI$_3$ and MAPbBr$_3$ crystals were grown with large dimensions (up to 1 cm^3) using this method [60]. In order to grow large-scale single crystals with high quality, another method was developed [46]. In this method, the solution is prepared inside a closed oven, and the solution is allowed to warm very slowly over the course of 20 days (around 2 °C per day) undisturbed. Single-crystalline MAPbBr$_3$ was grown around 26 cm^3 using the low-temperature gradient crystallization method [46]. Although solution-based methods have mainly been used for perovskite single-crystal growth, some groups have turned to melt growth methods (or Bridgman-Stockbarger method) for high-quality inorganic perovskite crystal growth (e.g., CsPbBr$_3$) [59]. In this method, precursors are added into a tube furnace and melted together to form a molten solution. Then the furnace is used to slowly cool the solution to form high-quality single crystals with the dimensions of the ampule used for growth. This method provides accurate control to produce high-quality single crystals. However, in the realm of hybrid perovskites, this method cannot be used. This is because the precursors mixed together must have similar melting temperatures, as is the case for CsBr and PbBr$_2$. However, the organic components and inorganic components have vastly different melting temperatures for hybrid perovskites. This is why solution-based methods have been at the forefront for high-quality single-crystal growth of hybrid perovskites.

3.3 Single-Crystal Characterization

Once single crystals are grown and harvested from solution, it is important to characterize them to ensure high-quality, proper composition, and desired properties. There are abundant characterization techniques developed in the literature to understand properties of hybrid perovskite single crystals, but the most important properties belong to structural, optical, and electronic categories for semiconducting device development.

When single crystals are first grown and harvested from solution, a well-utilized tool to initially characterize the quality of the growth is visual inspection. As shown in Fig. 3d, when the bromide- and chloride-based crystals are grown, they are transparent. When visually inspecting the single crystal, it is important to look for various signs. The most common visual inspection for these types of single crystals among the community relies on transparency of the crystals without distortion of imagery or graphical lines placed underneath the crystal. However, if the crystal is not transparent, cloudy, or even slightly distorts the image below, one can speculate on the quality of the crystal growth through a relative concentration of defects or micro-cracks in the bulk of the crystals. In the case for nontransparent single crystals, like the black, iodide-based single crystals, transparency cannot be checked for quality. Instead, the quality inspection here comes from smoothness and mirror-like finishes of the crystal faces. Other visual inspections include checking for smooth parallel sides without curvature for cubic crystals and sharp edges.

The visual inspection is a first, simple step to initially speculate on the quality of the bulk material; however, phase and defect concentration are very important properties in these materials for semiconductor applications and must be quantified properly. Powder X-ray diffraction (XRD) is most commonly used to study the crystal structure of single crystals. In this method, the single crystal is finely ground into powder and an XRD pattern is collected. The data is then matched to the pure compound from a crystallographic database to ensure that the material is in the proper phase (i.e., $MAPbBr_3$ = cubic, $MAPbCl_3$ = cubic, $MAPbI_3$ = tetragonal for room temperature structures) [61]. Other X-ray diffraction techniques can be used to understand imperfections in the crystallographic structure throughout the material, such as Laue diffraction patterns to characterize the alignment of the crystal faces and high-resolution XRD (such as rocking curves) to see any deviations in the crystal structure [62]. Temperature-dependent XRD can also be used to understand structure transitions such as in $MAPbI_3$ which undergoes a transition from cubic to tetragonal phase around 330 K and again transition to the orthorhombic phase around 160 K [63].

The next common structural characterization deals with quantifying defect concentration and characterizing the types of defects that directly play an important role in semiconductor optoelectronic applications. There are a wide variety of experimental techniques including secondary ion mass spectrometry, positron annihilation spectroscopy, X-ray measurements, electrical measurements, optical measurements, and more [64]. Electrical measurements such as space-charge-limited current (SCLC) and Hall measurements can be also used to quantify defect concentrations [33, 63]. Meanwhile, other common techniques include time-of-flight and temperature-modulated techniques to determine the levels of traps within the band gaps, which help to understand the roles of traps in charge transport processes [65, 66].

The next techniques to continue characterizing as-grown hybrid perovskite single crystals use optical spectroscopy. Absorption is used to directly measure the band gap and states that below the gap, photoluminescence can be a close substitute to estimating the band gap and trap density of the material. When they are excited by an absorbable wavelength, the excited states relax from the conduction band to the valence band, and the corresponding energy is released via light such as the photoluminescence shown in Fig. 3f.

4 Electronic Properties

Unlike commonly used semiconductors, such as Si, Ge, and CZT, halide perovskite materials are ionic compounds, which give rise to unique electrical properties as well as challenges. The electronic band structure, charge transport, traps, and defects depend on the perovskite compositions and also the crystal growth conditions. Here,

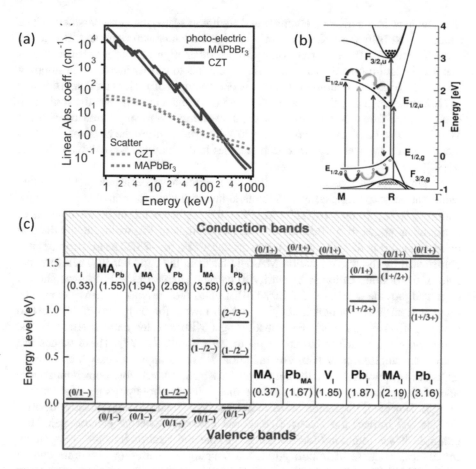

Fig. 4 Perovskite properties. (**a**) Calculated linear absorption cross-sectional coefficient of photo-electric effect and scatter process as a function of gamma-ray photon energy of CZT and MAPbBr$_3$ crystals. (**b**) Schematic of calculated electronic band diagram of high-temperature cubic phase of MAPbI$_3$ considering spin-orbit coupling [78]. (**c**) Calculated transition energy of defects in MAPbI$_3$ [79]

we discuss the general electronic properties of perovskites relevant to radiation detection applications. Lead-halide perovskites have a strong radiation stopping power due to high effective atomic number (consists of heavy elements like Pb, Cs, halides, etc.) and high density. Figure 4a shows the linear attenuation coefficients of hybrid perovskite materials, (MAPbI$_3$, CsPbBr$_3$, 2D perovskite, and double perovskite) which are comparable to conventional CZT and much higher than Si. Therefore, the quantum efficiency of a MAPbI$_3$ crystal with 2 mm thickness (density 4.15 g/cm^3) for a radiation photon with energy of 100 keV is about 99.9% (assuming the photon deposits all its energy in the crystal and all the electron-hole pairs generated are collected). Even at an energy of 300 keV, photoelectric absorption is the dominant interaction mechanism and Compton scattering is much smaller. The inorganic $CsPbBr_3$ perovskite has an even higher attenuation coefficient. Due to the tunable nature of the material compositions, other high atomic number elements can also be incorporated in the perovskite materials to further increase the stopping power. $Cs_2AgBiBr_6$ double perovskite is one example of high detection efficiency [67]. Moreover, using the optical band gap (E_g), one can estimate the ionization energy or the free electron-hole pair creation energy (W) using empirical relation $W \sim 3*E_g$. The band gap for common perovskites falls within the range of 1.1 eV to 2.5 eV, which translates to ionization energies in the range of 4.6 eV–6.83 eV [49, 51, 68].

Electronic band structures of halide perovskites have been thoroughly studied using various experimental and theoretical investigations for optoelectronic applications [69–73]. In the perovskite $MAPbI_3$, the valence band maximum consists of antibonding states of 5p orbitals of I and 6s orbitals of Pb, while the conduction band minimum is formed mostly by 6p orbitals of Pb [69–73]. The electronic states contributed by the organic cation MA^+ lie far from the band edges. Therefore, the role of the organic cation is primarily to balance charge in the PbI_6^{4-} octahedral cage and provide structural stability [74]. Due to the presence of heavy elements, there is significant spin-orbit coupling which lowers the band gap of $MAPbI_3$ to a value of ~1.62 eV [74]. By substituting I with smaller halides (Br, Cl), the band gap can be further increased up to 2.9 eV [69, 70, 75]. These values fall within the suitable band gaps for radiation detection applications which is large enough to suppress thermally generated background noise for room temperature operation, yet small enough to produce many electron-hole pairs per high-energy photon. However, the halide substitution has been found to significantly increase defects and impact the electrical response under an external electric field. For example, it was found in $MAPbBr_3$ that the hole and electron transports are widely imbalanced because of impurity-induced trapping or scattering. Accurate doping of Cl ($MAPbBr_{3-x}Cl_x$, where $x \leq 0.15$) in the perovskite structure can create more balanced charge transport and reduction of dark current, leading to enhanced radiation detection performance [50, 53, 76]. Cl doping in $MAPbI_3$ has also been demonstrated ($MAPbI_{3-x}Cl_x$) to enhance photovoltaic performance via tuning of transport and structural properties [77].

X-ray detection using double perovskite $Cs_2AgBiBr_6$ single crystals, which consists of heavy elements like Cs, Ag, and Bi and has an indirect band gap of 2.1 eV, has also been reported [67]. Therefore, halide perovskites offer incredible flexibility to tune the band gap, increase the ionizing radiation stopping power, and deliver an impressive detection limit of 59.7 nGy_{air} s^{-1}.

4.1 Charge Transport Properties

Due to strong spin-orbit coupling, the electron and hole effective mass in $MAPbI_3$ is reduced to 0.1 m_0−0.29 m_0 [80–82]. These effective mass values are comparable to Si. However, the mobilities of halide perovskites are lower than that of Si. Depending on the measurement technique used and the quality of the single crystal, the reported electron and hole mobility values for $MAPbI_3$ are in the range of 24–800 cm^2/Vs [32, 33, 56]. Moreover, temperature dependence of mobility measurements indicates that these modest mobility values are mainly due to the scattering of electrons with longitudinal-optical phonons [83]. Notably, the reported electron and hole mobilities are in the same range, which is consistent with effective mass [81]. A lower hole mobility in CZT is a major problem limiting the device performance [84]. Therefore, ambipolar charge transport in lead-halide perovskites is ideal for efficient extraction of both electrons and holes.

On the other hand, the perovskites have long carrier lifetimes and diffusion lengths [32, 33, 56, 85]. The reported lifetimes of $MAPbI_3$ are around 10–17 μs [51], while a record long lifetime of $CsPbBr_3$ that is more than 25 μs is reported [51]. From first principle calculations, the proposed mechanism for these long carrier lifetimes is attributed to Rashba splitting arising from strong spin-orbit coupling which results in spin-forbidden transitions [86]. The long lifetimes also make up for the modest mobility values and result in mobility-lifetime ($\mu\tau$) product comparable to CZT materials. The reported $\mu\tau$ product for $MAPbI_3$, $MAPbBr_3$, and $CsPbBr_3$ are 0.8×10^{-3} cm^2/V [51], 1.2×10^{-2} cm^2/V [49], and $1.33–1.69 \times 10^{-3}$ cm^2/V [51, 87], respectively. In addition, high $\mu\tau$ product of 1.8×10^{-2} cm^2/V and resistivity of 3.6×10^9 Ω cm [50] for mixed-halide $MAPbBr_{2.94}Cl_{0.06}$ have been obtained by doping $MAPbBr_3$ with Cl.

4.2 Defects

Low trap state densities in the range 10^9 cm^{-3}–10^{10} cm^{-3} has been reported for solution-grown halide perovskite single crystals [32, 33, 88]. Using Bridgman melt growth method, centimeter-sized $CsPbBr_3$ crystals with impurity levels below 10 p.p.m. for total 69 elements have also been obtained [51]. The absorption spectra of perovskites show sharp onsets with Urbach energy as small as 15 meV even for polycrystalline $MAPbI_3$ films [89, 90]. Single-crystalline Si for comparison has an Urbach energy of 11 meV [89]. These experimental results are consistent with theoretical studies which indicate that dominant intrinsic defects are shallow traps, while deep defects located near the middle of the band gap are unlikely to form due to high enthalpy of formation [91, 92]. Figure 4c shows the formation energy and energy level of possible defect states in $MAPbI_3$ [79]. It can be seen that defects with low formation energy such as interstitials (I_i, MA_i), vacancies (V_{MA}, V_{Pb}, V_I), and antisites (MA_{Pb}, Pb_{MA}, MA_I) have shallow transition levels. Similarly, defects with higher formation energies such as I_{MA}, I_{Pb}, Pb_I, and Pb_i have higher transition levels. This "defect tolerance" in perovskites has been attributed to the strong lone pair in Pb s orbital, antibonding coupling in I p orbital, and the high iconicity [91, 92]. Due to the low density of electrically active defects in halide perovskites, charge carriers excited by radiation can be extracted without being trapped. Moreover, the defect tolerance nature of perovskites also gives rise to radiation hardness [93, 94]. Yang et al. have demonstrated perovskite solar cells which retain more than 96% of their initial efficiency under continuous irradiation of gamma-rays from ^{137}Cs benchtop irradiator for accumulated dose of 2.3 Mrad [94].

4.3 Surface and Interface Effects

Different surface termination and facets can have different electronic levels and stability. Density-functional theory (DFT) studies have predicted that the energetically favorable (110)/(001) facets and MAI termination is more stable than the PbI$_2$ termination in MAPbI$_3$ [95, 96]. MAI-terminated surface also has higher valence and conduction band edges than PbI$_2$ [96]. However, the different surfaces retain the defect tolerant nature of the bulk and hence have low mid-gap defect states [95]. Similar to bulk defects, surface defects are also electronically benign. Using transient reflectance spectroscopy, the surface recombination velocity in perovskite single crystal was measured to be ~10^3 cm s^{-1} [97, 98]. This value is orders of magnitude lower than that of conventional inorganic semiconductors.

Uratani et al. used first principle calculations to investigate possible surface defects in MAPbI$_3$ under different growth conditions (I-rich, Pb-rich, and moderate) [99]. Under the I-rich condition, excessive I atoms on flat and vacant surfaces can trap charge carriers [99]. Therefore, many passivation strategies such as using organic polymer interlayers have been successfully used to suppress recombination at the surface [100].

5   Solid-State Detector

5.1   Introduction

Most of the reported perovskite solid-state radiation detectors (PSSRDs) have similar structures to photodetectors where a *p-i-n* device configuration is often chosen. It comprises a highly resistive intrinsic semiconductor and is sandwiched between *p* and *n* contacts, as shown in Fig. 5a [49]. For applications such as X-ray detectors, the working principles of PSSRD are like photodetector operation probing in either photocurrent or photovoltage mode, where free carriers are excited by multiple X-ray photons and are subsequently collected by the electrodes and analyzed by an external circuit [49]. Due to the high absorption coefficient of lead-halide perovskites, thin film devices with thicknesses from microns to millimeters are often used [101].

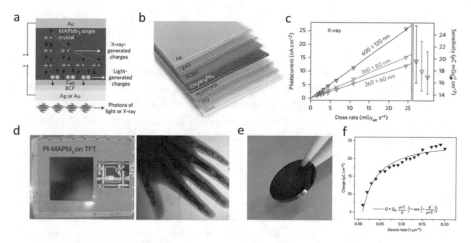

Fig. 5 X-ray detectors based on lead-halide perovskites. (**a**) Single-crystal MAPbBr₃ radiation detector structure. The charge generation regions are located close to the surface for visible light excitation and deeper inside the single crystal for X-ray excitation, respectively. (**b**) Schematic of layer stacking of the MAPbI₃-based *p-i-n* photodiode, the MAPbI₃ layer is 260 ± 60 nm. (**c**) Averaged short-circuit X-ray photocurrent as function of dose rate. Inset: Sensitivity normalized to the active volume for MAPbI₃ layers with different thicknesses. (**d**) The left panel shows an optical image of spin-cast polyimide-MAPbI₃ composite on an a-Si:H TFT backplane. The inset shows a single-pixel structure of TFT (scale bar 30 μm) in which the collection electrode (white outline) is connected to the drain contact of the TFT through a circular pad. The right panel shows a hand phantom X-ray image obtained (using 100 kVp and 5 mGy$_{air}$ s^{-1} for 5 ms exposure, resulting in a dose of 25 μ Gy$_{air}$ and a bias voltage of 50 V). (**e**) Image of a freestanding sintered MAPbI₃ wafer (1/2 inch × 1 mm). (**f**) Extracted charge versus electric field from a MAPbI₃ wafer-based X-ray detector shown in (**e**). (**a**) is adapted with permission from Ref [49]. (**b**) and (**c**) are adapted with permission from Ref [48]. (**d**) is adapted with permission from Ref [101]. (**e**) and (**f**) are adapted with permission from Ref [68]

For other applications, such as photon counting and gamma-spectroscopy, the PSSRDs need to count single X- or γ-ray photons, that is, free carriers excited by each photon are collected by the electrodes and are converted into an electrical pulse signal measured by an external circuit. The corresponding charge density is proportional to the height of the pulse. Specifically, PSSRDs require several important characteristics to successfully construct the energy-resolved spectrum. First, the dark current of the detector needs to be minimized so that the electronics can resolve pulse signals at all amplitudes for spectrum construction. Second, to collect all the ionized charges caused by a single incident gamma-ray photon, a large electric field is required during detector operation. Third, the semiconductor must consist of heavy elements to efficiently interact with X- and γ-ray photons, while maintaining excellent transport properties (i.e., mobility-lifetime ($\mu\tau$) product) in order to maximize the photoelectric peak that contributes to the spectrum [59].

5.2   X-Ray Detectors

Several advantages including high absorption coefficients, long carrier diffusion lengths [85], high mobilities, and solution-processible techniques [13, 19, 35, 102, 103], described in the Electronic Properties section, make lead-halide perovskites promising candidates for next-generation SSRD. Table 1 presents the detector characteristics of several lead-halide perovskites and traditional semiconductors, such as amorphous selenium (a-Se) and mercury iodide (HgI).

Yakunin et al. first reported X-ray detector based on solution-processed MAPbI$_3$ thin films (Fig. 5b) [48]. The sensitivity of 25 μC Gy$_{air}$$^{-1}$ cm^{-2} is commensurate with the current solid-state detectors (Fig. 5c) [101]. The sensitivity of MAPbI$_3$ detectors is further improved to over 2000 μC Gy$_{air}$$^{-1}$ cm^{-2} by fabricating high-quality polycrystalline MAPbI$_3$ wafers with millimeter thickness using a mechanical sintering process (Fig. 5e and f) [68]. Moreover, further developments have been made in X-ray imaging based on lead-halide perovskites. Benefiting from the high sensitivity, the X-ray dose can be reduced by at least an order of magnitude compared with conventional selenium detectors (Fig. 5d) [101, 108]. Due to the cubic crystal shape of the MAPbBr$_3$ single crystal, it has been commonly used and is also proven as excellent X-ray detector. Wei et al. demonstrated the first single-crystalline MAPbBr$_3$ X-ray detector with a sensitivity of 80 μC Gy$_{air}$$^{-1}$ cm^{-2} and a high $\mu\tau$ product of 1.2×10^{-2} cm^2 V^{-1} [49]. The sensitivity is further improved up to 529 μC Gy$_{air}$$^{-1}$ cm^{-2} by tailoring the interface state density [104].

While 3D perovskites in the ABX$_3$ form have been extensively studied, another group of perovskite materials, namely, Ruddlesden-Popper (RP) layered 2D perovskites, also attracted a lot of interest. These 2D perovskites are more environmentally stable due to the existence of protecting organic spacers [17, 18]. Recently, Tsai et al. demonstrated high sensitivity and robust thin film X-ray detectors using 2D RP perovskites [42]. Different from previous detectors that usually require high electric field to collect X-ray-induced free carriers, these detectors generate both short-circuit current (primary current) and open-circuit voltage from X-ray photons, providing an alternative detecting mechanism.

5.3 γ-Ray Detectors

Besides low-energy X-ray detection below 100 keV, lead-halide perovskites have also proven to be feasible for high-energy γ-ray spectroscopy in the range of 0.1 to 1 MeV. Unlike X-ray detectors that focus more on the sensitivity and detection limit, γ-ray detectors are typically used for energy-resolved spectroscopy. It also requires large-volume (several cm^3 to hundreds cm^3) single crystal to improve the efficiency of photoelectric absorption [50, 51]. Table 2 summarizes device performance of conventional and halide perovskite detectors.

Table 1 Device performance of X-ray detectors based on different materials

Material	Atomic number	Thickness (μm)	Resistivity (Ω cm)	Electric field (V cm^{-1})	Attenuation coefficient at 10 keV (cm^{-1})	Sensitivity (μCGy$_{air}^{-1}$ cm^{-2})	$\mu\tau$ product (cm^2 V^{-1})	Detection low limit (μGy$_{air}$ s^{-1})	References
MAPbI$_3$	53, 82	830	10^8–10^9	240	567	10^4	8×10^{-4}	–	[51, 101]
MAPbBr$_3$	35, 82	2000	1.7×10^7	50	291	529	1.2×10^{-2}	0.5	[49, 104]
(BA)$_2$(MA)$_2$Pb$_3$I$_{10}$	53, 82	0.47	–	0	500	13	–	10	[42]
a-Se	34	200	$10^{14\text{-}15}$	10^4	200	20	10^{-6}	5.5	[105, 106]
HgI	53, 80	300	10^{13}	2×10^3	922	10^3	1.5×10^{-5}	–	[106, 107]

Table 2 Device performance of γ-ray detectors based on different materials

Material	Atomic number	Resistivity (Ω cm)	Thickness (mm)	Electric field (V cm^{-1})	Attenuation coefficient at 662 keV (cm^{-1})	Band gap (eV)	Energy resolution (%)	μτ product (cm^2 V^{-1})	References
MAPbBr$_3$	35, 82	1.7×10^7	5	1.8	0.07	2.2	6.5 (662 keV)	1.8×10^{-2}	[50]
MAPbI$_3$	53, 82	10^8–10^9	1.5	46	0.08	1.5	6.8 (122 keV)	8×10^{-4}	[51]
CsPbBr$_3$	35, 55, 82	10^9–10^{11}	1.24	726	0.09	2.3	3.8 (662 keV)	10^{-3}	[59, 87]
CdZnTe	30, 48, 52	10^{10}	10	200	0.04	1.6	0.8 (662 keV)	1.2×10^{-2}	[109, 110]
CdTe	48, 52	10^9	0.5	600	0.05	1.4	12 (662 keV)	3×10^{-3}	[84, 111]
TlBr	35, 81	10^{10}	1	200	0.23	2.7	3.3 (662 keV)	3×10^{-5}	[112–114]

Several initial studies have demonstrated prototype perovskite detectors. For example, Yakunin et al. reported the first energy-resolved gamma-ray spectrum from ^{241}Am at 59.6 keV with 35% energy resolution using solution-grown FAPbI$_3$ single crystal [115].Wei et al. used dopant-compensated MAPbBr$_{3-x}$Cl$_x$ crystals to increase the bulk resistivity and achieved below 10% energy resolution at 662 keV from ^{137}Cs radiation at a field of 18 V/cm [50]. He et al. achieved energy resolution of 12% at 59.6 keV from ^{241}Am and 6.8% at 122 keV from ^{57}Co with MAPbI$_3$ single crystals at −2 °C using asymmetric Schottky-type electrodes [51].

Despite the rapid progress, the organic-inorganic perovskites suffer from the long-term environmental instability due to the hydroscopic nature of the organic cations. On the contrary, pure inorganic perovskite CsPbBr$_3$ provides better stability due to the lack of organic molecules. He et al. have shown CsPbBr$_3$ single-crystal gamma-ray detector with energy resolution approaching the industrial standard CZT detectors when operating under 7×10^3 V/cm [59]. While these demonstrations are highly promising, their working conditions and reported energy detection ranges are very different despite all claimed high-energy resolution <10%.

Recently, Liu et al. conducted a detailed study of the working and design principle of perovskite γ-ray detector [116]. It has been found that the dark current is mainly contributed by electrons and the slow pulse generation is caused by ion migration. Therefore, creating p-type contact by using high work function contact and reducing the ion migration by lowering the temperature or using inorganic cation are the key to improve the detector performance (Fig. 6).

5.4 Scintillators

Apart from solid-state detectors, scintillators are also an important category of X-ray and γ-ray detectors. Unlike solid-state detectors which directly convert radiation into currents, scintillators convert high-energy photons into ultraviolet-visible photons which can be further detected by conventional photodetectors, usually photomultiplier tubes (PMTs) [117]. Conventional scintillator materials, such as Bi$_4$Ge$_3$O$_{12}$ (BGO), YAlO$_3$:Ce, CsI:Tl, and LaCl$_3$:Ce, usually require high-temperature synthesis, and their emission energies are not tunable to match the response of PMTs. Lead-halide perovskites, on the other hand, have several advantages including low-temperature solution process, tunable emission, and high sensitivity. Recently, there has been significant progress in scintillators based on lead-halide perovskites. Table 3 summarizes the properties of common scintillator materials and lead-halide perovskites.

Two groups, Birowosuto et al [118] and Mykhaylyk et al [119], studied the scintillation properties of organic-inorganic lead-halide perovskite single crystals, MAPbI$_3$, MAPbBr$_3$, and 2D perovskite crystal (EDBE) PbCl$_4$ (EDBE = 2,2′-(ethylenedioxy)bis(ethylammonium)). They found these perovskite crystals exhibit high light yield of over 1.5×10^6 photon/MeV for MAPbI$_3$ and MAPbBr$_3$ at cryogenic

Table 3 Material properties of different scintillators

Material	Atomic number	Emission (nm)	Light yield (10–77 K) (photon/MeV)	Light yield (293 K) (photon/MeV)	Absorption coefficient at 662 keV (cm^{-1})	Decay time (ns)	References
MAPbBr$_3$	35, 82	550	152000 @ 10 K, 90000 @ 77 K	<1000	0.07	0.1, 1	[118, 119]
MAPbI$_3$	53, 82	750	296000 @ 10 K	<1000	0.08	4.3	[118]
CsPbBr$_{3-x}$Cl$_x$-QD	35, 55, 82	420–700	–	~30000	0.09	44.6	[120]
CsI-Tl	55, 81	550	–	~60000	0.05	600, 3500	[117]
BGO	32, 83	480	–	8200	0.22	300	[117]

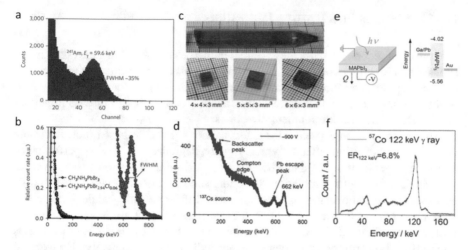

Fig. 6 γ-Ray detectors based on single-crystal lead-halide perovskites. (**a**) Energy-resolved spectrum of ^{241}AM recorded with a FAPBI$_3$ single crystal. (**b**) ^{137}Cs energy spectrum obtained by CH$_3$NH$_3$PbBr$_{2.94}$Cl$_{0.06}$ with energy resolution of 6.5%. (**c**) As-grown CsPbBr$_3$ single-crystal ingot with a diameter of 11 mm and the single-crystal wafers with different sizes. (**d**) Energy-resolved spectrum of ^{137}Cs γ-ray source from CsPbBr$_3$ detector with a shaping time of 0.5 μs. The dimension of the detector is $4 \times 2 \times 1.24$ mm^3. (**e**) Schematic diagram of Schottky-type MAPbI$_3$ detector (left) and corresponding energy level diagram (right) with asymmetrical electrode design. (**f**) Energy-resolved spectrum by Schottky-type MAPbI$_3$ detector at 2 °C and −70 V bias under ^{57}Co γ-ray source at 2 °C with a characteristic energy of 122 keV. The dimension of the detector is $4 \times 3 \times 1.52$ mm^3. (**a**) is adapted with permission from Ref [115]. (**b**) is adapted with permission from Ref [50]. (**c**) and (**d**) are adapted with permission from Ref [59]. (**e**) and (**f**) are adapted with permission from Ref [116]

temperature of 10 K. The fast decay of a few nanoseconds enables fast timing measurements.

Comparing with bulk crystals, perovskite nanocrystals have many advantages such as stability, tunable emission, bottom-up synthesis, and large-area fabrication. Chen et al. synthesized CsPbBr$_{3-x}$Cl$_x$ nanocrystals with average size of 9.6 nm and tunable emission from 420 nm to 700 nm and demonstrated first report of X-ray scintillator based on perovskite nanocrystals [120]. The nanocrystal thin film with thickness of 0.1 mm demonstrated optical sensitivity comparable (60%) to CsI:Tl scintillator at 10 keV photon energy. They also demonstrated X-ray imaging by integrating CsPbBr$_3$ nanocrystals to a pixelated α-silicon photodiode array. Thanks to the high sensitivity, the required X-ray dose was 400 times lower than typical medical imaging doses.

6   Challenges and Future Directions

The previous sections discuss the essential properties of the halide perovskite materials for high-energy radiation detectors, including structure characterization, crystal growth methods, and electronic properties, which led to pioneer demonstrations of X-ray sensors and gamma-spectroscopy. In this section, we will discuss important challenges and future development directions.

The main challenge recognized for halide perovskite-based optoelectronic devices is polarization effects, which rapidly degraded the device performance overtime under high external electric fields. Polarization is a known issue in other semiconductors used for radiation detection, such as TlBr detector [121]. Halide perovskites are ionic materials, and ions can diffuse (mostly halides) via defect sites under large electric fields [122, 123] which presents an essential and fatal problem in radiation detectors. Unlike photovoltaic devices (e.g., no external bias is applied), high-energy radiation detectors generally required relatively high external voltages to extract ionized carriers efficiently. In particular, the detectors for gamma-spectroscopy described in the Solid-State Detector section necessitate thick crystals with large cross section to fully stop high-energy gamma-ray photons, and single photon counting operation mode is required to construct energy-resolved gamma-ray spectrum. Therefore, collecting every ionized charge in solid-state gamma-ray detector becomes critical, and large external voltages (above 500 volts across a 5-mm-thick crystal or 1000 V/cm) are needed. However, recent works by Liu et al. [116] and Tisdale et al. [53] have identified the problems of the increasement in background noise and baseline fluctuation when the single-crystal MAPbBr$_3$-based detectors were operated under electric fields higher than 250 V/cm. This is far from the required electric fields to collect all the ionized charges, and the limitation in applied electric field is partially contributed to the poorly energy-resolved spectra. After careful and detailed detector characterizations, Tisdale et al. attributed such instability under high electric field to the field-induced ion migration that created conducting channels through the device, causing a significant increase in dark current. This effect can be circumvented by constantly switching the voltage polarity applied on the gamma-ray detector while collecting both of the positive and negative pulses. As a result, the detector could be operated under 500 V/cm and output an energy resolution of ~30% for the 662 keV gamma-ray photon emitter from a ^{137}Cs source. However, the intrinsic material instability induced by the electric field still needs to be addressed and will require further improvements such as material structure and composition optimization and interface passivation approaches. Fortunately, suppressing ion migration in perovskite materials has been widely investigated in the field of perovskite photovoltaics. For instance, since the ion migration often occurs via defect sites, bulky organic cations used for fabricating 2D perovskites have been incorporated into 3D perovskites to passivate the defects and facilitate crystal growth [103, 124]. This strategy has proven to be effective for photovoltaic devices and light-emitting diodes. Furthermore, surface passivation strategies have been proposed to suppress the interfacial ion migration that can cause device polarizations [125, 126]. Utilizing potassium iodide on the perovskite surface has been shown useful for maximizing and stabilizing the photoluminescence [127]. These surface passivation approaches are all promising strategies to be employed for radiation detector development.

References

1. Fu, Y., Zhu, H., Chen, J., Hautzinger, M. P., Zhu, X. Y., & Jin, S. (2019). *Nature Reviews Materials, 4*, 169.
2. Tennyson, E. M., Doherty, T. A. S., & Stranks, S. D. (2019). *Nature Reviews Materials, 4*, 573.
3. Wei, H., & Huang, J. (2019). *Nature Communications, 10*, 1066.
4. Leijtens, T., Bush, K. A., Prasanna, R., & McGehee, M. D. (2018). *Nature Energy, 3*, 828.
5. Li, Z., Klein, T. R., Kim, D. H., Yang, M., Berry, J. J., van Hest, M. F. A. M., & Zhu, K. (2018). *Nature Reviews Materials, 3*, 18017.
6. Chen, Y., et al. (2020). *Nature, 577*, 209.
7. Hou, Y., et al. (2020). *Science, 367*, 1135.
8. Kim, D., et al. (2020). *Science, 368*, 155.
9. Lei, Y., et al. (2020). *Nature, 583*, 790.
10. Xu, J., et al. (2020). *Science, 367*, 1097.
11. Zhao, X., & Tan, Z.-K. (2020). *Nature Photonics, 14*, 215.
12. Rose, G. (1839). *Annalen der Physik, 124*, 551.
13. Lee, M. M., Teuscher, J., Miyasaka, T., Murakami, T. N., & Snaith, H. J. (2012). *Science, 338*, 643.
14. Goldschmidt, V. M. (1926). *Naturwissenschaften, 14*, 477.
15. Kieslich, G., Sun, S., & Cheetham, A. K. (2015). *Chemical Science, 6*, 3430.
16. Li, Z., Yang, M., Park, J.-S., Wei, S.-H., Berry, J. J., & Zhu, K. (2016). *Chemistry of Materials, 28*, 284.
17. Smith, I. C., Hoke, E. T., Solis-Ibarra, D., McGehee, M. D., & Karunadasa, H. I. (2014). *Angewandte Chemie International Edition, 53*, 11232.
18. Cao, D. H., Stoumpos, C. C., Farha, O. K., Hupp, J. T., & Kanatzidis, M. G. (2015). *Journal of the American Chemical Society, 137*, 7843.
19. Tsai, H., et al. (2016). *Nature, 536*, 312.
20. Soe, C. M. M., et al. (2019). *Proceedings of the National Academy of Sciences, 116*, 58.
21. Ramos-Terrón, S., Jodlowski, A. D., Verdugo-Escamilla, C., Camacho, L., & de Miguel, G. (2020). *Chemistry of Materials, 32*, 4024.
22. Mao, L., Ke, W., Pedesseau, L., Wu, Y., Katan, C., Even, J., Wasielewski, M. R., Stoumpos, C. C., & Kanatzidis, M. G. (2018). *Journal of the American Chemical Society, 140*, 3775.
23. Ke, W., Mao, L., Stoumpos, C. C., Hoffman, J., Spanopoulos, I., Mohite, A. D., & Kanatzidis, M. G. (2019). *Advanced Energy Materials, 9*, 1803384.
24. Soe, C. M. M., et al. (2017). *Journal of the American Chemical Society, 139*, 16297.
25. Mao, L., Stoumpos, C. C., & Kanatzidis, M. G. (2019). *Journal of the American Chemical Society, 141*, 1171.
26. Dou, L., et al. (2015). *Science, 349*, 1518.
27. Ning, Z., et al. (2015). *Nature, 523*, 324.
28. Protesescu, L., Yakunin, S., Bodnarchuk, M. I., Krieg, F., Caputo, R., Hendon, C. H., Yang, R. X., Walsh, A., & Kovalenko, M. V. (2015). *Nano Letters, 15*, 3692.
29. Li, L., et al. (2015). *Nature Nanotechnology, 10*, 608.
30. Fedorovskiy, A. E., Drigo, N. A., & Nazeeruddin, M. K. (2020). *Small Methods, 4*, 1900426.
31. Maculan, G., et al. (2015). *The Journal of Physical Chemistry Letters, 6*, 3781.

32. Saidaminov, M. I., et al. (2015). *Nature Communications, 6*, 7586.
33. Shi, D., et al. (2015). *Science, 347*, 519.
34. Liu, Y., Zhang, Y., Yang, Z., Yang, D., Ren, X., Pang, L., & Liu, S. (2016). *Advanced Materials, 28*, 9204.
35. Nie, W., et al. (2015). *Science, 347*, 522.
36. Tsai, H., Nie, W., Cheruku, P., Mack, N. H., Xu, P., Gupta, G., Mohite, A. D., & Wang, H.-L. (2015). *Chemistry of Materials, 27*, 5570.
37. Nie, W., et al. (2016). *Nature Communications, 7*, 11574.
38. Tsai, H., Nie, W., Lin, Y.-H., Blancon, J. C., Tretiak, S., Even, J., Gupta, G., Ajayan, P. M., & Mohite, A. D. (2017). *Advanced Energy Materials, 7*, 1602159.
39. Nie, W., et al. (2018). *Advanced Materials, 30*, 1703879.
40. Tsai, H., et al. (2018). *Advanced Materials, 30*, 1704217.
41. Tsai, H., Liu, C., Kinigstein, E., Li, M., Tretiak, S., Cotlet, M., Ma, X., Zhang, X., & Nie, W. (2020). *Advanced Science, 7*, 1903202.
42. Tsai, H., Liu, F., Shrestha, S., Fernando, K., Tretiak, S., Scott, B., Vo, D. T., Strzalka, J., & Nie, W. (2020). *Science Advances, 6*, eaay0815.
43. Tsai, H., et al. (2018). *Nature Communications, 9*, 2130.
44. Schlipf, J., & Müller-Buschbaum, P. (2017). *Advanced Energy Materials, 7*, 1700131.
45. Chen, A. Z., Shiu, M., Ma, J. H., Alpert, M. R., Zhang, D., Foley, B. J., Smilgies, D.-M., Lee, S.-H., & Choi, J. J. (2018). *Nature Communications, 9*, 1336.
46. Liu, Y., et al. (2019). *Materials Today, 22*, 67.
47. Zhang, Y., et al. (2020). *Nature Communications, 11*, 2304.
48. Yakunin, S., et al. (2015). *Nature Photonics, 9*, 444.
49. Wei, H., et al. (2016). *Nature Photonics, 10*, 333.
50. Wei, H., DeSantis, D., Wei, W., Deng, Y., Guo, D., Savenije, T. J., Cao, L., & Huang, J. (2017). *Nature Materials, 16*, 826.
51. He, Y., et al. (2018). *ACS Photonics, 5*, 4132.
52. Tisdale, J. T., et al. (2018). *CrystEngComm, 20*, 7818.
53. Tisdale, J. T., Yoho, M., Tsai, H., Shrestha, S., Fernando, K., Baldwin, J. K., Tretiak, S., Vo, D., & Nie, W. (2020). *Advanced Optical Materials, 8*, 2000233.
54. Su, C.-H., & Lehoczky, S. L. (2008). NASA/Marshall Space Flight Center. Huntsville.
55. Liu, Y., et al. (2015). *Advanced Materials, 27*, 5176.
56. Dong, Q., Fang, Y., Shao, Y., Mulligan, P., Qiu, J., Cao, L., & Huang, J. (2015). *Science, 347*, 967.
57. Dang, Y., Zhou, Y., Liu, X., Ju, D., Xia, S., Xia, H., & Tao, X. (2016). *Angewandte Chemie International Edition, 55*, 3447.
58. Konstantakou, M., Perganti, D., Falaras, P., & Stergiopoulos, T. (2017). *Crystals, 7*, 291.
59. He, Y., et al. (2018). *Nature Communications, 9*, 1609.
60. Su, J., Chen, D. P., & Lin, C. T. (2015). *Journal of Crystal Growth, 422*, 75.
61. Dang, Y., Ju, D., Wang, L., & Tao, X. (2016). *CrystEngComm, 18*.
62. Tisdale, J. T. (2018). PhD dissertation, University of Tennessee.
63. Whitfield, P. S., Herron, N., Guise, W. E., Page, K., Cheng, Y. Q., Milas, I., & Crawford, M. K. (2016). *Scientific Reports, 6*, 35685.
64. Alkauskas, A., McCluskey, M. D., & V, C. G. (2016). d. Walle. *Journal of Applied Physics, 119*, 181101.
65. Pospisil, J., Zmeskal, O., Nespurek, S., Krajcovic, J., Weiter, M., & Kovalenko, A. (2019). *Scientific Reports, 9*, 3332.
66. Musiienko, A., et al. (2019). *Energy & Environmental Science, 12*, 1413.
67. Pan, W., et al. (2017). *Nature Photonics, 11*, 726.
68. Shrestha, S., et al. (2017). *Nature Photonics, 11*, 436.
69. Mosconi, E., Amat, A., Nazeeruddin, M. K., Grätzel, M., & De Angelis, F. (2013). *The Journal of Physical Chemistry C, 117*, 13902.
70. Tanaka, K., Takahashi, T., Ban, T., Kondo, T., Uchida, K., & Miura, N. (2003). *Solid State Communications, 127*, 619.

71. Even, J., Pedesseau, L., Jancu, J.-M., & Katan, C. (2013). *The Journal of Physical Chemistry Letters, 4*, 2999.
72. Whalley, L. D., Frost, J. M., Jung, Y. K., & Walsh, A. (2017). *The Journal of Chemical Physics, 146*, 220901.
73. Yin, W.-J., Shi, T., & Yan, Y. (2015). *The Journal of Physical Chemistry C, 119*, 5253.
74. Brivio, F., Butler, K. T., Walsh, A., & van Schilfgaarde, M. (2014). *Physical Review B, 89*.
75. Caputo, M., et al. (2019). *Scientific Reports, 9*, 15159.
76. Rybin, N., et al. (2020). *Chemistry of Materials.*
77. Colella, S., et al. (2013). *Chemistry of Materials, 25*, 4613.
78. Even, J., Pedesseau, L., & Katan, C. (2014). *The Journal of Physical Chemistry C, 118*, 11566.
79. Yin, W. J., Shi, T., & Yan, Y. (2014). *Advanced Materials, 26*, 4653.
80. Giorgi, G., Fujisawa, J., Segawa, H., & Yamashita, K. (2013). *Journal of Physical Chemistry Letters, 4*, 4213.
81. Umari, P., Mosconi, E., & De Angelis, F. (2014). *Scientific Reports, 4*, 4467.
82. Miyata, A., Mitioglu, A., Plochocka, P., Portugall, O., Wang, J. T.-W., Stranks, S. D., Snaith, H. J., & Nicholas, R. J. (2015). *Nature Physics, 11*, 582.
83. Herz, L. M. (2017). *ACS Energy Letters, 2*, 1539.
84. Watanabe, T. T. S. (2001). *IEEE Transactions on Nuclear Science, 48*, 950.
85. Stranks, S. D., Eperon, G. E., Grancini, G., Menelaou, C., Alcocer, M. J. P., Leijtens, T., Herz, L. M., Petrozza, A., & Snaith, H. J. (2013). *Science, 342*, 341.
86. Zheng, F., Tan, L. Z., Liu, S., & Rappe, A. M. (2015). *Nano Letters, 15*, 7794.
87. Stoumpos, C. C., et al. (2013). *Crystal Growth & Design, 13*, 2722.
88. Suarez, B., Gonzalez-Pedro, V., Ripolles, T. S., Sanchez, R. S., Otero, L., & Mora-Sero, I. (2014). *Journal of Physical Chemistry Letters, 5*, 1628.
89. De Wolf, S., Holovsky, J., Moon, S. J., Loper, P., Niesen, B., Ledinsky, M., Haug, F. J., Yum, J. H., & Ballif, C. (2014). *Journal of Physical Chemistry Letters, 5*, 1035.
90. Samiee, M., et al. (2014). *Applied Physics Letters, 105*, 153502.
91. Yin, W.-J., Shi, T., & Yan, Y. (2014). *Applied Physics Letters, 104*, 063903.
92. Kim, J., Lee, S. H., Lee, J. H., & Hong, K. H. (2014). *Journal of Physical Chemistry Letters, 5*, 1312.
93. Lang, F., Shargaieva, O., Brus, V., Neitzert, H. C., Rappich, J., & Nickel, N. H. (2018). *Advanced Materials, 30*.
94. Yang, S., Xu, Z., Xue, S., Kandlakunta, P., Cao, L., & Huang, J. (2019). *Advanced Materials, 31*, e1805547.
95. Haruyama, J., Sodeyama, K., Han, L., & Tateyama, Y. (2014). *Journal of Physical Chemistry Letters, 5*, 2903.
96. Quarti, C., De Angelis, F., & Beljonne, D. (2017). *Chemistry of Materials, 29*, 958.
97. Yang, Y., Yan, Y., Yang, M., Choi, S., Zhu, K., Luther, J. M., & Beard, M. C. (2015). *Nature Communications, 6*, 7961.
98. Selig, O., Sadhanala, A., Muller, C., Lovrincic, R., Chen, Z., Rezus, Y. L., Frost, J. M., Jansen, T. L., & Bakulin, A. A. (2017). *Journal of the American Chemical Society, 139*, 4068.
99. Uratani, H., & Yamashita, K. (2017). *Journal of Physical Chemistry Letters, 8*, 742.
100. Han, T. H., Tan, S., Xue, J., Meng, L., Lee, J. W., & Yang, Y. (2019). *Advanced Materials, 31*, e1803515.
101. Kim, Y. C., Kim, K. H., Son, D.-Y., Jeong, D.-N., Seo, J.-Y., Choi, Y. S., Han, I. T., Lee, S. Y., & Park, N.-G. (2017). *Nature, 550*, 87.
102. Saliba, M., et al. (2016). *Science, 354*, 206.
103. Wang, Z., Lin, Q., Chmiel, F. P., Sakai, N., Herz, L. M., & Snaith, H. J. (2017). *Nature Energy, 2*, 17135.
104. Li, L., Liu, X., Zhang, H., Zhang, B., Jie, W., Sellin, P. J., Hu, C., Zeng, G., & Xu, Y. (2019). *ACS Applied Materials & Interfaces, 11*, 7522.
105. Kasap, S. O. (2000). *Journal of Physics D: Applied Physics, 33*, 2853.

106. Kasap, S. O., Zahangir Kabir, M., & Rowlands, J. A. (2006). *Current Applied Physics, 6*, 288.
107. George, Z., et al. (2003). In *Proceedings of SPIE*.
108. Wei, W., et al. (2017). *Nature Photonics, 11*, 315.
109. Szeles, C., Cameron, S. E., Soldner, S. A., Ndap, J.-O., & Reed, M. D. (2004). *Journal of Electronic Materials, 33*, 742.
110. Zhang, F., He, Z., & Seifert, C. E. (2007). *IEEE Transactions on Nuclear Science, 54*, 843.
111. Watanabe, S., et al. (2001). In *2001 IEEE nuclear science symposium conference record (Cat. No.01CH37310)* (pp. 2434).
112. Owens, A. (2006). *Journal of Synchrotron Radiation, 13*, 143.
113. Hitomi, K., Kikuchi, Y., Shoji, T., & Ishii, K. (2009). *Nuclear Instruments and Methods in Physics Research Section A: Accelerators, Spectrometers, Detectors and Associated Equipment, 607*, 112.
114. Hitomi, K., Tada, T., Kim, S., Wu, Y., Tanaka, T., Shoji, T., Yamazaki, H., & Ishii, K. (2011). *IEEE Transactions on Nuclear Science, 58*, 1987.
115. Yakunin, S., Dirin, D. N., Shynkarenko, Y., Morad, V., Cherniukh, I., Nazarenko, O., Kreil, D., Nauser, T., & Kovalenko, M. V. (2016). *Nature Photonics, 10*, 585.
116. Liu, F., et al. (2020). *Materials Today*.
117. Li, Y., & Chueh, W. C. (2018). *Annual Review of Materials Research, 48*, 137.
118. Birowosuto, M. D., Cortecchia, D., Drozdowski, W., Brylew, K., Lachmanski, W., Bruno, A., & Soci, C. (2016). *Scientific Reports, 6*, 37254.
119. Mykhaylyk, V. B., Kraus, H., & Saliba, M. (2019). *Materials Horizons, 6*, 1740.
120. Chen, Q., et al. (2018). *Nature, 561*, 88.
121. Hitomi, K., Kikuchi, Y., Shoji, T., & Ishii, K. (2009). *IEEE Transactions on Nuclear Science, 56*, 1859.
122. Li, C., Tscheuschner, S., Paulus, F., Hopkinson, P. E., Kießling, J., Köhler, A., Vaynzof, Y., & Huettner, S. (2016). *Advanced Materials, 28*, 2446.
123. Xing, J., Wang, Q., Dong, Q., Yuan, Y., Fang, Y., & Huang, J. (2016). *Physical Chemistry Chemical Physics, 18*, 30484.
124. Cho, Y., et al. (2018). *Advanced Energy Materials, 8*, 1703392.
125. Chen, B., Rudd, P. N., Yang, S., Yuan, Y., & Huang, J. (2019). *Chemical Society Reviews, 48*, 3842.
126. Zhou, Y., et al. (2020). *Chemistry of Materials, 32*, 5104.
127. Abdi-Jalebi, M., et al. (2018). *Nature, 555*, 497.

Modelling Spectroscopic Performance of Pixelated Semiconductor Detectors Through Monte-Carlo Simulation

Kjell A. L. Koch-Mehrin, Sarah L. Bugby, and John E. Lees

1 Introduction

When choosing or designing a detector for a particular application, a complete understanding of the expected response is vital. Parameters such as quantum efficiency, spatial resolution, signal-to-noise ratio, count rate capability and spectral resolution all play a part in the suitability of a detector. These parameters are not necessarily independent, and in many cases an improvement in one parameter can only be achieved with deterioration of another.

In order to understand the complex interplay between these parameters, predict detector response and inform design decisions, models that employ the Monte-Carlo method are commonly used. The Monte-Carlo approach allows the stochastic nature of photon detection to be simulated through a series of randomly determined outcomes.

For solid state detection, pixelation allows imaging and improves photon counting as simultaneous, but spatially separated events can be differentiated. Additionally, if the pixel pitch is small compared to sensor thickness, the 'small pixel effect' can mitigate the effects of poor charge carrier mobilities and improve spectroscopic response [1]. For these reasons, it is often desirable to decrease pixel size, but doing so increases the impact of electronic noise thresholding, pile up, fluorescence and the sharing of charge from a single event across multiple pixels. A Monte-Carlo model of a pixellated solid state detector must therefore incorporate all these effects since each has an impact on the detector response.

In this chapter we describe such a Monte-Carlo detector model. The model is general and applicable to a range of sensor materials and detector architectures but is

K. A. L. Koch-Mehrin (✉) · S. L. Bugby · J. E. Lees
Leicester University, Leicester, UK
e-mail: kalkm1@le.ac.uk

© Springer Nature Switzerland AG 2022
K. Iniewski (ed.), *Advanced X-ray Detector Technologies*,
https://doi.org/10.1007/978-3-030-64279-2_3

demonstrated here through comparison of simulated and experimental results from observations using a 1-mm-thick CdTe sensor coupled to the HEXITEC ASIC [2]. The model accurately predicts the proportion of charge sharing events as a function of incoming photon energy, and this is used to explore the influence of internal fluorescence on charge sharing. More broadly, the model can be used to predict ideal spectroscopic response of a given sensor and understand how a detector's design can impact performance.

2 Monte-Carlo Detector Model

The Monte-Carlo model that we present here has been developed in the Python programming language [3] and is appropriate for use for photon energies where pair production is negligible. For pair production to occur, the energy of the photon needs to be at least equal to the combined resting mass of an electron and positron. Since this threshold is at 1.022 MeV, which is above the X-ray energies typically detected directly with semiconductor radiation detectors, attenuation by pair production is not included in this model. Although applicable to any pixellated semiconductor detector if the appropriate material properties are entered, throughout this chapter we will use a CdTe detector as an example.

Figure 1 shows a schematic of the simulated detector, detecting a single photon, and defines variables that will be used to describe the model. The model assumes that the detector has a planar cathode and a pixelated anode. The z-axis is defined

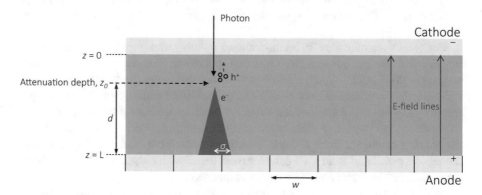

Fig. 1 Schematic of the detector as simulated by the model. Electrodes are positioned above and below the sensor material. The entire sensor is assumed to be depleted by the electric field, produced by applying a bias voltage to the electrodes. The cathode is planar, and the anode is pixelated. Although the cathode is drawn here, its attenuation is assumed to be negligible by the model. The thickness of the sensor is given by L, and d is the drift length of the electrons to the anode. The electron charge cloud will expand as it drifts towards the anode (represented by the triangle with its apex at the photon attenuation depth z_0). The width of the Gaussian charge cloud (i.e. standard deviation) at the anode is given by σ. The width of the pixel electrodes is given by w

to be from the cathode to the anode, measured from the surface of the sensor on the cathode side. In practice the model is run with a large number of input photons in order to ensure a smooth simulated response with good statistics. For photon counting, the error on the number of detected events, σ_{n_E}, is derived from Poisson statistics, i.e.

$$\sigma_{n_E} = \sqrt{n_E}, \tag{1}$$

where n_E is the number of detected events. If, for example, 1000 events are recorded in an energy bin, the associated counting error is $\sim$3%.

The following sections describe the path of a single photon and its resulting charge cloud as it steps through the model for a CdTe detector that ultimately produces a signal at the pixelated anode.

2.1 Photon Generation and Attenuation

The energy, E, of the incoming photon is specified by the user. Alternatively, a radioactive isotope such as Cd^{109}, Am^{241} or Co^{57} can be selected, and the source's possible emission energies are randomly sampled from known rates to determine the emitted photon energy E.

Uniform illumination across the entire collecting area is modelled, with the photon given a random incident position on the surface of the detector. The model assumes a back-illuminated detector, with the photon incident perpendicular to the detector surface through the cathode.

2.1.1 Photon Attenuation

Any attenuation by the cathode is considered negligible as electrode thicknesses are typically of the order of nanometres.

The depth, z_0, into the sensor from the cathode side at which the photon is attenuated is found using the probability distribution $P(x)$ obtained from the Beer–Lambert law [4]

$$P(x) = e^{-\mu(E)x}, \tag{2}$$

where $\mu(E)$ is the total linear attenuation coefficient of CdTe for the photon energy E and x the distance the photon travels before being attenuated. For the incident photon, the attenuation depth z_0 is simply the position along the z-axis after distance x. Values for $\mu(E)$ are taken from the National Institute of Standards and Technology (NIST) XCOM database [5], interpolated on a log–log scale. If the calculated attenuation depth z_0 is greater than the thickness of the sensor L, the photon is not detected by the sensor and instead passes through unattenuated. In this case, the photon is removed from the simulation, and the model proceeds with the computation of the next photon.

2.1.2 Scattering

Since the probability distribution from Eq. (2) uses a total linear attenuation coefficient, we do not yet know if attenuation occurred via the photoelectric effect, Compton scattering or Rayleigh scattering, whose separate attenuation coefficients make up the total $\mu(E)$ [5]. The type of attenuation that occurred at position z_0 is determined by random sampling from the relative likelihoods of each attenuation type. For example, the probability of attenuation by Compton scattering $P_{\mu_{cs}(E)}$ at photon energy E is

$$P_{\mu_{cs(E)}} = \frac{\mu_{cs}(E)}{\mu(E)}, \tag{3}$$

where μ_{cs} is the attenuation coefficient for Compton scattering only.

If attenuation occurs via the photoelectric effect, the primary photon is assumed to be completely absorbed at depth z_0 and deposits all of its energy at that location. If the photon is scattered, any energy transfer at position z_0 depends on the scattering that occurred. Once the new direction of the scattered photon has been determined, we return to Eq. (2) to calculate whether it interacts at a new position within the sensor or escapes the sensor entirely in which case it is removed from the simulation. For every interaction position of the scattered photon that is within the sensor, the relative likelihoods of the attenuation types are used to determine if the photon is again scattered or finally absorbed.

For compounds, it is also necessary to determine which type of atom the interaction was with. The probability, P_{atom}, of the photon interacting with a Cd atom in CdTe, for example, is [6]

$$P_{atom} = W_{Cd} \frac{\left(\dfrac{\mu(E)}{\rho}\right)_{Cd}}{\left(\dfrac{\mu(E)}{\rho}\right)_{CdTe}}, \tag{4}$$

where W_{Cd} is the weight fraction of Cd in CdTe (0.468) and ρ the density of the sensor material [5].

If Rayleigh scattering occurs, since it is a form of coherent scattering, the scattered photon does not deposit any energy at the interaction position. Instead, only the direction of travel of the photon changes. The possible scattering angles are defined by a probability distribution that depends on the energy E of the photon being scattered and the number of the atom involved in the interaction Z [7]. The probability distribution is randomly sampled to determine the scattering angle θ.

When Compton scattering occurs, the primary photon imparts some of its energy to the electron it interacts with and is scattered in a new direction. The energy of the scattered photon, E', is defined by [8]

$$E' = \frac{E}{1 + \frac{E}{m_e c^2}(1 - \cos \theta)}, \qquad (5)$$

where θ is again the scattering angle of the photon, m_e the electron mass and c the speed of light. Compton scattered photons are typically assumed only to interact with free electrons. Under this assumption, the Klein–Nishina formula [9] can be used to determine θ as it gives the probability distribution function of all possible scattering angles from a single free electron for a photon with energy E. The energy deposited at the interaction position z_0 where the photon scatters is

$$E_e = E - E', \qquad (6)$$

where E_e is the energy transferred to the electron, in this case the Compton recoil electron.

The direction of the scattered photon relative to the photon's initial direction of travel is calculated using spherical coordinates (θ, ϕ) where the scattering in the polar direction is described by the scattering angle θ and the azimuthal direction ϕ of the scattered photon is isotropic and therefore obtained by uniformly sampling in the interval $[0, 2\pi]$.

2.1.3 Absorption

At this stage, the primary photon was either directly absorbed via the photoelectric effect or scattered before being absorbed at a new position and depth z_0. During absorption, the photon transfers its energy to a bound electron, ejecting it from the atom and leaving the atom in an excited state. The ejected electron is referred to as the photoelectron and leaves behind a vacancy in the shell that it was ejected from, which may lead to the emission of characteristic fluorescence photons as the atom relaxes.

For compounds, the atom the photon interacts with is determined using Eq. (4). Which shell the photon interacts with will firstly depend on the binding energy of the shells relative to the energy of the photon. If the energy of the photon does not exceed the binding energy E_b of a shell, absorption by that shell is not possible. The binding energy of the shells is taken from Thompsen et al. [10]. Where $E > E_b$, the shell the photon is absorbed by is determined by comparison of the mass attenuation coefficients of the individual shells. Electrons bound to K, L_1, L_2 and L_3 shells are considered in the model. The mass attenuation for an individual shell can be calculated from the total mass attenuation of the atom and its jump factor, J, [6]

$$\left(\frac{\mu}{\rho}\right)_{shell} = \frac{J-1}{J} \left(\frac{\mu}{\rho}\right)_{tot}. \tag{7}$$

The jump factor is the fraction of the total photoelectric absorption coefficient due to absorption by electrons of the respective shell.

2.1.4 Fluorescence

When a vacancy is created in a shell, its fluorescence yield is used to determine if a fluorescence photon is generated as the vacancy is filled. If fluorescence does not occur, the complete energy of the primary photon is assumed to be absorbed by the photoelectron such that

$$E_e = E. \tag{8}$$

Equation (8) is a simplification that does not take into account non-radiative transitions (such as the Auger effect), which may result in new vacancies in higher energy shells. Given that fluorescence yield is lower for higher energy shells and that, if fluorescence does occur, its energy is lower, the likelihood of the energy escaping a pixel of realistic size via a fluorescence photon becomes very small. The absorbed energy can therefore be reasonably approximated using Eq. (8).

If a fluorescence photon is emitted from the excited atom, the relative radiative rates for each possible transition are used to determine the energy of the emitted fluorescence photon. Possible transitions, their line energies and their relative rates for both Cd and Te are shown in Table 1 [10].

A vacancy that forms in an L-shell of the same atom due to a $K\alpha$ transition may also fluoresce, producing an additional fluorescence photon. The energy of the photoelectron E_e is therefore calculated by

$$E_e = E - \sum_{i=K_{shell}}^{L_3} E_{XRF_i}, \tag{9}$$

where E_{XRF_i} is the energy of the X-ray fluorescence photon, if any, from the ith shell. Vacancies that arise in an M or N shell due to the cascading of electrons or otherwise are ignored due to the low energy of any potential fluorescence emissions, as in the case of Auger electrons. The energy from the ignored vacancies is therefore included in the photoelectron energy E_e (i.e. it is never subtracted from the incident photon energy E in Eq. (9)).

An emitted fluorescence photon may travel in any direction, determined by sampling a uniform distribution over all solid angles, and the distance it travels before interaction is again calculated using Eq. (2). The steps described so far for the primary incident photon, including scattering, are repeated for any secondary fluorescence photon and any further fluorescence photons that result from those.

Table 1 Possible transitions for Cd and Te atoms used in the model, along with line energies and relative radiative rates for each transition [10] rounded to three and two significant figures

	Cd		Te	
Transition	Line energy (keV)	Rate	Line energy (keV)	Rate
$K\alpha_1$ (KL$_3$)	23.174	0.55	27.473	0.53
$K\alpha_2$ (KL$_2$)	22.984	0.29	27.202	0.29
$K\beta_1$ (KM$_3$)	26.095	0.09	30.996	0.10
$K\beta_2$ (KM$_2$)	26.654	0.02	31.700	0.03
$K\beta_3$ (KM$_2$)	26.061	0.05	30.945	0.05
$L\alpha_1$ (L$_3$M$_5$)	3.134	0.77	3.769	0.75
$L\alpha_2$ (L$_3$M$_4$)	3.127	0.08	3.758	0.08
$L\beta_2$ (L$_3$N$_4$)	3.528	0.12	4.302	0.14
Ll (L$_3$M$_1$)	2.767	0.03	3.336	0.03
$L\beta_1$ (L$_2$M$_4$)	3.317	0.91	4.030	0.88
$L\gamma_1$ (L$_2$N$_4$)	3.717	0.09	4.571	0.12
$L\beta_3$ (L$_1$M$_3$)	3.401	1.00	4.120	1.00

2.2 Charge Transport

At this stage in the model, the primary photon and any secondary fluorescence photons have each transferred their energy to a photoelectron. The position of the photoelectrons in the sensor is known as these are the interaction positions of the photons at depth z_0. A photoelectron will lose its energy by ionization, creating electron–hole pairs along its path until it is stopped. The number of electron–hole pairs, N, is given by

$$N = \frac{E_e}{\varepsilon}, \tag{10}$$

where ε is the average pair creation energy, equal to 4.3 eV in CdTe [11]. The number of electron–hole pairs N generated for a given E_e will vary due to Fano noise. This noise captures the variance in electron–hole pairs generated as the photoelectron loses energy, which is assumed to follow Fano-adjusted Poisson statistics ($\sigma_N^2 = FN$). The Fano factor, F, is taken to be 0.1 for CdTe [12].

The model assumes a uniform electric field across the sensor material formed by the bias voltage applied to the electrodes. If a negative polarity bias is applied to the cathode, the field lines go from the anode to the cathode (parallel to the z-axis). Due to this field, charges will drift towards the electrodes, with the holes moving towards the cathode ($z = 0$) and the electrons towards the pixelated anode ($z = L$).

The motion of the charge carriers drifting towards the electrodes induces a current at the electrodes as described by the Shockley–Ramo theorem [13]. This induced current is what is read out by the detector electronics and signals that some energy deposited in the sensor by a photon has been detected. For a pixelated readout, the drifting charge carriers from a single photon energy deposition may induce a current over multiple pixels. The charge carriers form a cloud that grows due to thermal motion and electrostatic forces during drift. If the size of the charge cloud is calculated, it can be used to estimate over how many, and which, pixels the charge carriers induce a signal in.

2.2.1 Charge Cloud Size

The size of the electron charge cloud is calculated in the model, approximated by a symmetrical two-dimensional Gaussian distribution [14, 15] in the plane perpendicular to the electric field lines (i.e. the z-axis). The final size of the cloud will depend on the initial size, σ_i, due to the range of the photoelectron (over which it excites the electron–hole pairs) and the growth of the cloud, σ_d, from diffusion and electrostatic repulsion of the charges as the cloud drifts to the anode. The final size of the charge cloud at the anode, σ, is found by adding the two components in quadrature (under the assumption that their magnitudes are not correlated [16])

$$\sigma = \sqrt{\sigma_d^2 + \sigma_i^2}. \tag{11}$$

Initial cloud size σ_i has been estimated by Blevis and Levinson [17] for a number of photoelectron energies, and these can be used to model X-ray spectra from a CdTe detector [15]. The values determined in [17] are the average range (i.e. displacement) of the photoelectron and therefore the diameter of the initial charge cloud. For a Gaussian cloud, we can estimate its diameter as the FWHM [18] such that $diameter \approx 2.355\sigma$. Using this approximation, the values for σ_i determined for our model are $0\,\mu m$, $1\,\mu m$ and $4\,\mu m$ for photons of energy 0 keV, 35 keV and 100 keV, fit with a second-order polynomial to find σ_i for any E_e. The relationship between σ_i and E_e is visualized in Sect. 4.2.

The cloud growth during drift, σ_d, is calculated using the diffusion equation derived from Fick's second law [19]

$$\frac{\partial \sigma_d(t)^2}{\partial t} = 2D, \tag{12}$$

where D is the diffusion coefficient and $\sigma_d(t)^2$ is a function of time. The diffusion coefficient is given by the Einstein relation [20]

$$D = \mu_e \frac{k_b T}{e}, \tag{13}$$

where μ_e is the electron mobility, k_b the Boltzmann's constant, e the elementary charge and T the temperature of the sensor. For a uniform field, the total drift time t_d for the electron cloud to reach the anode is equal to

$$t_d = d\frac{L}{\mu_e V},$$ (14)

where L is the thickness of the CdTe sensor, d the drift length of the electron cloud to the anode (given by $L - z_0$) and V the applied bias voltage. The electrostatic repulsion between the electrons that occurs during drift is included by using an effective diffusion constant D' [14] in place of D in Eq. (12)

$$D' = D + \frac{\mu_e N q}{24\pi^{3/2}\epsilon}\frac{1}{\sigma(t)},$$ (15)

where ϵ is the permittivity in CdTe [21] and $\sigma(t)$ is the size of the electron cloud at any instantaneous point in time t. By combining Eqs. (12), (13) and (15), the differential equation that gives the size of the electron cloud due to diffusion and electron repulsion after drifting some period of time t is

$$\frac{\partial\sigma_d(t)^2}{\partial t} = \frac{2\mu_e k_b T}{e} + \frac{\mu_e N e}{12\pi^{3/2}\epsilon}\frac{1}{\sigma(t)}.$$ (16)

This equation must be solved numerically to get a well-approximated solution for σ_d after the total drift time. In our model we use the fourth-order Runge–Kutta method [22] with over 10^3 iterative time steps between $t = 0$ and $t = t_d$ with $\sigma(t) = \sigma_i$ at $t = 0$. Once $\sigma_d(t_d)$ is known, Eq. (11) is used to compute the final size of the electron cloud when it has arrived at the anode.

2.2.2 Signal Induction

It is well understood that the signal measured at an electrode is not the charge that arrives at the electrode but instead the induced charge from the charge carrier motion during drift [23]. From the Schokley-Ramo theorem, the charge induced ΔQ on a single electrode by a moving charge is [24]

$$\Delta Q = -q[\psi(\mathbf{r}_f) - \psi(\mathbf{r}_i)],$$ (17)

where $\mathbf{r}_i$ and $\mathbf{r}_f$ are the start and end positions of the moving charge q, respectively, and ψ the weighting potential of the corresponding electrode. A detector's weighting field depends on its geometry and electrode arrangement only and describes the electromagnetic coupling between moving charges and the conducting electrodes. Although the movement of the charges is controlled by the electric field, ΔQ is calculated more easily by use of the weighting potential, a dimensionless quantity of the weighting field for which $0 \leq \psi \leq 1$.

Ideally, to achieve the most accurate modelling of charge induction, Eq. (17) would be solved numerically for each electron and hole in motion over each pixel electrode, using a 3-dimensional weighting potential and electric field profile. In the model described here we use the approximation of a charge cloud of width σ to estimate how the induced charge is distributed over the pixelated anode (see Sect. 2.2.3) and so we need only to calculate how much of the original charge in the cloud is induced at the anode. In this case, simpler approximations exist, which are less computationally demanding but still provide valuable information.

The Hecht relation [25] can be used to calculate the induced charge ΔQ for a parallel-plate two-electrode configuration with a uniform electric field. By dividing the induced charge by the initial amount of charge in the cloud at creation, given by Ne, the charge collection efficiency (CCE) is found. Typically ΔQ is not equal to Ne because, as the charge carriers drift to the corresponding electrode, they can become trapped in defects and impurities in the sensor crystal, which results in incomplete charge collection. The trapped charges are no longer in motion and therefore do not induce a charge on the electrodes resulting in charge loss. The classical Hecht relation that takes into account this charge loss due to carrier trapping is given by

$$CCE = \frac{\Delta Q}{Ne} = \frac{\lambda_h}{L}\left(1 - e^{-\frac{z_0}{\lambda_h}}\right) + \frac{\lambda_e}{L}\left(1 - e^{-\frac{L-z_0}{\lambda_e}}\right), \tag{18}$$

where $\lambda_h = \mu_h \tau_h E_F$ and $\lambda_e = \mu_e \tau_e E_F$ are the mean drift lengths of holes and electrons, respectively, and are a product of the carrier lifetime τ, the carrier mobility μ and the electric field strength E_F.

The classical Hecht relation works for planar detectors where the pixel size is equal to or greater than the thickness of the sensor ($w/L \geq 1$). For such a detector geometry, the charge induced by some fixed carrier drift length is equal irrespective of the distance of the charge carriers to the electrode. However, for the case of a pixelated detector, if the pixel pitch is smaller than the sensor thickness, a more significant fraction of charge is induced by carriers drifting close to the pixel. This is what is referred to as the small pixel effect, and the weighting potential ψ describes how much charge is induced depending on the where the charge carriers are in motion relative to the electrode.

For our model, we present a slightly modified Hecht relation that employs a one-dimensional weighting potential along the z-axis

$$CCE = \frac{\lambda_h}{L}\left(1 - e^{-\frac{\psi(z_0/L)L}{\lambda_h}}\right) + \frac{\lambda_e}{L}\left(1 - e^{-\frac{L-\psi(z_0/L)L}{\lambda_e}}\right), \tag{19}$$

where $\psi(z_0/L)$ is the weighting potential at the normalized interaction depth z_0/L of the photon from the cathode. Figure 2 shows the one-dimensional weighting potential calculated by Duarte et al. [26] using a finite element simulation software for the small pixel $w/L = 0.25$ CdTe detector geometry used as an example here, compared with the weighting potential for a planar detector. We see from the figure that for a planar electrode configuration $\psi(z_0/L) = z_0/L$. Inserting this into Eq. (19), we obtain the classical Hecht relation as shown by Eq. (18). The weighting potential for the $w/L = 0.25$ case, on the other hand, exponentially approaches 1 at

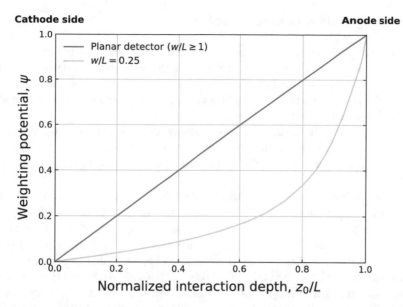

Fig. 2 Weighting potential for a planar detector ($w/L \geq 1$) and for the geometry of the HEXITEC CdTe sensor. A normalized interaction depth of 0 is at the cathode and 1 at the anode. The $w/L = 0.25$ weighting potential is from Duarte et al. [26], calculated using a finite element simulation software

the anode side. Therefore, charges moving close to the pixel anode induce a greater fraction of the charge signal. The weighting field in small pixel configurations is favourable for detectors with sensor materials that have significantly different carrier drift lengths, allowing the possibility to minimize the induced signal from the carrier that is more likely to trap. For CdTe, the mean drift length of holes is roughly a factor of 10 times smaller than that for electrons [27]. Therefore, by using a back-illuminated detector where most of the photon interactions take place closer to the cathode side, electrons will primarily traverse the more dominant regions of the weighting field and correspondingly induce a greater fraction of the charge signal as opposed to the holes. This results in a better CCE and thereby improves spectroscopic performance as charge loss due to the more easily trapped holes is reduced.

The CCE is calculated for each charge cloud by first sampling the weighting potential at the normalized interaction depth z_0/L and then using Eq. (19). The effective number of carrier pairs, equivalent to the induced current on the anode after consideration of the weighting field and trapping during carrier drift, N', is then

$$N' = CCE \times N. \tag{20}$$

2.2.3 Readout at the Pixelated Anode

For each charge cloud, we have a two-dimensional Gaussian distribution centred at the x and y position of the photon interaction point with standard deviation σ and containing N' charges. This is evaluated against the pixelated array. The number of charges distributed to a pixel are multiplied by the pair creation energy ε to obtain the signal induced on that pixel in energy (eV).

Many pixelated detectors will have inter-pixel gaps between the electrodes forming each pixel. It is expected that these will be small compared to pixel pitch— our CdTe detector example has a pixel pitch of $w = 250 \, \mu m$ with an inter-pixel gap of $50 \, \mu m$.

Charges drifting in close proximity to the inter-pixel gap may result in further incomplete charge collection due to changes in the electric and weighting field close to the gap [28]. Using the one-dimensional approximation for weighting potential only sensitive to the z position of the charge cloud in the sensor, it is not possible to model this effect. As a result, this model assumes that charges are always drifting perpendicular to the electrodes and charge loss in the inter-pixel gap is not considered. This is a limitation of the model, however; in practice, only a small fraction of the signal is induced by carriers drifting in the gaps as the majority of the weighting field has already been traversed. Although challenging, it is possible to use three-dimensional weighting and electric field maps to simulate the charge loss to the inter-pixel gap using the Shockley–Ramo theorem with numerical analysis. These more accurate field maps can, for example, be obtained using finite element analysis software. For our purposes, the benefits of this approach are outweighed by its complexity and computational requirements as results using the approximations outlined here are already very similar to those seen experimentally.

The signal induced on each pixel electrode is subject to electronic noise from the pixel readout and digitization process. Individual sources of electronic noise are not considered in the model; instead, we define a single parameter A that encapsulates all the noise to the signal from the readout electronics. The electronic noise is assumed to follow a Gaussian distribution with variance $\sigma^2 = A$, where A is equal to the equivalent noise charge (ENC). The ENC is a free parameter in the model that can be adjusted to match experimental data. Electronic noise is added to each pixel that contains an energy.

2.3 Spectral Response Reconstruction

Data is recorded in an event list format, and this must be reconstructed in order to distinguish individual photon events and obtain the detector response to the observed source. The model's event list format is similar to that produced by the HEXITEC ASIC (see Sect. 3.1 for ASIC details), where the charge recorded in each pixel for every frame in the observation is saved. The simulated event lists can therefore

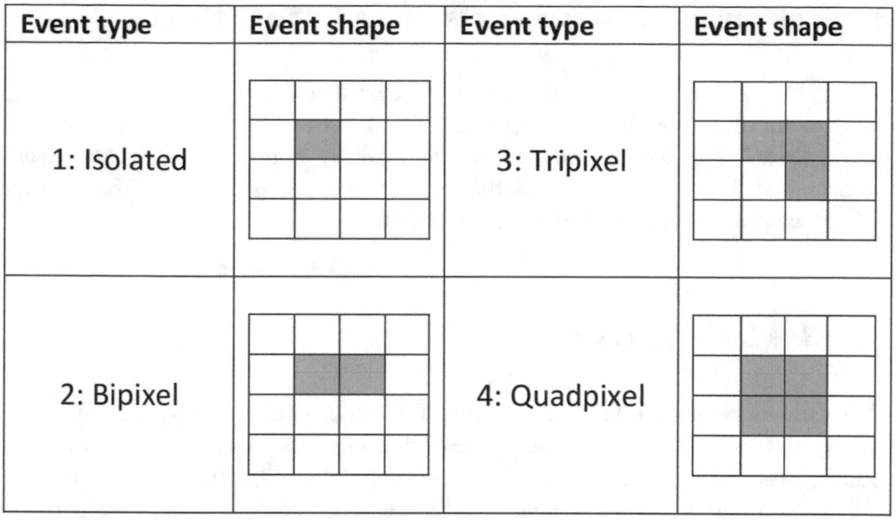

Event type	Event shape	Event type	Event shape
1: Isolated		3: Tripixel	
2: Bipixel		4: Quadpixel	

Fig. 3 Event shapes for isolated, bi-, tri- and quad-pixel events. The greyed-shaded squares correspond to pixels that have recorded an energy greater than the noise threshold

be processed using a reconstruction algorithm based on the one used in [29] for experimental HEXITEC detector event lists.

The reconstruction algorithm classifies the type of each event. Pixels that are adjacent to one another are said to be part of the same event. The event type then depends on the number of pixels the event contains—as shown in Fig. 3. The energy of an event is the sum of the energy in all pixels making up that event. A separate spectrum is reconstructed for each event type, where only events with a shape as shown in Fig. 3 are considered. This excludes some events (e.g. a line of four pixels) that are known to occur at a very low frequency and more likely due to coincident events. Different orientations for identically shaped events show no difference in their energy response and are therefore considered as the same event type. Diagonally adjoined events are counted as two separate events.

It is common practise to apply a noise threshold to the energy in each pixel to remove counts that are due to thermal charge leakage from the crystal and detector electronics, instead of energy deposited by a photon. The noise threshold is applied by the algorithm—removing all pixels in an event that have an energy lower than a user-defined threshold. The event type and its energy are determined after applying the noise threshold.

3   Experimental Measurements with a CdTe Detector

In this section we introduce the CdTe detector that is used as a case study to validate the model, with the comparisons discussed in Sect. 4. The experimental data collected with the detector is from the work by Bugby et al. [29] and Koch-Mehrin et al. [30]—a summary of those observations is given here. The detector calibration process is described in detail in [29].

3.1   The CdTe Detector

Experimental data were taken using the STFC High Energy X-ray Imaging Technology (HEXITEC) [2] detector system, developed to detect hard X-rays and gamma rays. The HEXITEC ASIC is a pixelated 80×80 array with a pitch of $250\,\mu m$ manufactured by a standard $0.35\,\mu m$ CMOS (complementary metal-oxide semiconductor) process, such that each ASIC pixel contains a charge amplifier, shaping amplifier and a peak track-and-hold circuit. Therefore, induced charge is converted to a voltage and digitized at the pixel level, allowing each pixel to provide a full energy spectrum.

The detector used consists of a platinum planar cathode of $\sim$200 nm thickness and pixelated aluminium anode on opposite sides of the CdTe semiconductor material as the sensor. The aluminium anode forms a Schottky (i.e. blocking) contact, which helps reduce charge leakage. The CdTe sensor has a thickness of 1 mm and a 20×20 mm collecting area. The anode is pixelated by an 80×80 array with a pixel electrode size of $200\,\mu m$ and an inter-pixel spacing of $50\,\mu m$. The ASIC and anode are attached by flip-chip assembly to gold studs placed on each of the ASIC bond pads.

3.2   Data Collection

The detector and ASIC module were mounted in a readout system with off-chip digitization, an FPGA (field programmable gate array) for ASIC clocking and a gig-ethernet connection to a PC. An ASIC frame rate of 1.6 kHz was used for all experiments. The readout system controlled the detector temperature to $28 \pm 1\,°C$ and applied a bias voltage of -500 V to the CdTe cathode. The bias voltage was periodically refreshed by switching the bias to 0 V for 2 s once every minute. This was done to prevent known time-dependent polarization effects in Schottky contact CdTe [31] and therefore help maintain a more uniform electric field across the detector.

Sources used were either a radioisotope that emits characteristic X-rays as it decays or an X-ray tube that produces X-rays via fluorescence and bremsstrahlung

Table 2 Experimental observations using the CdTe HEXITEC detector carried out by [29] for the radioisotope runs (top section of table) and [30] for the X-ray tube runs (bottom section of table). For the radioisotope photopeaks the errors depend on if multiple peaks (i.e. $K_{\alpha 1}$ and $K_{\alpha 2}$) have been combined into a single peak energy because these cannot be resolved, in which case their relative intensities are taken into account to determine the error; otherwise, a nominal error of 10 eV is used. For the X-ray tube acquired photopeaks, the error is determined from the photopeak FWHM since the peak is not from characteristic X-rays. Source activity or X-ray tube current, total acquisition time and average frame occupancy are also given for each observation

Source	Source activity (MBq)	Photopeak energies (keV)	Acquisition time (s)	Frame occupancy (%)
^{109}Cd	84.9	22.00 ± 0.10	1000	0.4
		24.90 ± 0.01		
		88.00 ± 0.01		
^{241}Am	351.7	59.54 ± 0.01	1000	0.5
^{57}Co	51.7	14.40 ± 0.10	7000	0.2
		122.10 ± 0.05		
		136.50 ± 0.10		
^{55}Fe	414.0	5.95 ± 0.01	3600	0.2
^{99m}Tc	188.3	140.50 ± 0.01	1800	0.4
Tube voltage (kV)	Tube current (μA)	Photopeak energy (keV)	Acquisition time (s)	Frame occupancy (%)
35	110	32.5 ± 3.5	1800	0.2
40	80	38.0 ± 3.5	1800	0.2
50	5	45.8 ± 3.5	1200	0.2

as electrons accelerated by an applied voltage collide into a target. Table 2 lists all the observation runs carried out, showing the radioisotope or X-ray tube voltage used for that run and the corresponding photopeak energies obtained from the source. The various sources used provide a range of photopeak energies from 5.95 keV to 140.5 keV.

Frame occupancy, defined as the percentage of pixels with an energy above the set noise threshold [32], was kept below 1% for each observation. At such a low frame occupancy, we can assume that multi-pixel events are indeed due to charge sharing and not photon pileup—this is important when accurately calculating the charge sharing rates due to the charge cloud size and X-ray fluorescence (Sect. 4.2).

3.3 Event Type Spectra and Charge Sharing

An energy spectrum is reconstructed from each source observation listed in Table 2 using the algorithm we described in Sect. 2.3 and analysed for each event type introduced in Fig. 3. Figure 4 shows the different event type energy spectra for

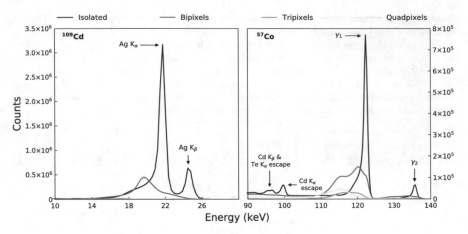

Fig. 4 Event type spectra for ^{109}Cd (LHS) and ^{57}Co (RHS) from experimental observations

the ^{109}Cd and ^{57}Co observations. By inspecting the individual event type spectra around the main source photopeaks, we get a good idea of the effect of charge sharing on a pixelated detector's response. The photopeaks of multi-pixel events are much broader than the same photopeak in the isolated event spectrum, and the peak position of the multi-pixel events also shifts to lower energies (indicating charge loss). For the ^{109}Cd source, the Ag K_α line was measured at 21.75 keV (expected at 22 keV) with isolated events but 19.65 keV for the bi-pixel events, and the FWHM of the peaks were 0.85 keV and 2.85 keV, respectively. Multi-pixel events exhibit worse spectral performance than isolated events and negatively impact the detector response.

Some multi-pixel events do not suffer from as much, if any, charge loss. This is easily seen in the 122-keV photopeak (^{57}Co) in the bi-pixel event spectrum that shows two components—a peak at ~121 keV followed by a flat shoulder at lower energies. This is due to two types of charge sharing—those created by fluorescence and those created by a single charge cloud inducing a charge on two neighbouring pixels. In the case of fluorescence charge sharing events, a fluorescence photon escapes into a pixel adjacent to the pixel of the incident photon and deposits its energy there directly. The charge sharing event is not created by charges drifting within the inter-pixel region and inducing a signal on a neighbouring pixel, and therefore does not suffer from charge loss. For the 22-keV photons in the ^{109}Cd bi-pixel spectrum, no double peak is visible. Since K-shell fluorescence does not occur at this energy, all of the counts in the ^{109}Cd bi-pixel photopeak are due to charges spreading from one cloud centre and exhibit some charge loss.

An effective way to visualize these two varieties of bi-pixel events and reveal the effects of charge loss is shown in Fig. 5. Figure 5 shows the energy split across the two pixels in each bi-pixel event, once for the 59.5-keV photopeak from ^{241}Am data and again for the 122-keV photopeak from the ^{57}Co data. Counts along the straight line have not suffered from any charge loss, with the energy in the two

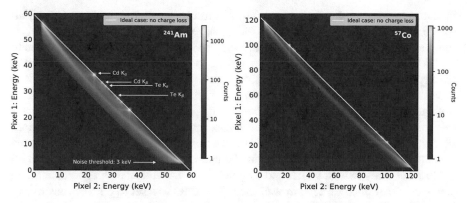

Fig. 5 Two-dimensional histograms showing the distribution of the energy split for all bi-pixels for the ^{241}Am data (LHS) in the energy range from 50 to 62 keV, and for the ^{57}Co data (RHS) within 110–125 keV. The white line represents the ideal case in which no charge loss occurs. The gaps on either axis from 0 to 3 keV correspond to the noise threshold. In the ^{241}Am distribution, the bright spots are annotated with the fluorescence photon causing the bi-pixel event (these correspond to the fluorescence spots in the ^{57}Co distribution). In total there are eight bright spots—only four are annotated because the remaining four are the same but with the energies in the pixels reversed

pixels adding up to the exact photopeak energy. The majority of the bi-pixel counts however appear below the straight line indicating charge loss. Along the line and not affected by charge loss are the bright spots, which correspond to the fluorescence charge sharing events. Bi-pixels due to fluorescence have characteristic energy splits where in one pixel the fluorescence photon energy is recorded and in the other pixel the escape peak energy (i.e. the photopeak energy minus the energy of the escaped fluorescence photon). This results in a large collection of counts at specific energies, forming the bright spots. Since fluorescence charge sharing events exhibit less charge loss, their energy resolution is typically better than that of the charge spreading events.

4   Comparison Between Model and Experiment

Both the Monte-Carlo model and the experimental data collected with the pixelated CdTe detector have been described. We now compare certain spectroscopic performance quantities as estimated by the model against values determined from the experimental observations. A detector response is simulated at multiple line energies covering the experimental photopeak energy range from 5.95 keV to 140.5 keV. Using the simulated responses, the rate of charge sharing as a function of incoming photon energies is calculated. The proportion of charge sharing events caused by fluorescence in bi-pixels and tri-pixels is also estimated by the model.

For each simulation, the model was run with 10^7 primary photons incident on the detector at a count rate of 1 photon per frame. This count rate was chosen to ensure that all observed multi-pixel events are due to charge sharing and not influenced by photon pileup. The thickness of the sensor was set to 1 mm with the sensor temperature fixed to 28 °C. Although in the experiments the detector was biased to −500 V, in the model the bias voltage was set to −425 V. Cola et al. [33] show that for Schottky contact CdTe, charge buildup causes a non-uniform electric field across the sensor, which decreases in strength with distance from the anode. An effective voltage of −425 V with a uniform field is suggested in [33], which approximates the average electric field strength felt by the charge carriers drifting in the non-uniform field under a −500 V bias. Since the model ignores the inter-pixel gap, the pixel size used in the simulations was the pixel electrode size of 200 μm on a 80 × 80 pixel array, which was found to give better agreement with experiment than when the pixel pitch of 250 μm was used. The ENC was set to 60 electrons rms in order to match the FWHM of the simulated photopeaks to that of the experimental photopeaks for isolated events. A noise threshold of 3 keV was applied to every pixel, and recorded energies were binned with a bin size of 0.3 keV (in order to match with the experimental data processing) to produce the response spectrum.

4.1 Simulated Detector Response

Figure 6 shows the simulated detector response to incoming photons with energy 59.5 keV, compared with the response obtained experimentally from the [241]Am source for isolated events only. The lower energy X-ray L-shell lines emitted from [241]Am were not observed experimentally due to the 0.5-mm-thick steel housing of the sealed source. This allowed us to obtain an experimental detector response to the 59.5-keV line only. The simulated and experimental responses are shown for isolated events only and were normalized by the maximum number of counts in an energy bin of the primary γ photopeak.

The detector response is, as we would expect, dominated by a photopeak at the energy of the incoming γ photons, 59.5 keV. However, multiple other smaller peaks are also visible in the response, despite the source not emitting those energies. All the peaks visible in the response and their measured positions are listed in Table 3. The escape peaks are due to K-shell fluorescence from either the Cd or Te atom escaping a pixel that absorbed a 59.5-keV photon, by either leaving the sensor completely or being reabsorbed beyond the nearest pixel neighbours. The escaped fluorescence photons that are reabsorbed within the sensor lead to the fluorescence peaks at ∼23 and ∼26 keV. The total number of isolated absorbed 59.5-keV events is the sum of counts in the primary gamma photopeak and all escape peaks. The proportion of counts in each of the peaks relative to this total sum is shown in Table 3, revealing good agreement between experiment and simulation. The multiple peaks in addition to the 59.5-keV photopeak are solely a product of the detector's response and do not describe the source that was observed. Furthermore, the use of pixels, and particularly as they become smaller, increases the proportion of counts in the fluorescence and escape peaks while decreasing counts in the

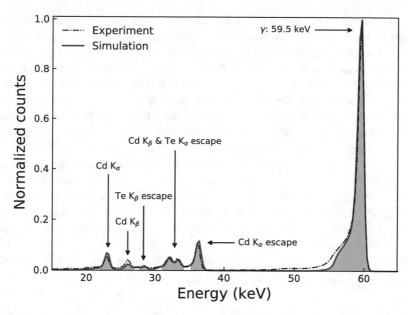

Fig. 6 Simulated detector response to 59.5-keV photons compared with the experimentally measured response to the same energy. The spectra were normalized to the maximum counts in an energy bin of the 59.5-keV photopeak and only include isolated events. The shaded regions highlight the energy windows used to calculate the total number of counts belonging to each peak

Table 3 All visible peaks in the detector response to 59.5-keV photons and their peak energy positions as measured from the simulated and experimental response shown in Fig. 6. The energy window corresponds to the shaded region underneath the respective peak in Fig. 6. The percentages are the proportion of counts in the respective peak from the total number of absorbed incident isolated events (i.e. γ photopeak and escape peak counts summed). The errors for the proportion of counts were determined by increasing the energy window by 0.5 keV at both the lower and upper bounds and calculating the difference

	Peak position (keV)			Proportion of counts (%)	
Peak	Experiment	Model	Energy window (keV)	Experiment	Model
γ	59.5	59.5	50–62	85.0 ± 1.0	84.0 ± 0.8
Cd K_α escape	36.2	36.4	35–38	6.6 ± 0.3	7.9 ± 0.2
Cd K_β escape	33.1	33.4	31–34.5	6.7 ± 0.3	6.7 ± 0.3
Te K_α escape	31.9	32.1			
Te K_β escape	28.4	28.5	27.5–29.5	1.8 ± 0.4	1.4 ± 0.3
Cd K_β	26.0	26.0	25–27	3.1 ± 0.4	1.9 ± 0.2
Cd K_α	23.0	23.0	22–24	3.6 ± 0.2	4.2 ± 0.1

photopeak actually corresponding to the source emission line. Understanding the detector's response therefore helps us not only determine which detected energies belong to the source but also to calculate spectroscopic performance indicators such as the spectral efficiency [15] (the proportion of counts recorded in the source photopeak relative to the total number of incident source photons at that energy).

The simulated response is somewhat able to reproduce the low energy tail seen in the experimental response on the LHS of the 59.5-keV photopeak. The tail arises due to two distinct mechanisms. The first is due to the noise threshold, which removes any energy belonging to the main event that is in an adjacent pixel and lower than the noise threshold. The second is a consequence of incomplete charge collection, which in the case of CdTe is due to the lower mobility hole charge carriers that are more likely to become trapped during drift. This results in what is known as hole tailing [12] as the charge from trapped holes is not induced on the anode. Both mechanisms are considered in the model, with the latter process taken into account by the modified Hecht relation (Eq. (19)) used with the appropriate weighting potential for the small pixel geometry of the detector. The discrepancy between the simulated and experimental tail is likely because the carrier mobilities will in practise vary across the sensor material, meaning that charge collection will differ slightly depending on the pixel. Whereas in the model the same carrier mobilities are used across the entire sensor.

4.2   Charge Sharing Rates

The proportion of charge sharing depends heavily on the energy of the source photons and the pixel size. Using the simulated detector responses, we can calculate the total amount of charge sharing and the proportion of different event types at specific photon energies. These are compared to experimentally determined values from the HEXITEC CdTe detector in Fig. 7. The proportion of a specific event type, for both simulation and experiment, was calculated by summing all counts within the primary photopeak energy window of the corresponding event type and dividing by the total number of counts for all event types within the same energy window [30]. The charge sharing proportion is then given by the proportion of all multi-pixel event types combined.

The proportion of charge sharing events as a function of energy is well estimated by the model, giving very good agreement to the experimentally determined values (Fig. 7a). At an incoming photon energy of ∼50 keV, 50% of all events already show charge sharing. Beyond this point, the amount of charge sharing increases gradually with photon energy, reaching 62.5% at 140.5 keV. The steep increase in charge sharing at lower energies is due to the noise threshold. At these energies, energy shared with adjacent pixels is more likely to be lower than the threshold, resulting in events that end up in the low energy tail of the isolated photopeak instead of being correctly identified as a multi-pixel event.

Above the K-shell binding energies of 26.7 keV and 31.8 keV for Cd and Te, respectively, there is a sudden jump in charge sharing due to the addition of fluorescence photons and the increase in attenuation due to the K-shell (i.e. the K-edge). The increase in charge sharing events above the Cd K-edge is larger due to

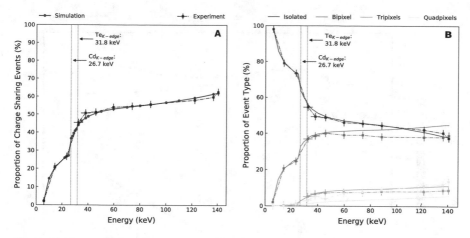

Fig. 7 (**a**) The proportion of charge sharing events as a function of incoming photon energy as predicted by the model and compared to experiment. (**b**) Proportion of each event type separately as a function of incoming photon energy. The solid lines are the model prediction (interpolated between the simulation points shown on the curve in **a**), and the dashed lines the experimental result. For both **a** and **b**, the data points mark the energies at which the calculations/simulations were performed. The experimental values have an uncertainty of ±2% due to the dependency on the position of the energy window. A low noise energy threshold of 3 keV was applied during event reconstruction

the Cd K-shell fluorescence energies (with photons having a mean attenuation range of around 100 μm in CdTe, compared with ~60 μm for the Te K-shell fluorescence). The Te K-shell fluorescence photons have energies (27–31 keV) above the binding energy of the K-shell in Cd. Therefore, if reabsorbed, they can cause additional K-shell fluorescence from the Cd atoms, resulting in more charge sharing events.

Above the absorption edges, the amount of charge sharing begins to level off as fluorescence charge sharing rates are constant and the noise threshold no longer has a significant impact. Charge sharing rates now depend primarily on the size of the charge cloud, gradually climbing as the final cloud size becomes increasingly dependent on the width of the initial excitation of charge carriers from the photoelectron, rather than on the cloud growth during drift.

Figure 7b shows that the majority of the charge sharing events within the investigated energy range are bi-pixels. Although the total charge sharing probability between the simulated and experimental data agrees very well, the relative contribution of the event types differs at higher energies. The simulated data predicts more bi-pixels and fewer quad-pixels than what is observed experimentally. Furthermore, the simulated results do not show a jump in the probability for quad-pixels after the Cd K-edge as suggested by experiment. These disagreements possibly reveal the limitations of our simplified signal induction model. As the charge cloud size becomes increasingly comparable to the pixel size, and charge sharing rates less dependent on the electronic noise threshold and fluorescence, the

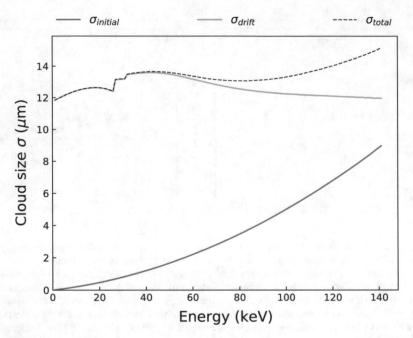

Fig. 8 The average charge cloud size at different energies due to the initial width and due to growth during drift to anode as predicted by the model. The combined average total size is also shown. The model was simulated between 1 and 141 keV with 1-keV intervals—only the curve is shown

Gaussian cloud approximation will predict a charge sharing event but not necessarily the correct multi-pixel event type.

Figure 8 shows the average contribution to the final cloud size at the anode in terms of the σ of the two distinct components (growth during drift and initial size), as predicted by the model. This suggests that the final cloud size is dominated by the growth due to diffusion and electron repulsion during drift towards the anode up to $\sim$100 keV. The average total charge cloud size is found to be relatively constant between 1 and 100 keV at $\sigma \approx 13 \, \mu m$. A charge cloud containing $\sim$99.7% (i.e. $\pm 3\sigma$) of all charges therefore extends $\sim$78 μm in diameter. Above the Cd and Te K-edges, photons are again absorbed closer to the cathode, increasing the drift time of the excited charges and therewith the cloud size. The relatively constant cloud size supports the argument that at low energies it is the noise threshold that prevents much larger proportions of charge sharing, although it will also depend somewhat on their being fewer charge carriers N for lower E_e.

4.3 Fluorescence Charge Sharing

We have seen that fluorescence from the sensor material itself can considerably impact the presence of charge sharing in a detector's response. Fluorescence charge sharing events suffer less from charge loss and as a result show better spectroscopic performance and can be recombined through simple addition to recover the original photon energy [29]. A Monte-Carlo model is therefore a useful tool to investigate the relationship between pixel size and fluorescence events.

Using the data in Fig. 5 from Sect. 3.3, the portion of bi-pixels that are only due to fluorescence was calculated by applying two-dimensional energy windows around the bright fluorescence spots in the image, summing the number of counts in those regions and normalizing by the total number of bi-pixels. The same analysis was performed using simulated data. Table 4 shows the results from these calculations.

In order to perform the same analysis for tri-pixel events, the tri-pixels themselves were further separated into four types. These are defined as:

- RA centre events—where the pixels make a right angle (RA) and the maximum energy is recorded in the centre/middle pixel
- RA edge events—where the pixels also make a right angle but the maximum energy is recorded in one of the outer end pixels
- Line centre events—where the pixels make a line and the maximum energy is in the central pixel
- Line edge events—where maximum energy is in an outer pixel for a line tri-pixel

The four tri-pixel event types are shown in Fig. 9 where the filled pixel signifies the pixel with the maximum recorded energy. The majority of tri-pixel events were found to be in the right-angle arrangement for both the 59.5- and 122 keV photopeak experimental data, at 93.0% and 91.5%, respectively, with the rest being line events. From simulated data, it was estimated that 92.0% of all tri-pixels are in the right-angle arrangement at both photopeak energies.

The location of the maximum energy pixel in a tri-pixel event is an indication of whether charge sharing occurred due to fluorescence or charge spreading, for both the right-angle or line pixel arrangement. This can be seen in Fig. 9, which shows the distribution of the energy in the remaining two non-maximum pixels for each of the four tri-pixel types, once from the experimental data and again from simulated data. The distributions from the modelled data show similar results compared with

Table 4 Percentage of bi-pixels and tri-pixels that are due to fluorescence as calculated experimentally and predicted by the model. The uncertainties were calculated from the difference when using a 1- and 2-keV energy window around the fluorescence spots/streaks

	Experiment (%)		Model (%)	
	59.5 keV	122 keV	59.5 keV	122 keV
Bi-pixels	12.4 ± 2.1	9.3 ± 1.9	11.8 ± 1.3	8.6 ± 1.2
Tri-pixels	61.1 ± 1.1	57.0 ± 2.5	65.4 ± 1.5	59.3 ± 2.0

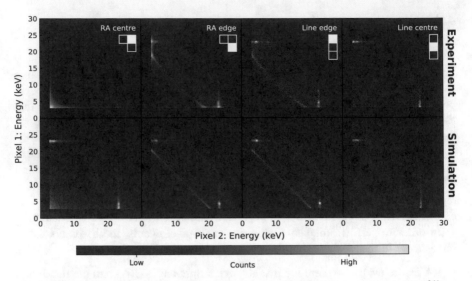

Fig. 9 Each image is a distribution of the energy in the two non-maximum pixels from the [241]Am data, including 59.5-keV photopeak tri-pixel events only. The two rows of images correspond to experiment (top) and simulation (bottom). Each column of images corresponds to one of the specific tri-pixel event types—RA centre, RA edge, line centre and line edge events. A drawing of the pixels making up the tri-pixel event type is shown in the first image of each column, with the filled pixel indicating the pixel containing the maximum energy. The energies in the unfilled pixels is therefore what is shown in the distributions

experiment. By visual inspection, it is clear the distributions for the edge and centre events are distinctly different. Edge events are mostly made up of events where summing the energy in the two non-maximum pixels equals the energy of Cd (~23 keV) or Te (~26 keV) K-shell fluorescence, for example at the bright streaks in the distribution at 20 and 23 keV. Counts along the downwards sloping curve are cases where the energy from the fluorescence photon spreads between the two non-maximum pixels. This leads to charge loss across the pixel gap, giving rise to the non-linear curve in the experimental data as observed previously in Fig. 5. The curves in the simulated edge event distributions are linear since the model does not account for charge loss across the pixel gap. Curves from the Cd K_β and Te K-shell fluorescence can therefore be resolved. The model agrees with experiment that edge events appear to be caused by fluorescence, where either an isolated or bi-pixel event becomes a tri-pixel due to the fluorescence photon.

The distribution for centre events in contrast show a large proportion of events with non-maximum pixel energies much lower around 3–5 keV—well below fluorescence energies. This suggests that centre events are predominately a result of charge spreading over to the adjacent pixels. The centre events distribution shows that some of the events are still due to fluorescence (i.e. the bright streaks), but at lesser rates than for edge events. The model predicts that the energy shared amongst the non-maximum pixels is less concentrated at lower energies than the experimental data shows—this again is likely a result of the simplifications made to the charge induction modelling.

By determining all edge events as resulting from fluorescence, and centre events from charge spreading (with the exception of the bright spots/streaks at 23 and 26 keV), the fraction of tri-pixels due to fluorescence was calculated, from both experimental and simulated data. Table 4 shows the portion of tri-pixels due to fluorescence from both experiment and model. The majority of tri-pixels appear to be a result of fluorescence, whereas bi-pixels are predominately due to charge spreading.

This comparison of modelled and experimental charge sharing events helps our understanding of the different processes and factors influencing charge sharing. This in turn has the potential to improve our understanding of inter-pixel charge loss and to inform methods of energy reconstruction for multi-pixel events, which will become increasingly important as detector designs for high-energy photon applications shift towards smaller pixels in a bid to achieve better spatial resolution and higher count rate imaging.

5 Summary

The detector response describes the spectrum recorded by the detector after effects such as electronic noise thresholding, photon pile up, fluorescence and charge sharing, which are all influenced by a detector's design. For high-energy photon detection such as of X-rays, design decisions such as pixilation, which offer the benefit of imaging and improved spectral resolution by means of the small pixel effect, come with some significant drawbacks. Relatively high-energy X-ray photons can produce charge clouds with sizes comparable to that of the pixel, resulting in detected signals in multiple pixels from a single photon event. The high Z materials often used to attain good X-ray detection efficiency emit fluorescence photons with mean absorption lengths large enough to escape into neighbouring pixels or escape the detector completely, meaning that the full energy of an incident photon is not detected by a single pixel. The magnitude of these effects largely depends on the pixel size, complicates a detector's response and therefore influences spectroscopic performance parameters such as spacial and spectral resolution, count rate capability and spectral efficiency.

Fortunately, the response of a detector can be modelled by use of the Monte-Carlo technique, such that the influence of a detector's design on its response can be studied. In this chapter we have described a possible Monte-Carlo model that successfully simulates the response of a pixelated detector. It is shown that although the model makes some simplifications, particularly to the charge induction processes, the rate of charge sharing can still be estimated well for a pixelated CdTe detector for multiple incoming photon energies. Additionally, the fraction of charge sharing events that are due to fluorescence shows good agreement. The fit between model and experimental charge sharing rates means that we can confidently extract information on the charge cloud sizes from the model. The use of the model in this chapter focused primarily on estimating the proportion of charge sharing due to the spreading of charges and fluorescence as we have demonstrated how charge sharing events negatively impact spectral resolution; however, it could equally be used to study other spectroscopic performance parameters such as the spectral efficiency and count rate capabilities of a detector as pixel size and incoming count rate are varied.

The limitations of our Monte-Carlo model were observed as it failed to accurately predict the correct experimental rates for the different multi-pixel event types at higher incoming photon energies. This loss in accuracy is likely a result of using the simplified charge induction approach employing the Hecht relation modified to work with a one-dimensional weighting potential to calculate induced charge, and using the computed cloud size to estimate the induced charge distribution over the pixelated anode. In contrast, a 3-dimensional weighting field and numerical solutions to the Shockley–Ramo theorem to calculate signals induced on multiple pixels for individual charge carrier can be used to achieve greater accuracy. This however is more complicated to implement and comes at the cost of greater computational demand than our model, which we have shown can nonetheless simulate a detector's response to good agreement.

References

1. Barret, H., & Eskin, J. (1995). Charge transport in arrays of semiconductor gamma-ray detectors. *Physical Review Letters, 75*(1), 156–159.
2. Jones, L., Seller, P., Wilson, M., & Hardie, A. (2009). HEXITEC ASIC-a pixellated readout chip for CZT detectors. *Nuclear Instruments and Methods in Physics Research A, 604*(1–2), 34–37.
3. Python. https://www.python.org/downloads/release/python-370/. Accessed on 7 April 2020.
4. Evans, R. (1955). *The atomic nucleus* (1st ed.). New York: McGraw-Hill Book Company, Inc.
5. Berger, M., Hubbell, J., Seltzer, S., Chang, J., Coursey, J., Sukumar, R., et al. (1998). *XCOM: Photon Cross Sections Database.* NIST, PML, Radiation Physics Division (pp. 87–3597).
6. Jenkins, R. (1995). *Quantitative X-ray spectrometry. Practical spectroscopy* (Vol. 20, 2nd ed.). New York: Marcel Dekker Inc.
7. Born, M. (1969). *Atomic physics* (8th ed.). Glasgow: Blackie and Sons.
8. Compton, A. (1923). A quantum theory of the scattering of X-rays by light elements. *Physical Review, 21*(5), 483–502.
9. Klein, O., & Nishina Y. (1929). Über die Streuung von Strahlung durch freie Elektronen nach der neuen relativistischen Quantendynamik von Dirac. *Zeitschrift für Physik, 52*, 853–868.
10. Thompson, A., & Attwood, D. (2009). *X-ray data booklet* (3rd ed.). Berkeley, CA: Center for X-ray Optics and Advanced Light Source.
11. Owens, A. (2012). *Compound semiconductor radiation detectors* (1st ed.). European Space Agency, Noordwijk: Taylor & Francis Group.
12. Redus, R., Pantazis, J., Pantazis, T., Huber, A., & Cross, B. (2009). Characterization of CdTe detectors for quantitative X-ray spectroscopy. *IEEE Transactions on Nuclear Science, 56*(4), 2524–2532.
13. Shockley, W. (1938). Currents to conductors induced by a moving point charge. *Journal of Applied Physics, 10*(9), 635.

14. Benoit, M., & Hamel, L. (2009). Simulation of charge collection processes in semiconductor CdZnTe γ-ray detectors. *Nuclear Instruments and Methods in Physics Research A, 606*(3), 508–516.
15. Trueb, P., Zambon, P., & Broennimann, C. (2017) Assessment of the spectral performance of hybrid photon counting X-ray detectors. *Medical Physics, 44*(9), 207–214.
16. Dreier, E. S., Kehres, J., Khalil, M., Busi, M., Gu, Y., Feidenhans'l, R., et al. (2018). Spectral correction algorithm for multispectral CdTe X-ray detectors. *Optical Engineering, 57*(05), 1.
17. Blevis, I., & Levinson, R. (2015). In J. Iwanczyk (Ed.), *Photon-counting detectors and clinical applications in medical CT imaging* (pp. 169–192). Oxfordshire/Oxford: Taylor & Francis Group/Oxford University Press.
18. Iniewski, H. (2007). Modeling charge-sharing effects in pixellated CZT detectors. In *Nuclear Science Symposium Conference Record* (Vol. 6).
19. Fick, A. (1855). Ueber Diffusion. *Annalen der Physik, 94*, 59–86.
20. Einstein, A. (1901). Über die von der molekularkinetischen Theorie der Wärme geforderte Bewegung von in ruhenden Flüssigkeiten suspendierten Teilchen. *Annalen der Physik, 332*(8), 549–560.
21. Strzalkowski, I., Joshi, S., & Crowell, C. R. (1976). Dielectric constant and its temperature dependence for GaAs, CdTe, and ZnSe. *Applied Physics Letters, 28*(6), 350.
22. Runge, C. D. T. (1895). Über die numerische Auflösung von Differentialgleichungen. *Mathematische Annalen, 46*(2), 167–178.
23. Ramo, S. (1939). Currents induced by electron motion. *Proceedings of the IRE, 27*(9), 584–585.
24. He, Z. (2001). Review of the Shockley-Ramo theorem and its application in semiconductor gamma-ray detectors. *Nuclear Instruments and Methods in Physics Research A, 463*(1–2), 250–267.
25. Hecht, K. (1932). Zum Mechanismus des lichtelektrischen Primärstromes in isolierenden Kristallen. *Zeitschrift für Physik A Hadrons and Nuclei, 77*, 235–245.
26. Duarte, D., Lipp, J., Schneider, A., Seller, P., Veale, M., Wilson, M., et al. (2016). Simulation of active-edge pixelated CdTe radiation detectors. *Nuclear Instruments and Methods in Physics Research A, 806*, 139–145.
27. Owens, A., & Peacock, A. (2004). Compound semiconductor radiation detectors. *Nuclear Instruments and Methods in Physics Research, 531*(1–2), 18–37.
28. Kalemci, E., & Matteson, J. L. (2002). Investigation of charge sharing among electrode strips for a CdZnTe detector. *Nuclear Instruments and Methods in Physics Research A, 478*(3), 527–537.
29. Bugby, S., Koch-Mehrin, K., Veale, M., & Lees, J. (2019). Energy-loss correction in charge sharing events for improved performance of pixellated compound semiconductors. *Nuclear Instruments and Methods in Physics Research A, 940*, 142–151.
30. Koch-Mehrin, K., Lees, J., & Bugby, S. (2020). A spectroscopic Monte-Carlo model to simulate the response of pixelated CdTe based detectors. *Nuclear Instruments and Methods in Physics Research A, 976*. https://doi.org/10.1016/j.nima.2020.164241
31. Cola, A., & Farella, I. (2009). The polarization mechanism in CdTe Schottky detectors. *Applied Physics Letters, 94*(10), 1–4.
32. Veale, M., Seller, P., Wilson, M., & Liotti, E. (2018). HEXITEC: A high-energy X-ray spectroscopic imaging detector for synchrotron applications. *Synchrotron Radiation News, 31*(6), 28–32.
33. Cola, A., & Farella, I. (2006). Electric field properties of CdTe nuclear detectors. In *IEEE Nuclear Science Symposium Conference Record*.

High-Z Pixel Sensors for Synchrotron Applications

Stergios Tsigaridas and Cyril Ponchut

1 Towards the 4th Generation of Synchrotrons

Radiation emitted from accelerated charges was predicted already in the beginning of the twentieth century [1]. However, the experimental observation of such radiation performed a few decades later thanks to the development of electron accelerators. The first indirect detection took place on 1945 [2], followed by the first visual observation 2 years later [3]. Both achieved by utilising the electron synchrotron hosted at the Research Laboratory of General Electric. Hence this type of radiation is called synchrotron radiation. Soon enough it was realised that the synchrotron light could provide a boost to science by overcoming some limitations of the conventional X-ray tubes. Thanks to their unique properties synchrotron beams are a powerful tool for probing the structure of matter at the molecular and the atomic scale.

In the early years and up to 1970, the only available synchrotron radiation sources were the electron synchrotrons. These sources, often called as the 0th generation, were able to provide high intensities; however, limitations were arising from the fluctuations per cycle of the source position, the spectrum and the intensity. The development of the storage rings for high energy physics applications led to sources with constant source position and spectrum which provide a long beam lifetime. The usage of synchrotron radiation was limited to parasitic operation in parallel with the high energy physics experiments and these kinds of sources referred to as the 1st generation. The beneficial use of synchrotron radiation became quickly evident which in turn initiated the discussions for the construction of storage rings specifically dedicated to the production of synchrotron beams.

S. Tsigaridas (✉) · C. Ponchut
European Synchrotron Radiation Facility (ESRF), Grenoble, France
e-mail: stergios.tsigaridas@esrf.fr; ponchut@esrf.fr

© Springer Nature Switzerland AG 2022
K. Iniewski (ed.), *Advanced X-ray Detector Technologies*,
https://doi.org/10.1007/978-3-030-64279-2_4

With an emerging synchrotron radiation community in terms of people and scientific output, storage rings designed for utilisation as light sources became a reality. These specialised sources established the 2nd generation and by 1980 there were many (1st and 2nd generation) sources in operation/construction all over the world, [4]. In their primary form, the radiation was emitted tangentially to the orbit of the accelerated particles by the bending magnets of the storage ring. The synchrotron light produced in this case covers a broad and continuous spectrum. Gradually in the mid-1980s, the free straight sections of the storage rings were equipped with wigglers and undulators, [5–7]. These structures are known as insertion devices and consist of magnet arrays. The alternating magnetic field in the insertion devices deflects the particle beam sinusoidally so that the emitted radiation is concentrated forward in a narrow cone and at specific spectral range.

Thanks to the insertion devices the synchrotron radiation sources were able to offer light with even higher intensity over a wide spectral range turning synchrotrons into multidisciplinary research centres. As a natural consequence, there was a great interest to enhance the source performance and the beam quality. The typical measure of the beam quality is its brightness or in other words the concentration of light per unit area and time. The brightness is measured in photons/s/mm^2/$mrad^2$/0.1%BW and depends on the emittance of the particle beam, which is a measure of the collimation of the beam. The emittance of a particle beam is given by the product of its angular divergence and its transverse size. The 2nd generation sources with their design at that time were able to reach a beam brightness of about 10^{15}, see Fig. 1. The further increase of the brilliance would require a refined design in order to achieve lower particle beam emittance.

In the early 1990s, the European Synchrotron Radiation Facility (ESRF) [8], the Advanced Photon Source (APS) [9] and the Super Photon ring-8 GeV (SPring-8) [10] became operational and were the first 3rd generation sources. These sources implement a high energy storage ring (6 GeV, 7 GeV and 8 GeV, respectively) including many straight sections and an optimised design for insertion devices capable to deliver a brilliance up to 10^{20}, see Fig. 1. Thanks to the improved performance, the 3rd generation sources deliver stable X-ray beams over a wide spectral range with a high flexibility, enabling plenty of new applications. Nowadays there are more than 50 light sources worldwide [11], hosting a large number of users every year. The strong impact of synchrotron radiation is also confirmed by the rapid increase of industrial applications [12]. With such an extensive use of light sources from academia and industry a reasonable question would be whether a further increase in the performance would be feasible.

The endeavour to achieve increased performance is targeting the development of a new generation of synchrotron radiation sources which outperform their predecessors in terms of beam quality parameters such as the brightness or the coherence by an order of magnitude at least. As mentioned above, in order to maximise the brilliance it is essential to minimise the particle beam emittance which translates in reduced emittance for the photon beam. However, this is effective up to some extent where due to diffraction the lower limit on the photon beam emittance is determined by the wavelength λ. Reduction of the particle beam emittance below

Fig. 1 An illustration of the evolution of the peak brilliance delivered by the various generations of synchrotron radiation sources towards the diffraction-limited storage rings

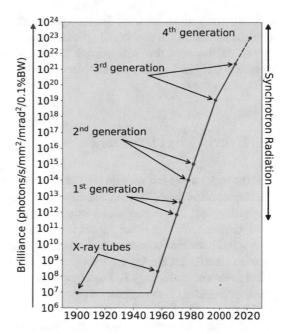

the diffraction limit may cause stability problems. Therefore, the guideline for new storage ring designs is that the particle beam emittance should not exceed the photon beam emittance, [13]. Such ultra-low emittance storage rings or diffraction-limited storage rings belong to the 4th generation of synchrotron radiation sources.

In the beginning of 2000s, new ideas for the design of 4th generation sources introduced based on a concept of multiple bend achromat lattice (MBA) proposed by Einfeld [14]. This concept implements a unit cell with an increased number of bending magnets compared to the double bend achromat lattices (DBA) present in 3rd generation sources. Such a design would be able to deliver photon beams more brilliant and coherent by two orders of magnitude and with an emittance approaching the diffraction limit. This would enable the focusing of X-rays to the nanoscale level, opening new opportunities for fundamental and applied research.

The first 4th generation source is the MAX-IV in Sweden [15, 16], implementing a MBA lattice in a 3 GeV storage ring. The ESRF in 2012 introduced a new lattice based on MBA, the hybrid multi-bend achromat (HMBA) or Raimondi lattice [17]. The new storage ring (currently under commissioning) known as the Extremely Brilliant Source (EBS) is expected to increase the brilliance and coherence of the X-ray source by a factor of 100. The HMBA-lattice has been adopted as the basis for the design of other future 4th generation facilities such as the upgraded Advanced Photon Source (APS-U) [18], the Advanced Light Source Upgrade (ALS-U) [19], the upgrade to PETRA-IV [20] and the update to Spring8-2 [21].

The modern synchrotron radiation sources will boost the brilliance and coherence of photon beams at unprecedented levels and bring photon science to a new era. Furthermore, such developments set new requirements for instrumentation. Besides the necessary advancements for the optical elements, there is a need for precise detectors, efficient over a broad spectral range and able to cope with high photon fluxes. In the next few sections we discuss the recent advances in hybrid pixel detectors (HPD), focusing on the materials used for the direct conversion of X-rays to electric charge.

2 Hybrid Pixel Detectors

For the various experiments taking place in the beamlines of synchrotron facilities, the high performance on the detection of X-rays is always an important factor. Depending on the experimental conditions, there is a wide range of detector technologies featuring either the direct conversion of X-rays into an electric charge or a scintillator that converts the X-rays into visible light first which is then turned into an electric charge by a photodetector. For the direct detection of X-rays the most common technology is a detector where a semiconductor is used as an active medium. Incident X-rays generate charge carriers (electron and holes) which can be collected by the readout electronics.

Since the 1960s, monocrystalline silicon (Si) wafers have been used extensively by the semiconductor industry for the fabrication of integrated circuits. The rapid growth in the silicon technology attracted soon the attention of the detector community. In the late 1970s, particle physics experiments were studying the production of short-lived particles with lifetimes in the order of picoseconds. In such experiments, there is a need for very precise tracking detectors to identify the secondary particles, the decay vertex and to measure decay distances. At that time the first silicon strip prototypes based on the surface barrier technology appeared. However, the availability of such detectors was limited by the complicated manufacturing technology. A step forward in the production of silicon detectors was taken by Kemmer [22] in 1980, with the production of single-sided silicon microstrip detectors by making use of the standard planar process of silicon wafers.

The early Si microstrip detectors were able to provide high precision; however, the readout electronics were occupying much more space than the actual active area. The solution was provided by the development of custom readout electronics using VLSI technology [23]. The critical breakthrough for detector community was the concept of hybridisation and the first pixel detectors. On 1984, Gaalema [24] introduced the design of a pixelated readout chip for focal plane imaging sensors. The readout chip could be coupled to an array of semiconductor diodes by means of bump-bonding for the detection of X-rays. This triggered the attention of CERN which performed first a feasibility study [25] and soon after initiated a collaboration for the development of silicon pixel detectors [26].

The performance of the first prototypes [27] demonstrated the great potential of the HPDs and motivated the first utilisation of such a detector in a synchrotron beamline [28]. The advances in the CMOS technology which enabled the design of readout chips with more functionality per pixel led to the realisation of HPDs with optimised designs for the detection of single photons [29]. This development was critical for photon science and a large number of synchrotron beamlines in 3rd generation facilities progressively replaced the traditional charged-coupled devices (CCDs) with HPDs.

2.1 Working Principle

Figure 2a shows the basic geometry of a HPD, which consists of two main components, the sensor and the readout chip. The pixelated sensor is made out of high resistivity semiconductor wafers and in fact is an array of reversed biased diodes or non-rectifying electrical junctions. The chip is a custom Application Specific Integrated Circuit (ASIC) produced with CMOS technology that contains a pixel matrix where each individual pixel is equipped with readout electronics. The most common technique used to connect the sensor and the readout chip is the bump-bonding technique, where each pixel of the sensor is connected to a pixel of the readout chip with a solder bump.

An absorbed X-ray photon in the sensor layer, generates by ionisation a certain number of electrons and holes which depends on the energy of the incident photon. The electric field applied between the back side metallisation and the pixels contacts separates the charge carriers and depending on its polarity the electrons (or the holes) drift towards the pixels. The signal induced in each pixel of the sensor layer is collected by the corresponding pixel of the readout chip and is then further processed by its circuitry.

Due to the separate development of the sensor and the readout electronics, the HPDs offer great flexibility on the design. This technology is not limited only to Si and is also compatible with other types of single-element or compound semiconductors. Thanks to the miniaturisation of electronics, advanced functionality can fit within a few tens of μm^2 which is the typical size of a pixel. Depending on the target application, there are many variants for the design of custom ASICs. For applications in X-ray science, new developments are focusing on ASICs for single photon-counting and recently on ASICs with charge-integrating capability.

2.2 Photon Counting

The HPDs equipped with ASICs capable for single photon-counting revolutionised the X-ray detection in synchrotron facilities. Thanks to their superior performance compared to traditional technologies, photon-counting detectors are offering improved performance over a large energy range. Moreover, they provide high spatial resolution due to the small pixel size and accurate measurement of the photon intensity over a large dynamic range, preserving at the same time the single photon

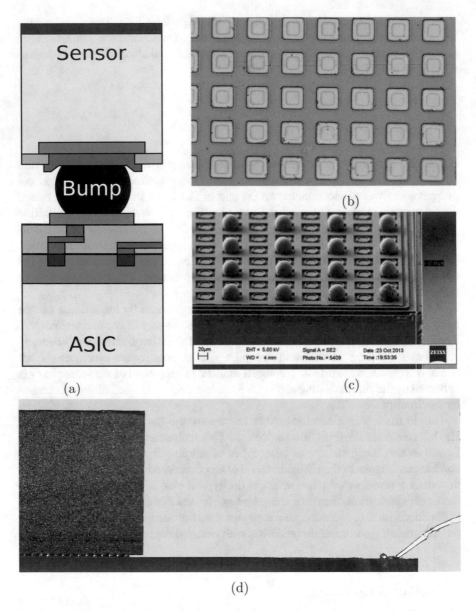

(a)

(b)

(c)

(d)

Fig. 2 (a) An illustration of a pixel cell which is the basic building block of a HPD. During the bump-bonding procedure, a solder bump is used to connect a pixel of the semiconductor sensor on top with a readout pixel of the ASIC at the bottom. (b) A photograph showing a small area of the pixelated side of a semiconductor sensor. (c) Solder bumps deposited on every second pixel of a readout ASIC with a pixel pitch of 55 μmm. The photograph is courtesy of ADVACAM. (d) A close look under microscope of a HPD after the bump-bonding. The solder bumps are visible in between the semiconductor sensor on top and the readout ASIC at the bottom. On the right hand side a wire bond is also visible

sensitivity. As a result, such detectors have become a standard for synchrotron beamlines specialised in crystallography, diffraction or small-angle scattering.

In photon-counting ASICs, the signal induced in each pixel is collected at the input pad and amplified by a charge sensitive amplifier. The amplified signal then goes through a band-pass filter, which acts as a pulse shaper. The output of the shaper is fed to a comparator and if the signal exceeds a user-defined threshold value, it is routed to the digital part of the pixel for further processing. The digital part implements a counter which is incremented every time the signal crosses the threshold. The main advantage of this technology is the noise-free operation for threshold values well above the amplifier noise, typically set at five times above. Moreover, photon-counting ASICs are less sensitive to leakage current thanks to dedicated circuitry within their front-end, which allows the compensation of leakage currents up to a certain range.

By implementing designs with more circuitry per pixel, photon-counting detectors with more functionality are now possible. One or more additional thresholds can be used to target photons in specific energy windows, useful for material discrimination. With more sophisticated counters, the ASICs can measure the time-over-threshold or the time-of-arrival of the signal, apply the on-chip correction of charge sharing between adjacent pixels and achieve higher frame rates. Examples of photon-counting ASICs are the PILATUS [30] and EIGER [31] developed by the detector group of the Paul Scherrer Institude (PSI) and the Medipix family of chips [32] developed by CERN. Large area systems based on these ASICS (or their next generations) operate in many synchrotron beamlines. The development of such systems performed either by synchrotron detector groups like the MAXIPIX [33], the Excalibur [34] and the LAMBDA [35] or by commercial companies.

Despite their versatility, the photon-counting detectors are count-rate limited due to the pile up effect. Coincident photons in a single pixel induce overlapping signals which are detected as one count. The pile up constraints the count rate of photon-counting systems for a given pixel size [36]. This is a well-known problem that should be addressed for future detectors that will be used in beamlines at 4th generation synchrotron sources. The expected increase of the coherent flux in diffraction-limited storage rings would require detectors with high count-rate capability. Going to smaller pixel sizes helps to deal with higher fluxes, however in order to extend the dynamic range of the count-rate alternative approaches in the ASIC design have to used.

2.3 Charge Integrating

Due to the count-rate limitations of the photon-counting HPDs, the photon science community is aiming to the development of new detector technologies in order to facilitate high-flux and fast time-resolved applications in 4th generation synchrotrons.[1] The most common technology under consideration is the HPDs with a charge-integrating front-end. In order to ensure the high performance of charge-integrating front-ends, it is crucial to keep the noise and the dark current at the lowest possible level. This technology offers high count-rate capabilities maintaining at the same time the single photon sensitivity.

In charge-integrating ASICs, the charge collected at the input pad of a single pixel is amplified by a charge sensitive amplifier in integrating configuration. In this approach the charge integration is implemented either in a scheme of multiple feedback capacitors (with increasing capacitances), which provides an adaptive gain scheme or with a single feedback capacitor combined with a charge removal circuit to avoid saturation. For high frame-rate applications, the charge-integrating ASICs are featuring on-chip storage of a certain number of frames temporarily, with the off-chip transmission taking place on a later stage.

Several charge-integrating ASICs for HPDs have been developed and tested the past few years. For example, the CS-PAD chip [37] and a few chips from the ePix series [38] implement a statically selected gain switching in view of low noise or larger dynamic range. Other ASICs implement the dynamic gain switching (or adaptive gain) approach such as the JUNGFRAU chip [39], the MÖNCH chip [40] and the AGIPD chip [41]. The alternative approach of a charge-integrating front-end with charge removal circuit is implemented by the MM-PAD [42] and the HDR-PAD [43].

3 High-Z Sensor Materials

For the majority of the HPDs used in synchrotron facilities, silicon sensors are used for the direct conversion of photons to electrical charge. The silicon technology has evolved over the years and today the fabrication of wafers containing high quality silicon pixel sensors is possible. HPDs coupled to silicon sensors are able to achieve low dark (leakage) current operation and show stability for long-term usage. In addition, thanks to absence of major crystal defects and the high spatial resolution down to a few tens of microns (μm), silicon sensors are suitable for many applications which require the detection of photons in the energy range 2–20 keV (soft X-rays).

However, the absorption efficiency of silicon drops significantly for photons with energies higher than 20 keV. This is limiting the use of silicon sensors for experiments using hard X-rays. In order to overcome this limitation, compound semiconductor materials of high atomic number (Z) are under investigation for use in HPDs. The most common materials used today are Gallium Arsenide

---

[1] This discussion also applies to another type of light sources, the X-ray Free electron lasers (XFELs). Such light sources are able to produce high instantaneous photon fluxes in short periods of time. However, this kind of sources is out of the scope of this chapter.

Table 1 Some characteristic properties of Si in comparison with high-Z compound semiconductors

Property	Si	GaAs.	CdTe	CdZnTe[a]
Z (average)	14	32	50	49.1
Density (g/ cm^3)	2.33	5.32	5.85	5.81
Energy gap (eV)	1.12	1.42	1.44	≥ 1.6
Intrinsic carrier concentration (1/ cm^3)	$\sim 10^{10}$	2×10^6	$\sim 10^7$	$\sim 10^7$
W-value (eV)	3.62	4.3	4.43	4.64

[a] For composition $Cd_{0.9}Zn_{0.1}Te$

(GaAs), Cadmium Telluride (CdTe) and Cadmium Zinc Telluride (CdZnTe). Table 1 summarises some of the properties of these materials in comparison with Si. Pixel sensors made from high-Z materials offer high absorption efficiency due to their high density and their wide band gap allows low dark (leakage) current operation at room temperature.

The major challenge for the development of these materials is the level of the crystal quality. Mostly due to their nature of being compounds, the fabrication of high-Z semiconductors is not trivial. These materials tend to be more brittle and need special care for handling and processing. Even though, reliable techniques developed in the past years allow the production of 3–4 inch wafers, which are then processed and diced for use in HPDs. During the crystal growth of such materials certain type of defects may appear due to dislocations, impurities, metallic precipitates and grain boundaries. Additional surface or bulk detects can be introduced during the processing of the material into pixel sensors.

Regardless of their origin, the defects have a strong impact on the uniformity and the performance of the sensor. The defects are randomly distributed over the crystal, being additional sources of leakage current and active centres for the trapping of charge carriers. The electric field in the close vicinity of the defects is distorted which in turn leads to certain areas of the sensor with reduced (or increased) sensitivity with a direct impact on the charge collection efficiency. Such variations though can be corrected in order to restore the uniformity of reconstructed images, provided that the count rate remains stable. However, in most cases, defects show a dynamic behaviour with irradiation and time. In this case, such a correction would be less effective which could be crucial for long-lasting experiments or high-flux applications. The advances in the crystal growth technology enable the fabrication of high-Z pixel sensors with an improved quality. In the next few subsections we report recent development for GaAs, CdTe and CdZnTe sensors.

3.1 Chromium Compensated Gallium Arsenide (GaAs:Cr)

GaAs is a well-known material for the semiconductor industry and is used widely for microwave integrated circuits and optoelectronic devices. With a composition of equal amounts of gallium ($Z = 31$) and arsenic ($Z = 33$) giving an average atomic number of 32 provides sufficient absorption up photons with energies up to 50 keV. In addition, thanks to its large band gap of 1.42 eV it can be operated at room temperature with low leakage current values. Shortly after the birth of hybrid pixel technology, the application of epitaxial or semi-insulating GaAs in pixel detectors was reported [44, 45]. However, it was difficult to achieve the fabrication of thick sensors with good transport properties mainly due to limitations in the growth technique.

On 2003 though, a research group demonstrated an alternative approach for the fabrication of GaAs sensors with improved quality [46]. In this technique, low resistivity n-type GaAs wafers obtained with liquid encapsulated Czochralski (LEC) growth are diffused with chromium (Cr) in a high-temperature procedure. The resulting material known as chromium compensated gallium arsenide (GaAs:Cr) is available in 3–4 inch wafers with a thickness up to 1 mm. The high resistivity of GaAs:Cr ($\sim 10^9$ Ω cm) and the improved charge transport properties make it suitable for use in X-ray detectors.

Within the past few years, GaAs:Cr attracted quite some attention from several research groups interested in X-ray imaging applications. HPDs using GaAs:Cr coupled to photon-counting or charge-integrating ASICs in a single or multi-chip configuration have been developed and tested. One of the main objectives for imaging applications is the quality of the reconstructed image. A basic qualitative measure of the sensor quality is the uniform illumination with a wide X-ray beam. Figure 3a shows a raw image recorded with a 500 μmm thick GaAs:Cr sensor irradiated using a Ag-anode X-ray tube. The image reveals pixel-to-pixel variations in the detector response which are associated with non-uniformities in the material. The granular pattern observed also in [47–49] and other works corresponds to dislocation networks arising during the crystal growth procedure. Such dislocations alter the electric field locally in the sensor resulting in deformations of the effective pixel shape as shown in Fig. 4b.

The stability observed in the detector response over time even after intense irradiation [50, 51] makes possible the flat-field correction in order to improve the image quality. Such a correction improves the signal-to-noise ratio (SNR) of the images by minimising the pixel-to-pixel variations and is stable over time [49]. However, one should be careful when applying the flat-field correction, since the effect of the pixel deformations on the spatial resolution is still present. The optimisation of the crystal growth of the n-GaAs wafers could lead to a more homogeneous material. Alternative growth techniques such as the vertical gradient freeze (VGF) are under consideration and could lead to lower dislocation density [52].

The charge transport properties of the material depend on the mobility and the lifetime of the free charge carriers. Their product known as $\mu\tau$-product is the parameter used to describe the dependence of the charge collection efficiency of a sensor with a certain thickness and for a given bias voltage. Several works reported the experimental determination of the $\mu\tau$-product of electrons in GaAs:Cr pixel sensors with typical values in the range $\mu\tau_e = (0.5–2) \times 10^{-4}$ cm^2/V. On a per-pixel basis the $\mu\tau_e$-product profile depends on the inhomogeneities and it follows the

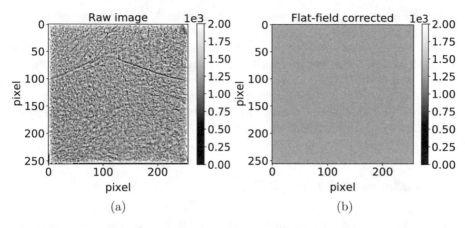

Fig. 3 (**a**) A raw image showing the response of a GaAs:Cr sensor with 500 μmm thickness and 55 μmm pixel pitch, under uniform illumination with X-rays provided from a Ag-tube operated at (35 kV, 25 mA). The image was taken with the sensor biased at −300 V and an energy threshold of 11 keV. (**b**) The corresponding image after flat-field correction

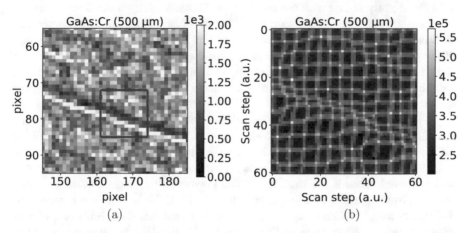

Fig. 4 (**a**) A close-up look of the raw image in Fig. 3a, focusing on a visible defect of the GaAs:Cr sensor. A monochromatic micro-focused X-ray synchrotron beam with 20 keV energy was aligned on the defect. The beam was used to perform a mesh (raster) scan with a finite step of 11 μmm within a 12×12 pixel area highlighted by the red rectangle. The energy threshold was set at 5 keV in order to get excessive counts in the inter-pixel areas due to charge sharing so as to better visualise the pixel borders and thus their effective shape. (**b**) The reconstructed map of pixel counts recorded during the mesh scan reveals deformations of the effective pixel shape due to the defect. The image was reproduced from reference [49]

same pattern [53]. For holes due to their short lifetime in GaAs:Cr, the measurement of the $\mu\tau$-product is not straightforward. Recently, it has been observed [54–56] that the short lifetime of holes could trigger the crater or halo effect. For photons absorbed close to the pixel electrode a signal with opposite polarity is induced in the neighbouring pixels. This is a sensor effect which could lead to lower count rates with increasing photon flux and energy.

Despite their imperfections, GaAs:Cr sensors offer stable performance, which could allow their use in synchrotron beamlines. In the photon energy range between 20 keV and 50 keV, GaAs:Cr provides high photon absorption, while it is less affected by Compton scattering or the fluorescence photons produced in the sensor. Therefore GaAs:Cr pixel detectors would make a good candidate for imaging applications in the medium energy range.

3.2   Cadmium Telluride (CdTe)

CdTe is a compound semiconductor composed out of equal amounts of cadmium (Cd) and tellurium (Te) with atomic numbers $Z = 48$ and $Z = 52$, respectively. Hence, its average atomic number $Z = 50$ offers sufficient absorption efficiency of photons with high energies even above 100 keV. The band gap of CdTe is about 1.44 eV, slightly higher than the band gap of GaAs. With typical values of $\mu \tau_e \approx 10^{-3}$ cm^2/V and $\mu \tau_h \approx 10^{-4}$ cm^2/V CdTe offers better charge transport properties with respect to GaAs:Cr. The larger drift lengths before trapping or recombination of the charge carriers allow the use in spectroscopic applications with high energy X-rays and γ-rays. However, the long ranged fluorescence photons emitted from Cd and Te in the range of 20–30 keV may escape or be re-absorbed leading to signal loss or image blurring, respectively.

The majority of pixelated CdTe sensors today are fabricated utilising the travelling heater method (THM) [57]. In this method, an ingot of polycrystalline CdTe is added in a furnace together with a Te-rich alloy doped with chlorine (Cl). In a high-temperature process, the Te-rich alloy melts and the CdTe ingot is slowly moved towards the melt zone. The polycrystalline ingot is dissolved and re-crystallised resulting in a single crystal material. CdTe is available in wafers of 3–4 inches with a thickness up to a few mm; however, it is more brittle than GaAs which is crucial during the hybridisation process, especially during bump-bonding.

Concerning the uniformity of CdTe, several types of defects have been observed which may occur during the crystal growth or the post-processing into pixelated sensors. Figure 5 shows a raw image taken with a 1 mm thick CdTe sensor with Ohmic contacts, illuminated with a uniform wide X-ray beam. The most characteristic type of defects that is observed is a network of lines all over the sensor area. Such lines show up due to dislocations along grain boundaries, usually emerging during the crystal growth. The electric field close to such lines is distorted which leads to an increased count rate measured by the corresponding pixels. In Fig. 5 other defects appear in the form of blobs, with a size of several pixels that may increase with the bias voltage. The blobs are associated with Te inclusions close to the surface and act as leakage current sources. Due to the increased current, the corresponding pixels are constantly above threshold and therefore measure one

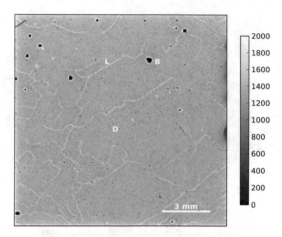

Fig. 5 A flood image recorded with a 1 mm thick CdTe sensor bonded to a Timepix chip [58]. The image obtained after uniform illumination of the sensor with X-rays using a Ag-anode tube (characteristic energy of 22 keV). The characteristic defects of type (L) appearing as network of lines correspond to dislocations at grain boundaries. The type (B) defects are large blobs covering several pixels and are associated with Te inclusions. Defects of type (D), appearing as isolated tiny black spots, correspond to single pixels that were not connected during the bump-bonding procedure. The image was reproduced from reference [59]

count. Randomly distributed isolated dots with no counts correspond to unconnected pixels.

One major challenge for the fabrication of CdTe sensors is their stability over time due to polarisation effects. Due to the trapping of the charge carriers at the sensor defects, the electric field shows a time-dependent behaviour with irradiation. The space charge build-up gradually leads to the degradation of the electric field and consequently of the count rate. The polarisation effects seem to depend on the contact type used. CdTe devices with Schottky contacts operate at lower leakage current regimes; however, they tend to suffer from polarisation effects [57]. Devices with Ohmic contacts are less susceptible to polarisation and therefore suited for high-flux applications, though polarisation has been observed after irradiation over time [60]. In both cases, the long-term operation would require the reset of bias voltage or other solution [61].

Figure 6a shows a flood image recorded with a 1 mm CdTe sensor with Ohmic contacts. In the first place, the image reveals less blob defects compared to the one in Fig. 5, presumably due to lower density of Te inclusions. With both images recorded at similar operating conditions, the difference could be related to improvements in the crystal growth method since this sensor was fabricated a few years later. Nonetheless, the characteristic network of line defects is still present. Figure 6b shows a flood image recorded after 3 h of continuous irradiation. As illustrated in Fig. 6c, the sensor exhibits along the line defects an increased count rate in the order of 25–30% due to the charge build-up at the grain boundaries. In this case it is practically difficult to apply the flat-field in a real time experiment.

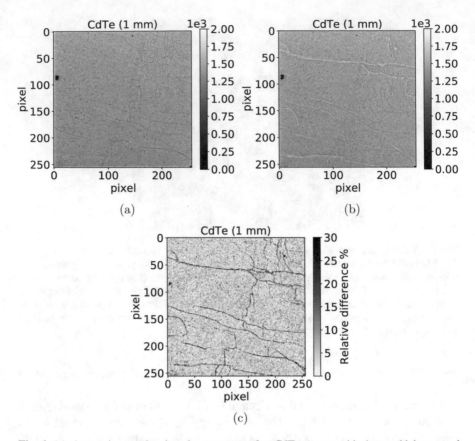

Fig. 6 (**a**) A raw image showing the response of a CdTe sensor with 1 mm thickness and 55 μmm pixel pitch, under uniform illumination with X-rays provided from a Ag-tube operated at (35 kV, 25 mA). The image was taken shortly after the ramp up to a bias voltage of −500 V with the energy threshold set at 11 keV. (**b**) A raw image of the same sensor recorded after the continuous illumination for 3 h at the same settings. (**c**) The relative percentage difference between the two images

CdTe thanks to its high absorption efficiency and the ability to withstand relatively high fluxes finds application in many synchrotron beamlines around the world [62]. Stability issues are posing certain limitations at the moment; however, with improved crystal quality, in the future such limitations would be less effective.

3.3 Cadmium Zinc Telluride (CdZnTe)

A more recent high-Z compound semiconductor alternative to CdTe is the CdZnTe. In the early 2000s CdZnTe has attracted some attention mostly for spectroscopic applications [63]. A small fraction of Cd is replaced by zinc (Zn), which is a metal with atomic number $Z = 30$, resulting in a crystal with a composition of $Cd_{1-x}Zn_x Te$, with $x = 0.08 - 0.15$. With the addition of Zn an increased band gap of $\sim 1.6\,eV$ (assuming a composition of $Cd_{0.9}Zn_{0.1}Te$) can be achieved which yields a resistivity larger by one order of magnitude with respect to CdTe. This in turn allows lower leakage currents at room temperature operation and therefore improved spectroscopic capability. CdZnTe with an average atomic number slightly smaller than the one of CdTe is also suited for the detection of high energy X-rays.

The growth of CdZnTe crystals was performed initially by making use of variants of the Bridgman technique [64]. In this method, a polycrystalline material is heated up beyond its melting point and brought into contact with a single crystal seed. A translation towards a cold section of the container gradually solidifies the melt at the interface with the seed, forming a crystal which follows in principle the crystallographic orientation of the seed. The Bridgman method, however, provides polycrystalline ingots which often makes difficult to find large single crystal areas. Sensors produced with Bridgman variants may contain grain boundaries responsible non-uniform charge transport properties, limiting the performance in spectroscopic and imaging applications.

In the past few years, it was demonstrated that high quality CdZnTe crystals can be fabricated utilising the THM method [65, 66]. The optimisation of the growth method by adding a few processing steps more enables the improvement of the $\mu\tau$ for holes [67]. An improvement by one order of magnitude has been reported in [68], mainly due to the larger lifetime of holes. The improvement is associated with the lower concentration of hole traps and therefore leads to a material less susceptible to polarisation. As a result, the use of THM-grown CdZnTe sensors in imaging applications with flux up to 200×10^6 counts/ mm^2/s is now possible [67]. The fabrication of CdZnTe is now performed on a wafer level and detector-grade material is available in thicknesses up to a few mm.

Figure 7a shows a raw flood image recorded using a 2 mm thick THM-grown CdZnTe sensor that was fabricated recently. The image confirms the high uniformity achieved using the THM growth method. A network of line defects is also observed; however, in this case the lines seem to be thinner with respect to the CdTe sensors in Figs. 5 and 6a. Besides some edge defects no other major defect show up, such as the blobs observed in CdTe sensors. Several isolated dark or bright dots that appear in the image correspond to pixels with different sensitivity. All the pixels are connected which is consistent with a high bonding yield, although the difference in the sensitivity could be related either to micro-defects in the crystal bulk or imperfections of the calibration procedure.

The major improvement of recent CdZnTe sensors with respect to CdTe sensors is their stability. Figure 7b shows a raw flood image of the sensor recorded after the continuous illumination with a uniform X-ray beam for 3 h. The count rate along the line defects is stable within a few percent. Figure 8 shows the monitoring of the total leakage current of a CdZnTe sensor over irradiation time, in comparison with the CdTe and GaAs:Cr sensors that are discussed in the previous sections. For CdZnTe and GaAs:Cr sensors the leakage current remains stable over irradiation

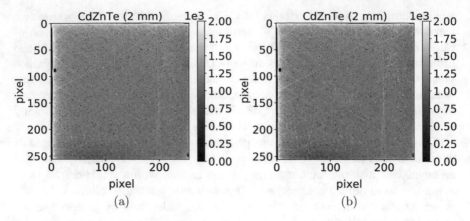

Fig. 7 (**a**) A raw image showing the response of a CdZnTe sensor with 2 mm thickness and 55 μmm pixel pitch, under uniform illumination with X-rays provided from a Ag-tube operated at (35 kV, 25 mA). The image was taken shortly after the ramp up to a bias voltage of −1000 V with the energy threshold set at 11 keV. (**b**) A raw image of the same sensor recorded after the continuous illumination for 3 h at the same settings

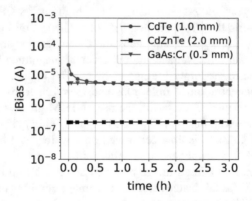

Fig. 8 Monitoring of the leakage current over irradiation time, for CdTe, CdZnTe and GaAs:Cr sensors with a thickness of 2 mm, 2 mm and 500 μmm, respectively. The pixel pitch is 55 μmm for all the sensors. The irradiation performed with a uniform X-ray beam provided from a Ag-tube operated at (35 kV, 25 mA). The sensors were biased with −500 V, −1000 V and −300 V, while the energy threshold was set at 11 keV

time, while a drift is observed for the CdTe sensor. The stable performance of CdZnTe sensors was also confirmed experimentally by monitoring of the leakage current and the count rate, in [51]. Both remain remarkably stable with irradiation time for a measured output count rate up to 1.25×10^6 counts/ mm^2/s. These results confirm the assumption of the reduced polarisation effects in this material. Consequently, long-lasting experiments could benefit from the application of the flat-field correction in order obtain improved data quality.

The optimisation of the CdZnTe growth method in view of high-flux capabilities is possible, at the expense of the spectroscopic performance though. The halo/crater effect observed for GaAs:Cr is present also in THM-grown CdZnTe. However, improved performance can be achieved by applying a depth-of-interaction correction [56]. In addition, CdZnTe sensors coupled to modern readout ASICs enable high resolution imaging thanks to the small pixel size, typically a few tens of μm. High spatial resolution was demonstrated for 2 mm thick CdZnTe in [51, 69]. Nowadays, CdZnTe sensors find wide use in X-ray imaging applications. Large area photon-counting systems have been developed for applications in spectral computed tomography as well as in baggage scanning [67]. Moreover, due to the low leakage current CdZnTe sensor is suitable for use in charge-integrating HPDs. Several groups are considering the use of CdZnTe for high-flux applications in diffraction-limited storage rings.

3.4 Other Materials

A special material which belongs to the high-Z family is germanium (Ge), with atomic number $Z = 32$. Benefiting from its elemental composition, the production of thick single crystals with extremely low concentration of defects is possible. However, due to the low band gap of 0.7 eV, Ge detectors generate large leakage currents at room temperature operation and therefore require cooling. Thanks to its excellent charge transport properties, Ge offers superior energy resolution with respect to other semiconductor and is mainly used for spectroscopic applications. Recently, HPDs featuring 1 mm thick Ge sensors have been developed in view of hard X-ray imaging applications [70]. Such prototypes show high image uniformity with a cooling down to $-100\,^\circ$C to ensure a sufficiently good performance.

Today, from the commercially available high-Z compound semiconductors, CdTe, CdZnTe and GaAs are the most commonly used materials. Several other high-Z compounds have been proposed as alternatives of Si and Ge in spectroscopic applications. Such compounds can be derived from elements in groups II to VI of the periodic table. Most of the elements that belong in these groups, being soluble within each other, enable the formation of binary, ternary or even quaternary compounds. A comprehensive discussion concerning compound semiconductors used for radiation detection can be found in [71].

4 Outlook

The photon science is entering in a new era with 4th generation synchrotron facilities already in operation or under commissioning phase. Diffraction-limited storage rings will be able to deliver X-rays more brilliant and coherent by an order of magnitude with respect to their predecessors. Therefore, the instrumentation scientists have to come up with novel ideas in order to meet the requirements set by the new generation of synchrotrons and facilitate this transition.

The HPD is a mature technology that is used for the direct detection of X-rays. Existing hybrid pixels detectors using Si as active medium offer high performance for soft X-ray applications up with photon energies up to 20 keV. The detection of X-rays with higher energy though, due to the poor absorption efficiency of Si, would require the use of other materials with high atomic number. Over the past 20 years compound semiconductors such as GaAs:Cr, CdTe and CdZnTe have attracted attention for use in HPD schemes. GaAs:Cr is a promising material for X-ray imaging applications with moderate energies up to 50 keV. CdTe and CdZnTe offer high absorption efficiency for high photon energies up to 100 keV. Moreover, recent developments in the crystal growth techniques enable the fabrication of detector-grade material with high-flux capabilities.

Pixelated sensors made from GaAs, CdTe and CdZnTe have been tested extensively over the past few years with promising results. Despite some imperfections observed, the sensor quality at the moment is sufficiently good to allow their use in synchrotron beamlines. The optimal choice of the detector material, however, depends on the requirements set by the target application. With the constant improvement that is observed in the crystal quality, high-Z sensors with lower concentration of defects, more stable and able to cope with high fluxes, would be available in the near future. There is always room for improvement and the future is bright.

References

1. Blewett, J. (1998). *Journal of Synchrotron Radiation, 5*(3), 135. https://doi.org/10.1107/S0909049597043306
2. Blewett, J. (1946). *Physical Review, 69*, 87. https://link.aps.org/doi/10.1103/PhysRev.69.87
3. Elder, F., Gurewitsch, A., Langmuir, R., & Pollock, H. (1947). *Physical Review, 71*, 829. https://link.aps.org/doi/10.1103/PhysRev.71.829.5
4. Winick, H., & Bienenstock, A. (1978). *Annual Review of Nuclear and Particle Science: Synchrotron Radiation News, 28*(1), 33. https://doi.org/10.1146/annurev.ns.28.120178.000341
5. Motz, H. (1951). *Journal of Applied Physics, 22*(5), 527. https://doi.org/10.1063/1.1700002
6. Hofmann, A. (1978). *Nuclear Instruments and Methods, 152*(1), 17. http://www.sciencedirect.com/science/article/pii/0029554X78902318
7. Winick, H., Brown, G., Halbach, K., & Harris, J. (1981). *Physics Today, 34*(5), 50 (1981). https://doi.org/10.1063/1.2914568
8. The ESRF website. https://www.esrf.eu/
9. The APS website. https://www.aps.anl.gov/
10. The SPring website. http://www.spring8.or.jp/en/
11. The lighsources.org community. https://lightsources.org/
12. Ishikawa, T. (2019). *Philosophical Transactions of the Royal Society A, 377*(2147). ID:2018.0231. https://royalsocietypublishing.org/doi/abs/10.1098/rsta.2018.0231
13. Wiedemann, H. (2014). In E. Jaeschke, S. Khan, J. Schneider, & J. Hastings (Eds.) *Synchrotron light sources and free-electron lasers: Accelerator physics, instrumentation and science applications* (1–41). Cham: Springer International Publishing, Cham. https://doi.org/10.1007/978-3-319-04507-8_1-1

14. Einfeld, D. (1993). *Nuclear Instruments and Methods A, 335*(3), 402. http://www.sciencedirect.com/science/article/pii/016890029391224B
15. The MAX-IV website. https://www.maxiv.lu.se
16. Tavares, P., Leemann, S., Sjöström, M., & Andersson, Å. (2014). *Journal of Synchrotron Radiation, 21*(5), 862. https://doi.org/10.1107/S1600577514011503
17. Raimondi, P. (2016). *Synchrotron Radiation News, 29*(6), 8. https://doi.org/10.1080/08940886.2016.1244462
18. Borland, M., Abliz, M., Arnold, N., Berenc, T., Byrd, J., Calvey, J., et al. (2018). In *Proceedings of the 9th International Particle Accelerator Conference (IPAC'18), Vancouver, BC, April 29–May 4, 2018* (Vol. 9, pp. 2872–2877). Geneva: JACoW Publishing. http://jacow.org/ipac2018/papers/thxgbd1.pdf
19. White, A., et al. (2019). *Synchrotron Radiation News, 32*(1), 32. https://doi.org/10.1080/08940886.2019.1559608
20. Schroer, C., et al. (2018). *Journal of Synchrotron Radiation, 25*(5), 1277. https://doi.org/10.1107/S1600577518008858
21. Tanaka, H., et al. (2016). In *Proceedings of the International Particle Accelerator Conference (IPAC'16), Busan, May 8–13, 2016* (Vol. 7, pp. 2867–2870). Geneva: JACoW. http://jacow.org/ipac2016/papers/wepow019.pdf
22. Kemmer, J. (1980). *Nuclear Instruments and Methods in Physics Research, 169*(3), 499. http://www.sciencedirect.com/science/article/pii/0029554X80909489
23. Walker, J., Parker, S., Hyams, B., & Shapiro, S. (1984). *Nuclear Instruments and Methods A, 226*(1), 200. http://www.sciencedirect.com/science/article/pii/0168900284901918
24. Gaalema, S. (1985). *IEEE Transactions on Nuclear Science, 32*(1), 417. https://doi.org/10.1109/TNS.1985.4336866
25. Heijne, E., Jarron, P., Olsen, A., & Redaelli, N. (1988). *Nuclear Instruments and Methods A, 273*(2), 615. http://www.sciencedirect.com/science/article/pii/0168900288900654
26. Heijne, E., et al. (1994). *Nuclear Instruments and Methods A, 348*(2), 399. http://www.sciencedirect.com/science/article/pii/0168900294907684
27. Heijne, E., et al. (1996). *Nuclear Instruments and Methods A, 383*(1), 55. Development and Application of Semiconductor Tracking Detectors. http://www.sciencedirect.com/science/article/pii/S0168900296006584.
28. Manolopoulos, S., et al. (1999). *Journal of Synchrotron Radiation, 6*(2), 112. https://onlinelibrary.wiley.com/doi/abs/10.1107/S0909049599001107
29. Campbell, M., et al. (1998). *IEEE Transactions on Nuclear Science, 45*(3), 751. https://doi.org/10.1109/23.682629
30. Broennimann, C., et al. (2006). *Journal of Synchrotron Radiation, 13*(2), 120. https://doi.org/10.1107/S0909049505038665
31. Dinapoli, R., et al. (2011). *Nuclear Instruments and Methods A, 650*(1), 79. International Workshop on Semiconductor Pixel Detectors for Particles and Imaging 2010. http://www.sciencedirect.com/science/article/pii/S0168900210027427
32. Campbell, M. (2011). *Nuclear Instruments and Methods A, 633*, S1. 11th International Workshop on Radiation Imaging Detectors (IWORID). http://www.sciencedirect.com/science/article/pii/S0168900210012969
33. Ponchut, C., et al. (2011). *Journal of Instrumentation, 6*(01), C01069. https://doi.org/10.1088/1748-0221/6/01/c01069
34. Tartoni, N., et al. (2012). In *2012 IEEE Nuclear Science Symposium and Medical Imaging Conference Record (NSS/MIC)* (2012) (pp. 530–533). https://doi.org/10.1109/NSSMIC.2012.6551164
35. Pennicard, D., et al. (2011). *Journal of Instrumentation, 6*(11), C11009. https://doi.org/10.1088/1748-0221/6/11/c11009
36. Ballabriga, R., et al. (2016). *Journal of Instrumentation, 11*(01), P01007. https://doi.org/10.1088/1748-0221/11/01/p01007
37. Herrmann, S., Boutet, S., Duda, B., Fritz, D., Haller, G., Hart, P., et al. (2013). *Nuclear Instruments and Methods A, 718*, 550. Proceedings of the 12th Pisa Meeting on Advanced Detectors. http://www.sciencedirect.com/science/article/pii/S0168900213001496.

38. Dragone, A., Caragiulo, P., Markovic, B., Herbst, R., Reese, B., Herrmann, S. C., et al. (2014). *Journal of Physics: Conference Series, 493*, 012012 (2014). https://doi.org/10.1088/1742-6596/493/1/012012

39. Mozzanica, A., Andrä, M., Barten, R., Bergamaschi, A., Chiriotti, S., Brückner, M., et al. (2018). *Synchrotron Radiation News, 31*(6), 16. https://doi.org/10.1080/08940886.2018.1528429

40. Dinapoli, R., Bergamaschi, A., Cartier, S., Greiffenberg, D., Johnson, I., Jungmann, J. H., et al. (2014). *Journal of Instrumentation, 9*(05), C05015. https://doi.org/10.1088/1748-0221/9/05/c05015

41. Henrich, B., Becker, J., Dinapoli, R., Goettlicher, P., Graafsma, H., Hirsemann, H., et al. (2011). *Nuclear Instruments and Methods A, 633*, S11. http://www.sciencedirect.com/science/article/pii/S0168900210012970

42. Tate, M. W., Chamberlain, D., Green, K. S., Philipp, H. T., Purohit, P., Strohman, C., & Gruner, S. M. (2013). *Journal of Physics: Conference Series, 425*(6), 062004. https://doi.org/10.1088/1742-6596/425/6/062004

43. Shanks, K. S., Philipp, H. T., Weiss, J. T., Becker, J., Tate, M. W., & Gruner, S. M. (2016). *AIP Conference Proceedings, 1741*(1), 040009. https://aip.scitation.org/doi/abs/10.1063/1.4952881

44. Via, C. D., Bates, R., Bertolucci, E., Bottigli, U., Campbell, M., Chesi, E., et al. (1997). *Nuclear Instruments and Methods A, 395*(1), 148. Proceedings of the Fourth International Workshop on GaAs Detectors and Related Compounds. http://www.sciencedirect.com/science/article/pii/S0168900297006311

45. Tlustos, L., Campbell, M., Fröjdh, C., Kostamo, P., & Nenonen, S. (2008). *Nuclear Instruments and Methods A, 591*(1), 42. Radiation Imaging Detectors 2007: Proceedings of the 9th International Workshop on Radiation Imaging Detectors. http://www.sciencedirect.com/science/article/pii/S0168900208003975

46. Tyazhev, A. V., Budnitsky, D. L., Koretskay, O. B., Novikov, V. A., Okaevich, L. S., Potapov, A. I., et al. (2003). *Nuclear Instruments and Methods A, 509*(1), 34. Proceedings of the 4th International Workshop on Radiation Imaging Detectors. http://www.sciencedirect.com/science/article/pii/S0168900203015456.

47. Tlustos, L., Shelkov, G., & Tolbanov, O. (2011). *Nuclear Instruments and Methods A, 633*, S103. 11th International Workshop on Radiation Imaging Detectors (IWORID). http://www.sciencedirect.com/science/article/pii/S0168900210013276

48. Hamann, E. (2013). *Characterization of High Resistivity GaAs as Sensor Material for Photon Counting Semiconductor Pixel Detectors*. Ph.D. thesis, Albert-Ludwigs-University, Freiburg, Germany.

49. Ponchut, C., et al. (2017). *Journal of Instrumentation, 12*(12), C12023. https://doi.org/10.1088/1748-0221/12/12/C12023

50. Veale, M. C., Bell, S. J., Duarte, D. D., French, M. J., Schneider, A., Seller, P., et al. (2014). *Nuclear Instruments and Methods A, 752*, 6. http://www.sciencedirect.com/science/article/pii/S0168900214003271

51. Tsigaridas, S., & Ponchut, C. (2019). *Journal of Instrumentation, 14*(04), C04009. https://doi.org/10.1088/1748-0221/14/04/C04009

52. Rudolph, P. (2005). *Crystal Research and Technology, 40*(1–2), 7. https://onlinelibrary.wiley.com/doi/abs/10.1002/crat.200410302

53. Billoud, T., Leroy, C., Papadatos, C., Pichotka, M., Pospisil, S., & Roux, J. S. (2018). *Journal of Instrumentation, 13*(04), P04002. https://doi.org/10.1088/1748-0221/13/04/p04002

54. Becker, J., Tate, M. W., Shanks, K. S., Philipp, H. T., Weiss, J. T., Purohit, P., et al. (2018). *Journal of Instrumentation, 13*(01), P01007. https://doi.org/10.1088/1748-0221/13/01/p01007

55. Greiffenberg, D., Andrä, M., Barten, R., Bergamaschi, A., Busca, P., Brückner, M., et al. (2019). *Journal of Instrumentation, 14*(05), P05020. https://doi.org/10.1088/1748-0221/14/05/p05020

56. Veale, M. C., Jones, L. L., Thomas, B., Seller, P., Wilson, M. D., & Iniewski, K. (2019). *Nuclear Instruments and Methods A, 927*, 37. http://www.sciencedirect.com/science/article/pii/S016890021930083X

57. Funaki, M., et al. (2013). Development of CdTe detectors in Acrorad. Published online by Acrorad Co., Ltd.
58. Llopart, X., Ballabriga, R., Campbell, M., Tlustos, L., & Wong, W. (2007). *Nuclear Instruments and Methods A, 581*(1), 485. VCI 2007: Proceedings of the 11th International Vienna Conference on Instrumentation. http://www.sciencedirect.com/science/article/pii/S0168900207017020
59. Ruat, M., & Ponchut, C. (2012). *IEEE Transactions on Nuclear Science, 59*(5), 2392. https://doi.org/10.1109/TNS.2012.2210909
60. Ruat, M., & Ponchut, C. (2014). *Journal of Instrumentation, 9*(04), C04030. https://doi.org/10.1088/1748-0221/9/04/c04030
61. Becker, J., Tate, M. W., Shanks, K. S., Philipp, H. T., Weiss, J. T., Purohit, P., et al. (2016). *Journal of Instrumentation, 11*(12), P12013. https://doi.org/10.1088/1748-0221/11/12/p12013
62. Förster, A., Brandstetter, S., & Schulze-Briese, C. (2019). *Philosophical Transactions of the Royal Society A, 377*(2147), 20180241. https://royalsocietypublishing.org/doi/abs/10.1098/rsta.2018.0241
63. Takahashi, T., & Watanabe, S. (2001). *IEEE Transactions on Nuclear Science, 48*(4), 950. https://doi.org/10.1109/23.958705
64. Sordo, S. D., Abbene, L., Caroli, E., Maria Mancini, A., Zappettini, A., & Ubertini, P., et al. (2009). *Sensors, 9*(5), 3491. https://www.mdpi.com/1424-8220/9/5/3491
65. Chen, H., Awadalla, S. A., Iniewski, K., Lu, P. H., Harris, F., Mackenzie, J., et al. (2008). *Journal of Applied Physics, 103*(1), 014903. https://doi.org/10.1063/1.2828170
66. Prokesch, M., Soldner, S., & Sundaram, A. (2018). *Journal of Applied Physics, 124*(4), 044503. https://doi.org/10.1063/1.5041006
67. Iniewski, K. (2016). *Journal of Instrumentation, 11*(12), C12034. https://doi.org/10.1088/1748-0221/11/12/c12034
68. Thomas, B., Veale, M. C., Wilson, M. D., Seller, P., Schneider, A., & Iniewski, K. (2017). *Journal of Instrumentation, 12*(12), C12045. https://doi.org/10.1088/1748-0221/12/12/c12045
69. Tsigaridas, S., Ponchut, C., Zanettini, S., & Zappettini, A. (2019). *Journal of Instrumentation, 14*(12), C12009. https://doi.org/10.1088/1748-0221/14/12/c12009
70. Pennicard, D., Struth, B., Hirsemann, H., Sarajlic, M., Smoljanin, S., Zuvic, M., et al. (2014). *Journal of Instrumentation, 9*(12), P12003. https://doi.org/10.1088/1748-0221/9/12/p12003
71. Owens, A., & Peacock, A. (2004). *Nuclear Instruments and Methods A, 531*(1), 18. Proceedings of the 5th International Workshop on Radiation Imaging Detectors. http://www.sciencedirect.com/science/article/pii/S0168900204010575

CdTe Sensors for Space-Based X-ray Detectors

Oliver Grimm

1 Introduction

Cadmium telluride (CdTe) is a semiconductor frequently used for X-ray detectors covering the energy range from a few keV up to hundreds of keV. The high density ($5.85 \, g/cm^3$) and high average atomic number (50) result in a high photoelectric absorption coefficient [1]. For example, 63% of all photons at 100 keV interact within a thickness of 1 mm [2]. In addition, the band gap of 1.5 eV [3] makes operation near room temperature possible with only moderate cooling.

These characteristics are favourable for space-based X-ray instruments which require compact, lightweight instruments with low power consumption because of high launch costs, limited electric power availability from solar cells or radioisotope thermoelectric generators, and difficulty in heat removal by radiators. One further important requirement of space experiments is reliability, which semiconductor detectors provide by being relatively simple, both electrically and mechanically.

A related material is cadmium zinc telluride (CZT). This report focusses on CdTe, but provides some information on CZT as well, as it is used similarly often in active space instruments. Manufacturing and further application details for both materials are given in [4].

O. Grimm (✉)
Institute for Particle Physics and Astrophysics, ETH, Zürich, Switzerland
e-mail: oliver.grimm@phys.ethz.ch

© Springer Nature Switzerland AG 2022
K. Iniewski (ed.), *Advanced X-ray Detector Technologies*,
https://doi.org/10.1007/978-3-030-64279-2_5

1.1   Cadmium Telluride

1.1.1   Material Characteristics

CdTe is a direct band gap semiconductor, used predominantly in thin-film photo-voltaics, where it provides about 5% of the worldwide solar energy production. Research into its application for X-ray detectors started in the 1960s, requiring improvements of the manufacturing process to obtain high enough crystal qualities that allow efficient carrier transport.

Undoped CdTe is typically strongly p-type due to the presence of a large number of acceptor-like cadmium vacancies in the crystal lattice. To compensate the vacancies, thus to increase the resistivity and to improve the charge transport, the material can be doped with chlorine that acts as a donor. The resulting crystal remains slightly p-type after this procedure [5, 6]. The electron affinity is 4.4 eV and the relative dielectric constant about 11.

The material is brittle and requires care when handling to avoid mechanical damage. Substantial variations in crystal quality are possible between samples, and extensive testing is required to select sensors that provide the highest energy resolution. Only a few places in the world supply material suitable for X-ray instruments.

Fluorescence photons from Kα and Kβ transitions in Cd and Te are emitted at energies 23 keV, 26 keV, 27 keV, and 31 keV [7]. Their range might be non-negligible in detectors with small pixels and they can contribute to signal cross-talk, in addition to charge sharing due to carrier diffusion. Escape lines appear in the spectra if the fluorescence photons are not detected. This is particularly pronounced at X-ray energies that interact near the surface, where the probability of escape is larger.

1.1.2   Carrier Lifetime, Schottky Electrodes

Compared to silicon or germanium, a binary compound as CdTe is significantly more difficult to produce with high crystal quality. One reason is that any departure from the correct stoichiometric ratio of the components during crystal growth (usually by the travelling heater method) will result in lattice defects from misplaced atoms. Since vapour pressures and solubility of Cd and Te are not equal, this is difficult to achieve during growth, and invariably the defect density is comparatively high.

Such defects can affect the charge transport properties and reduce the carrier lifetimes. The average distance a carrier can drift in an electric field before being lost, usually called drift length, might become comparable to the sensor thickness. For higher photon energies, where absorption occurs throughout the crystal, the measured signal amplitudes then depend on the depth of interaction. If no correction is applied, the spectra of photon lines show a characteristic tailing.

Methods to mitigate the tailing exist, for example, single carrier sensing or bi-parametric analysis [8], however, such more complicated schemes might be at odds with the reliability and simplicity requirements of a space instrument.

Another way to reduce the tailing is to increase the electric field, so that the charge collection time is well below the carrier lifetime. With the advent of good quality Schottky electrodes on CdTe, it became possible to apply bias voltages up to several 100 V per mm, without increasing the leakage current substantially, which limits tailing to negligible levels.

1.1.3 Bias Polarization

A consequence and disadvantage of the Schottky electrode configuration is bias polarization, i.e., a change of the electric field as function of time under bias due to increasing space charge from ionizing deep acceptors [9, 10].[1] At room temperature the time scale of the process is only minutes, whereas it becomes weeks or months when the sensors are cooled to $-20\,°C$. Since short bias resets of a few minutes are sufficient to restore the initial sensor behaviour, the operational detriment of polarization in cooled instruments is small.[2]

Polarization increases the electric field strength near the Schottky contact and, via barrier lowering, the leakage current. Figure 1 illustrates the polarization and bias reset effects on the leakage current for two temperatures. Also seen on the left plot is that too short depolarization results in incomplete recovery: polarization proceeds faster afterwards. Generally, longer depolarization is expected to be required at lower temperatures since electron release from ionized, localized deep acceptors depends on the presence of nearby holes, and thus their thermal movement.

There are indications that the polarization time scale of proton irradiated sensors becomes substantially shorter even when cooled (see Sect. 3). Another disadvantage of Schottky electrodes is their delicate nature, preventing efficient thermal annealing of sensors degraded by displacement damage.

Further information about CdTe, its band structure, and explicit derivations of the signal generation principles can be found in [11]. For reference, typical room temperature carrier transport parameters for CdTe are:

---

[1]This effect has to be distinguished from rate polarization, which is a change of the electric field at high event rates, when the average signal current is no longer negligible compared to the dark current.

[2]The ionized acceptors can quickly release their electrons to the valence band once the hole density is large. Contrary to the ionization process, there is no energy barrier involved, but the process is still not instantaneous, as the hole density is finite.

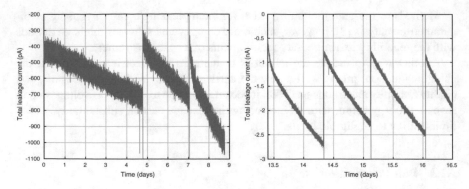

Fig. 1 Leakage current measurement with a $10 \times 10 \times 1\,\text{mm}^3$ CdTe sensor at $-200\,\text{V}$ bias. Temperature $-6\,°\text{C}$, bias resets ($0\,\text{V}$) for 1 min (left), and $4\,°\text{C}$, bias resets for twice ten and once 2 min (right)

Carrier type	Electron	Hole
Mobility ($\text{cm}^2/(\text{V s})$)	1100	100
Lifetime (μs)	3	1
Mobility-lifetime product (cm^2/V)	3.3×10^{-3}	1×10^{-4}
Drift length @ 200 V/mm (mm)	66	2
Drift time for 1 mm (ns)	45	500

The mobility scales with absolute temperature T approximately as $T^{-3/2}$ for the temperatures of interest.

1.2 Cadmium Zinc Telluride

CZT is an alloy of CdTe and ZnTe. The resulting band gap can be adjusted by the alloy blending ratio, and for a typical value of 10% ZnTe is $\approx 0.05\,\text{eV}$ higher than in CdTe. The resistivity of CZT is about an order of magnitude higher and allows room temperature operation, although cooling still improves spectral performance. Explicit Schottky contacting, for example, with indium, is possible, and in [12] also platinum was found to form a Schottky barrier, in contrast to its ohmic behaviour on CdTe (raising the question of polarization).[3]

Being a ternary semiconductor, CZT is even more difficult to produce with high crystal quality compared to CdTe and naturally has more defects, resulting in generally worse carrier transport properties. The electron mobility was found to be identical to the value for CdTe in [13], but the electron lifetime was shorter,

---

[3]The characteristics of contacts depend sensitively on surface layers and defects. Fermi level pinning might alter the behaviour from what would be expected by the bulk material parameters.

in the range 400 ns to 1000 ns for different samples. Mobility-lifetime products of 1.4×10^{-3} cm^2/V for electrons and 1×10^{-5} cm^2/V for holes are reported in [14]. For sensor thicknesses up to a few millimetres, electrons are still nearly completely collected, whereas most holes will be lost after one drift length, before reaching the cathode. As a result, the measurement signal consists of a fast rising electron part with amplitude roughly proportional to the interaction depth, and a small, slow rising, nearly constant contribution from holes. To obtain high spectral resolution in this case requires some depth sensing capability.

1.3 Heritage of CdTe and CZT in Space Experiments

1.3.1 Integral

An array of 16384 ohmic (platinum electrode) CdTe:Cl crystals of size $4 \times 4 \times 2$ mm^3 is used in the coded-aperture *IBIS* imager of the Integral satellite, launched in 2002 and still operational [15]. They are cooled to about 0 °C. The lower photon energy limit of the instrument is about 15 keV, the full-width-half-maximum (FWHM) resolution at 59.6 keV is 6.3 keV. The on-board calibration source is sodium-22.

A custom-developed 4-channel ASIC is used in the front-end, measuring both pulse height and rise time, such that an estimate of the interaction depth and correction for hole carrier loss is possible. The power consumption is 2.8 mW per channel.

1.3.2 Swift

The coded-aperture *Burst Alert Telescope* (BAT) on the Swift Gamma-Ray Burst Mission, launched in 2004 and which is also still active, uses 32768 CZT sensors ($4 \times 4 \times 2$ mm^3) operating at 20 °C and at about -200 V bias. It achieves a lower energy limit of 15 keV (individual detectors 10 keV) [16]. The FWHM resolution at 59.6 keV is 3.3 keV. Americium-241 is used as on-board calibration source.

The front-end ASIC is a 128-channel, self-triggering XA1.23 (Ideas, Norway, 2.8 mW per channel), each channel containing a charge sensitive pre-amplifier, shaping amplifier, and discriminator.

1.3.3 Chandrayaan-1

The *HEX* spectrometer on-board the Chandrayaan-1 moon mission, launched in 2008 and operational for about half a year, used nine 4×4 cm^2 CZT crystals of 5 mm thickness, each with 256 pixels of area 2.5×2.5 mm^2 and XAIM3.2

ASICs (Ideas, 3.2 mW per channel) [17]. Ground testing indicated a 12 keV FWHM resolution at 59.5 keV.

1.3.4 NuSTAR

The Nuclear Spectroscopic Telescope Array (NuSTAR), launched in 2012, has two separate telescopes, each with four $2 \times 2 \, cm^2$ CZT crystals of 2 mm thickness in the focal plane. The crystal anode is subdivided into 1024 pixels [18]. The FWHM resolution at 68 keV is 900 eV and the low-energy threshold 3 keV, with the detectors cooled to about 5 °C and at −400 V bias. Europium-155 is used as on-board calibration source.

Due to the small pixel size, pixels neighbouring the primary interaction will register a signal of opposite polarity that results from incomplete cancellation of electron and hole transient signals (unless substantial charge sharing occurs). The NuCIT read-out ASIC (Caltech, 100 μW per channel) samples signals of both polarities, and ground processing can then estimate the depth of interaction and correct for the effects of charge loss. In addition, the small pixel size of 0.6 mm provides sensitivity predominantly to the electron signal due to weighting field concentration.

1.3.5 AstroSat

The AstroSat mission, launched in 2015, uses 64 CZT crystals, 5 mm thick, having in total 16384 pixels of area $2.46 \times 2.26 \, mm^2$ in the *CZTI* coded-mask imager to cover an energy range from 20 keV to 150 keV [19]. The energy resolution is better than 7.2 keV FWHM at 60 keV for most pixels with the detectors at about 10 °C. Americium-241 is used as on-board calibration source.

1.3.6 Hitomi

A single $34 \times 34 \times 0.75 \, mm^3$ double-sided strip CdTe sensor (128 strips, pitch 250 μm) with an aluminium Schottky electrode was used for the *HXI* instrument of the short-lived Hitomi mission, launched in 2016 [20]. HXI was a combined silicon/CdTe imager, with CdTe covering the photon energies above about 30 keV. The FWHM resolution at 60 keV was found in orbit to be 2 keV with the sensor temperature kept below −25 °C. VATA461 ASICs (Ideas, 1.2 mW per channel) were used.

CdTe pixel sensors (pitch 3.2 mm) with the same thickness and similar wafer size, but using indium as Schottky electrode, were used for the *SDG* Compton camera in Hitomi.

A similar double-sided strip detector as for HXI, but with pitch 60 μm, was also used for two flights of the sounding rocket experiment *FOXSI* [21], obtaining a FWHM resolution of 1.3 keV at 60 keV in laboratory tests.

1.3.7 ASIM

The Atmosphere Space Interaction Monitor (ASIM), launched in 2018 and installed on the ISS, contains the coded-aperture *MXGS* instrument [22]. Its low-energy detector uses 256 CZT sensors of size $20 \times 20 \, \text{mm}^2$, 5 mm thick, each one with 64 pixels, operating at 10 °C. The FWHM resolution at 60 keV is 6 keV and the low-energy threshold is 15 keV. Cadmium-109 is used as in-flight calibration source.

The front-end ASIC is a 128-channel XA1.82 (Ideas), with about 500 μW power consumption per channel.

1.3.8 Spektr-RG

The Spektr-RG (SRG) X-ray observatory was launched in 2019. The *ART-XC* imaging X-ray telescope uses seven $30 \times 30 \times 1 \, \text{mm}^3$ double-sided strip CdTe sensor (48 strips on either side, pitch 595 μm) with aluminium Schottky electrodes [23]. The energy range is 4 keV to 30 keV, the upper limit due to the reflectivity of the Wolter mirror system, the energy resolution about 1.3 keV at 14 keV with the detectors cooled to −22 °C. Iron-55 and americium-241 are the in-flight calibration sources.

Readout is with VA64TA1 ASICs (Ideas, 0.9 mW per channel).

2 CdTe in STIX, Space Design Considerations

The *Spectrometer/Telescope for Imaging X-rays* (STIX) [24] is an instrument onboard the Solar Orbiter mission of the European Space Agency [25], launched on 10 February 2020. The detector employs 1 mm thick chlorine-compensated CdTe sensors of area $10 \times 10 \, \text{mm}^2$ for imaging of solar thermal (above 10 MK) and non-thermal X-ray sources. The aluminium Schottky anode[4] is segmented into 8 large $(9.6 \, \text{mm}^2)$ and 4 small $(1 \, \text{mm}^2)$ pixels, the platinum cathode is monolithic. The energy range of the instrument is 4 keV–150 keV, the FWHM energy resolution has been measured in flight to be 1 keV at 31 keV for the small pixels and 1.2 keV for the large ones.

---

[4]The anode is made from three layers: gold, titanium, and aluminium. Gold is used to ease external bonding and for surface protection, the intermediate titanium layer for improved mechanical contact of the gold layer to the Schottky contact forming aluminium.

Fig. 2 Moire pattern generated by X-rays vertically impinging on two grids with pitches 666 µm/690 µm and angles 60°/64°. The black rectangle indicates the grid size, the blue lines the pixel structure on the detectors. Figure from [26], © IOP Publishing Ltd and SISSA Medialab Srl. Reproduced by permission of IOP Publishing. All rights reserved

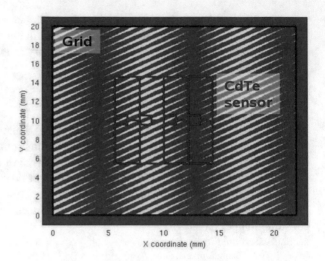

STIX uses a Fourier imaging technique. Pairs of X-ray opaque tungsten grids cast a Moiré patterned shadow on the 32 detectors. The orientation and pitch of the grids and the phase of the pattern on the detectors encode a Fourier component of the source distribution, as illustrated in Fig. 2. The low telemetry allocation of 700 bits per second average allows to transmit only energy-binned count rates to ground for image reconstruction. As the image information is contained in the count rate differences between pixels, a precise energy calibration of the bin boundaries is required to minimize image blurring. Since the solar X-ray sources have steeply falling thermal spectra at lower energies, a high instrument resolution is required to allow deduction of the source temperature from the slope with acceptable uncertainty. Weak on-board barium-133 radioactive sources are used for energy calibration.

The instrument development was accelerated significantly by the availability, at project start, of a read-out ASIC and a space-qualified hybrid design which could be adapted to the particular requirements of STIX. The modified design resulted in the Caliste-SO hybrid [27] with a power consumption of about 1 mW per channel. To maintain the space qualification, in particular to protect the crystal bonding that is potentially sensitive to humidity, most ground testing and all storage was either in vacuum or in a dry nitrogen atmosphere. Also the spacecraft provided nitrogen flushing until launch.

2.1 Sensor Selection

The quality of individual CdTe samples often varies substantially, even within a single production batch. For STIX it was found necessary to judge each sample based on three selection criteria [28]: (a) type, number, and location of microscopi-

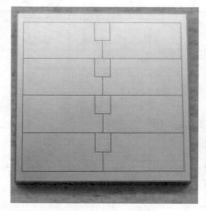

Fig. 3 View of the pixelized Schottky anode side of a STIX CdTe detector (left) and the CdTe interface board used for sensor selection testing (right). The sensor size is $10 \times 10\,mm^2$. Visible are the large and small pixels, and the surrounding guard ring. The interface board is connected to an ASIC board underneath. Figures from [28], © IOP Publishing Ltd and SISSA Medialab Srl. Reproduced by permission of IOP Publishing. All rights reserved

cally visible surface defect, (b) total and pixel leakage current, and (c) spectroscopic resolution. Leakage current and resolution were measured both at room temperature and when cooled to about $-20\,°C$.

Visual inspection was performed first as it provided the quickest removal of unsuitable samples. Besides obvious damage, surface defects at certain locations, which otherwise might not have had a performance penalty, had to be avoided to guarantee good bonding to the Caliste-SO hybrid.[5]

Since CdTe is a brittle material, handling should be minimized. An interface board was designed for the electrical characterization, so that a sensor had to be mounted only once (Fig. 3, right). The board uses low-force spring contacts coated with a soft electroconductive polymer and a thin Gold layer for electrical contact to the pixels. A cut-out in the top copper cover allows X-rays from radioactive sources to pass. A soft carbon foil between the cover and the crystal prevents damage to the platinum cathode.

The fabrication of the sensors started with oversized raw crystals with monolithic electrodes $(14 \times 14 \times 1\,mm^3)$, procured from the Japanese company Acrorad. Patterning was performed at the Paul Scherrer Institute in Switzerland using photolithography and a plasma etching process, followed by dicing to the final size of $10 \times 10\,mm^2$. The pixel structure (Fig. 3, left) consists of four stripes to sample the Moiré pattern, further subdivided into two large and one small pixel to address

---

[5]There are more bond locations than needed for the electrical contacts. Additional bonds ensure mechanical stability and vibration resistance. Frequent surface defects were found to be aluminium delaminations, likely due to underlying imperfections in the crystal.

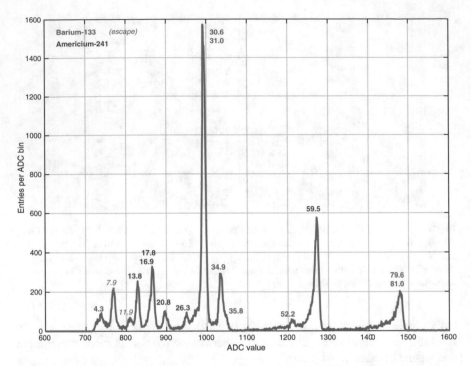

Fig. 4 Measured laboratory spectrum resulting from exposing a STIX CdTe crystal, cooled to −20 °C, to ^{133}Ba and ^{241}Am X-rays. Line source isotope and energies in keV are indicated

the large dynamic range in count rates. A guard ring along the crystal border protects the pixels from edge-related leakage currents.

A laboratory spectrum obtained with a STIX sensor exposed simultaneously to ^{133}Ba and ^{241}Am radioactive sources is shown in Fig. 4. In addition to the source lines, also escape lines due to fluorescence photons leaving the detector are seen.

The tailing due to short hole lifetime is clearly visible for the line at 81 keV. In addition, the largest line at 31 keV also shows some tailing towards the left, despite its signal being almost entirely from electrons with a very long drift length. This is attributed to a near-surface damage layer, reported in [29] and investigated for its effect on X-ray spectra in [30]. Photon lines in the energy range from about 20 keV to 40 keV are most affected because conversions occur all through the affected surface layer, but only rarely in the bulk.

2.2 Thermal Constraints

Cooling of the sensors to at least −20 °C is a key requirement in order to obtain good spectral resolution and to manage the bias polarization time scale. Due to the low instrument power allocation of 8 W, no active (e.g., Peltier) cooling was feasible. Instead, passive cooling via a spacecraft-provided radiator is employed. Size constraints of this radiator limit the allowable heat load to about 3 W, thus the thermal design was of high importance. Maintaining such a low heat load (which includes 1.2 W of electrical power) while the spacecraft environment is at nearly 50 °C was achieved by designing a *cold unit*, an enclosure thermally isolated by multi-layer insulation (MLI) and minimum-cross section mechanical support elements. Other than the Caliste-SO hybrids, only a small amount of analogue front-end electronics is contained in this enclosure. Digital electronics and power supplies are mounted separately, thermally connected to the warm environment, electrically connected to the front-end with Kapton cables of low heat conductivity. It was necessary to manufacture the MLI enclosure, despite its complicated shape, as a single piece with little stitching. No openings in the field-of-view of the detectors were possible, therefore the X-ray attenuation of the MLI had to be carefully considered.

Thermal design is an area that is often initially neglected in space detector systems. The constraints of the vacuum environment and the limited heat rejection only by inefficient radiators to deep space are unfamiliar from ground-based instruments. Late thermal re-design can be especially costly and disturbing as it affects the basic mechanical layout. It is advisable to invoke expert advice in this field early on in a project.

2.3 Reliability and Redundancy

Due to the non-serviceability of almost all space instruments and their typically extended operation period, long-term reliability is one of the prime design goals.

On component, material, and process levels, rigorous space qualification is attempted if no prior space heritage can be demonstrated. Qualification typically involves applying stresses (e.g., vibration levels or temperatures) well beyond those expected in flight, and demonstrating that no unacceptable degradation occurred. This is often a lengthy and costly process, with the risk of not being able to successfully qualify, thus for a science instrument usage of not yet qualified components might jeopardise schedule or budget. Sometimes suitably de-rated, standard commercial components can be used instead.

Redundancy is a further important contribution to reliability. Due to the typically tight mass constraint, complete redundancy, which maintains the full instrument performance in case of a failure, is not always chosen. Instead, redundancy is implemented at various levels in the design, depending on the added complexity that it entails, attempting to achieve a graceful degradation in case of a failure, so that the scientific output of the instrument is reduced, but not eliminated.

As example from STIX, the bias distribution system uses two independent high-voltage DC-DC converters. Cross coupling of each detector to both converters would have been component and space intensive (due to the required isolation distances). Instead, each converter supplies half of the detectors which are connected such that if one converter fails, the remaining detectors still provide data for the full Fourier range, i.e., maintaining imaging capability, but at reduced performance with only half as many Fourier components.

On the contrary, connecting two cold-redundant low voltages or signals to a single component is relatively straightforward, therefore a top level redundancy is available for the low voltage power supplies and the digital control. For similar reasons, each of the two redundant 28 V spacecraft power supply lines is connected to only one of the two redundant STIX power supply units, whereas the two SpaceWire communication lines are cross-coupled to either side.

In implementing redundancy, it has to be considered on a case-by-case basis if the extra complexity that is inevitably required, and which contains a risk for failure itself, is worth the effort, or might potentially even lower the overall reliability.

A further important step towards reliability for space instruments is acceptance testing of the completed instruments, and of sub-components earlier, with the main purpose to identify workmanship problems. This typically involves performance tests, but also mechanical and thermal-vacuum tests with levels reduced compared to the qualification process, but which still demonstrate margin to the expected in-flight conditions.

A particularity of space instruments is that one significant design driver, resistance to vibration and mechanical stress, is due to conditions that almost exclusively occur before instrument operation, i.e., during the few minutes of launch. This affects testing, as pre-stressing components for this one-off launch condition should be minimized.

3 Displacement Damage by Energetic Protons

The relatively poor carrier transport properties of CdTe are further degraded by exposure to energetic charged particles or to neutrons, both of which can generate crystal defects by atomic displacements. Environments having significant fluxes of such particles are more likely found in space than in terrestrial applications, except for some particle and nuclear physics experiments, therefore considering sensor degradation due to this effect is an important step in the design of a space-based X-ray detector.

The relevant quantity to estimate the severity of a given particle spectrum is the non-ionizing energy loss (NIEL). To explain the significance of NIEL for evaluations of detector systems, first the ionizing energy loss is briefly considered, helping also in clarifying the terminology.

3.1 Ionizing Energy Loss

Effects of shielding and energy deposition (dose) in materials for protons and heavier charged particles can be described well with the *Bethe energy loss formula*. The Bethe formula is derived from considering the energy transfer to shell electrons of the target material by a passing heavy charged particle, therefore yields by construction the ionizing energy loss. Integrating the Bethe formula for a given particle fluence yields the total ionizing dose (TID, energy per unit mass).

The Bethe formula for the ionizing energy loss per unit distance, dE/dx, is in SI units and for the momenta that are of interest here[6] [31, Section 33]

$$-\frac{dE}{dx} = \frac{4\pi}{m_e c^2} \times \frac{n_e z^2}{\beta^2} \left(\frac{e^2}{4\pi\epsilon_0} \right)^2 \left(\ln \frac{2m_e c^2 \beta^2 \gamma^2}{I} - \beta^2 \right).$$

m_e is the electron mass and n_e the electron density in the target material. ze is the charge of the incoming particle, βc its speed, and $\gamma^{-1} = \sqrt{1 - \beta^2}$, thus $E = \gamma M c^2$ with M the incoming particle mass.

The mean excitation potential I depends on the target material and can be approximated to better than 10% accuracy by $I = Z \cdot 10\,\text{eV}$, where Z is the atomic number of the target material, for $Z > 15$. The electron density n_e follows from Z and the target material density ρ and (average) atomic mass Au (u being the atomic mass unit) as $n_e = \rho Z/(A\text{u})$.

The quantity $(dE/dx)/\rho$ is known as the linear energy transfer (LET) in dosimetry.[7] Multiplying LET with a particle fluence F yields the total deposited ionizing dose, which scales to first order with Z/A.

Since this equation describes the dominant energy loss mechanism, it can be used to calculate the modification of an incoming charged particle spectrum by shielding material. As an example, Fig. 5 gives both the predicted mission-integrated proton fluence spectrum for Solar Orbiter without shielding, and assuming an average spherical aluminium shielding of 4 mm thickness for an instrument located inside of the spacecraft.

For comparison with other radiation environments, the TID deposited by protons in the CdTe sensors, calculated with the Bethe formula for 0, 4 mm, and 10 mm spherical aluminium shielding, is 2170 Gy, 140 Gy, and 46 Gy, respectively.

---

[6]The formula is valid for $2\gamma m_e \ll M$ and $\beta\gamma > 0.1$. The particles relevant for the majority of sensor degradation in a space environment will have non-relativistic energies, $\gamma \approx 1$, and, for 50 MeV protons, $\beta \approx 0.3$.

[7]The nomenclature in [31] calls dE/dx the *linear stopping power* and LET the *mass stopping power*.

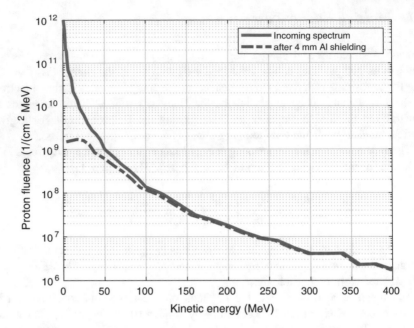

Fig. 5 Mission-integrated (10 years) incoming proton fluence spectrum for Solar Orbiter, and after modification by traversing 4 mm of aluminium (calculated using the Bethe formula and with data from [32])

3.2 Displacement Damage

Displacement damage refers to defects in the crystal created by displacing atoms from their correct lattice positions. The initial interaction of an energetic heavy particle may result in a so-called Primary Knock-on Atom (PKA), which, depending on energy, can displace further atoms. A displaced atom might migrate due to thermal motion for a while before forming a stable defect, or until re-settling into a correct lattice position. The latter process is the basis of thermal annealing. A PKA is the result of energy transfer to the whole atom, not to a shell electron (excitation or ionization), therefore the term non-ionizing energy loss is used.

The threshold energy that has to be imparted on a lattice atom to displace it is in the range 10 eV to 50 eV. As the collision of the incoming particles is with atoms at rest, the required kinetic energy for a displacement is much lower for protons and heavier ions than for electrons. For Cd or Te, the threshold kinetic energy of protons is 600 eV, but 700 keV for electrons.

NIEL is expressed in the same units as LET, for example, in MeV cm^2/g. There is no theory with a relatively simple, analytic result analogous to the Bethe formula that would allow to calculate NIEL from first principles, and experimental determinations of this quantity are difficult since NIEL typically amounts to only a small fraction of LET. Instead, numerical calculations of the energy deposited into atomic displacements are performed, using as input experimental cross-section values for elastic and inelastic interactions between the incoming particle and the target material. Results of such calculations are available in [33] for various particles and target materials. NIEL does not depend strongly on the target.

An extensive review of displacement damage and relevant literature is found in [34].

3.3 Equivalent Fluence Calculation

Sensor degradation by energetic charged particles can be studied in a controlled manner best at irradiation facilities that provide well defined beams of such particles. However, as facility beam time is relatively expensive, it is usually not feasible to precisely reproduce the shape of the expected in-flight spectrum. Instead, a compromise has to be found between representing the particle spectrum accurately, the preference for lower fluxes (which typically characterizes the actual instrument operation situation better), the desire for higher statistics, and the required time.

Experience shows that a linear relationship between NIEL and an observed, suitably quantified degradation effect is often a useful approximation. Assuming that such a relation holds, the actual particle spectrum seen by the instrument can be replaced for testing purposes by a single particle energy and a fluence that results in the same integrated NIEL. It is advisable to use the same particle type that dominates during operation also for laboratory testing.

A general recommendation for the case of using only a single beam energy is to choose it as the mean particle spectral energy weighted with the NIEL function [34]. For the case of the attenuated proton spectrum $\Phi(E)$ from Fig. 5 and NIEL data for silicon, the result is 38 MeV, whereas the mean energy of the spectrum without applying weights is 52 MeV.

The required equivalent fluence Φ_{eq} at energy E_{test}, assuming the validity of NIEL scaling, can be calculated from

$$\Phi_{eq}(E_{test}) = \frac{1}{NIEL(E_{test})} \int \Phi(E) \, NIEL(E) \, dE,$$

and, again for the previous example, yields $\Phi_{eq}(E_{test}) = 1.6 \times 10^{11} \, cm^{-2}$.

Consideration must also be given to neutrons, both in the primary beam of the test facility, and possibly generated by nuclear reactions in shielding material or collimators. A too large neutron flux, compared to the one expected in flight, would reduce the representativeness of the test. Estimating the neutron flux might require modelling the set-up in a particle tracking code, for example, Geant4 or FLUKA.

3.4 Literature

The available literature on the effects of energetic protons on CdTe performance is limited and scattered among different parameter sets. Some results for Schottky-CdTe are:

- Severe degradation in spectral performance was seen after exposure to a fluence of 10^{11} cm^{-2} with 200 MeV protons, and the detectors were found to polarize much faster [35]. Peak broadening and a decrease in the electron drift length by one order of magnitude were observed already at a fluence of 10^{10} cm^{-2}. The Schottky electrode was made of aluminium.
- Nearly no detectable degradation was found for protons in the energy range 10 MeV to 200 MeV up to a fluence of 6×10^9 cm^{-2} [36].[8] The authors investigated Schottky electrodes based on indium and aluminium.
- For an indium-Schottky electrode and 22 MeV proton energy, little degradation in spectral resolution was seen up to a fluence of 10^{10} cm^{-2}, whereas the resolution was nearly seven times worse at 10^{12} cm^{-2} and the electron mobility-lifetime product reduced by two orders of magnitude [37]. The leakage current at $-15\,°C$ was nearly unchanged. This study found CdTe unsuitable for the BepiColombo SIXS instrument due to this degradation by displacement damage.

Ohmic-contact CdTe sensors were investigated with 700 keV protons to a fluence of 1.2×10^{12} cm^{-2} [38]. The range of protons at this energy is very short, thus displacement damage is concentrated near the entrance electrode. The authors found the formation of a high density of electron traps.

The effects of 2 MeV protons (up to 2.1×10^{11} cm^{-2}), low-energy neutrons (to 10^{14} cm^{-2}), and high-energy neutrons (to 10^{13} cm^{-2}) on ohmic-contact CdTe were examined in [39]. The sensor degradation at the upper fluence level was substantial in all cases. This reference also provides band gap energies of carrier traps that were identified with photo-induced current transient spectroscopy.

Other studies are available for CZT detectors. The mobility-lifetime product for electrons, not for holes, is reported to be strongly affected by irradiation with 199 MeV protons at fluence 2.5×10^9 cm^{-2} [40]. The effect of neutrons has been investigated with no significant degradation seen for a fluence of 10^{10} cm^{-2} [41].

A comprehensive overview and many references about radiation effects in silicon, germanium, and GaAs are given in [42], which might help to elucidate the basic relations in CdTe as well.

---

[8] The five year in-flight fluence, calculated in the reference for the Hitomi mission, was found to be 10^{10} cm^{-2} without shielding and 2.6×10^9 cm^{-2} with 3 cm of bismuth germanate (scintillator) shielding. These numbers are representative for a satellite on a typical low-earth orbit.

3.5   Proton Irradiation Study with STIX Sensors

To predict the expected sensor degradation for STIX over the duration of the Solar Orbiter mission, and to evaluate possible mitigation options, e.g., shielding, increased bias voltage, or decreased temperature, irradiation studies were performed at the Proton Irradiation Facility of the Paul Scherrer Institute, operated in cooperation with ESA. Sensors were irradiated up to the above calculated fluence of 1.6×10^{11} cm^{-2} in steps of 20% and subsequently characterized by spectral measurements. Details of the set-up and analysis principles are given in [43], here only the main points are reproduced.[9] As there are no indications that the initial displacement damage depends on bias voltage or environmental conditions, irradiations were performed in air and without bias.

Barium-133 was used to obtain spectra of the irradiated crystals and performance parameters were extracted by applying the fitting procedure detailed in [30]. The procedure essentially combines the Hecht formula (giving the signal induced on the electrodes as function of photon absorption depth) with the exponentially attenuated photon beam intensity as function of depth, using the relevant absorption length, and this way simulated spectra are generated that can be fitted to the observed data. Electronic noise is considered by Gaussian spectral broadening. The free parameters of the fit are the peak heights of the main X-ray lines, the width of the Gaussian broadening, and the electron and hole drift lengths, λ_e and λ_h.

Example spectra and corresponding fits for different fluences are shown in Fig. 6. Since fluorescence, escape lines and charge sharing are not modelled, the reduced χ^2 is mostly well above unity. All shown measurements were performed under vacuum with the sensors cooled to $-20\,°C$ and at a bias of $-300\,V$.

Figure 7 shows the evolution of the parameters with fluence. Parameters are obtained for each pixel, however, the number of counts in the small pixel spectra is small and results in comparatively large parameter uncertainties. In addition, the *small pixel effect*[10] affects the extraction of bulk material parameters. Therefore, only results for large pixels are shown. Further, data points with an exceedingly large reduced χ^2 (above 20) are excluded.

No substantial annealing was observed by storing the irradiated sensors for 18 months at room temperature and repeating the spectral characterization, in line with [39]. Annealing at a temperature of $50\,°C$ for 6 days was attempted in a climate chamber, exposed to air with 10% relative humidity. The main result was, however, an increase in the leakage current of more than an order of magnitude and a spectral resolution much worse than before. This might be attributed to the delicate nature of the Schottky contact that is damaged by extended exposition to elevated temperatures.

---

[9]Figures in this section are also taken from [43] and are © Elsevier (2020).

[10]Since the transverse dimensions of the small pixels are comparable to the sensor thickness, carrier drift near the pixel anodes has a larger influence on the measured charge induction signals than drift further away due to weighting field concentration [1, Section 7.2].

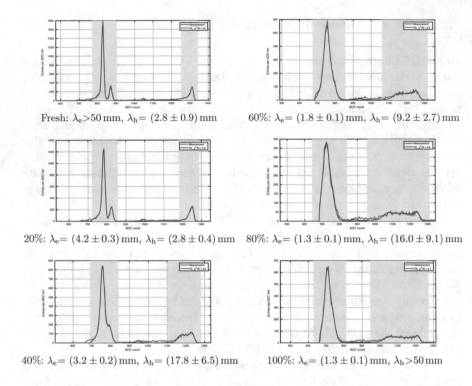

Fig. 6 ^{133}Ba spectra and fitting results for each fluence, measured at $-20\,°\text{C}$ and with a bias voltage of $-300\,\text{V}$. The highlighted bands indicate the count ranges used for fitting. 100% fluence corresponds to 1.6×10^{11} protons per cm^2

3.5.1 FWHM at 31 keV

The increase of the FWHM at 31 keV with fluence has been shown to be related to an increasing sensor capacitance and the resulting increase of electronic noise in the readout chip [43]. The capacitance C of a plane-parallel capacitor is inversely proportional to the electrode distance d. Within the depletion approximation, the depletion zone width W of a semiconductor is inversely proportional to the square root of the effective semiconductor dopant density N_{eff}. By equating W to d, it is found that $N_{\text{eff}} \sim C^2$, so an increasing capacitance can result from an increasing effective doping density.

The depletion approximation stipulates that all available acceptor and donor levels below the Fermi level are ionized within a distance W from the electrode. Due to band bending near the Schottky contact, those levels can include deep trap levels. The capacitance observations thus might be explained by an increase in deep acceptor traps with fluence, an effect which has indeed been described in some of the references given in Sect. 3.4.

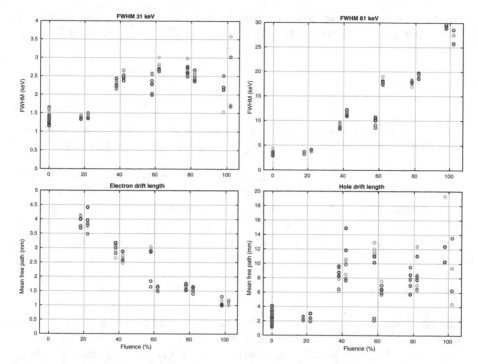

Fig. 7 Fluence dependency of the FWHM of the lines at 31 keV and 81 keV, and of the electron and hole drift lengths. The data points at zero fluence show the values for all large pixels of all crystals, colour differentiates the eight large pixels, and data points for the two crystals that were irradiated to each fluence percentage are slightly offset to the left and right

3.5.2 Electron Drift Length

A clear observation is the substantial, systematic reduction of the electron drift length with fluence. The value at zero fluence is too large to be extracted from the spectra, but is expected to be in excess of 50 mm from literature data, so even the smallest fluence of 20% affects this parameter strongly.[11] When λ_e approaches the thickness of the sensor, the average signal amplitude decreases, an effect that can be readily identified in Fig. 6 for the lines around 31 keV.

Assuming that the initial electric field, after bias application, does not change substantially with fluence, and that the carrier mobility is unaffected,[12] the electron lifetime is seen to be roughly inversely proportional to the fluence, $\tau_e \sim F^{-1}$ (for larger fluences). Minority carrier lifetime is inversely proportional to the trap density

---

[11] In [35], a decrease of the electron drift length by one order of magnitude was observed for 10% of the maximum fluence used here.

[12] For silicon, irradiations with 17 MeV protons left carrier mobilities largely unchanged [44], but affected the lifetime.

N_t, thus $N_t \sim F$. If deep acceptor traps are located above the Fermi level in the bulk of the sensor, they will mostly be empty and can easily capture electrons from the conduction band.

3.5.3 Hole Drift Length

The hole drift length shows a step-wise increase above 20% fluence and then large scatter, possibly indicating improved hole transport. There is presently no explanation for this behaviour.

The deep acceptor traps just mentioned cannot efficiently capture holes from the valence band since most traps are neutral and unable to release an electron to the valence band. Only ionized traps have electrons available, but since electrons are minority carriers, the effect on the majority holes is expected to be small. This at least provides an explanation why the fluence effect on λ_e is much stronger.

3.5.4 FWHM at 81 keV

The substantial width increase at 81 keV is predominantly a result of the decreasing λ_e and the stronger tailing that follows, with a small contribution from the increasing electronic noise. Due to the complicated shape of this line, especially for short electron drift lengths, its FWHM is not a very useful performance parameter on its own.

3.5.5 Leakage Current

The pixel leakage currents of the irradiated crystals as function of fluence are shown in Fig. 8. Because the current is very different for small pixels, large pixels, and the guard ring, the data is presented as ratio of current before and after irradiation (at zero fluence, all data points would thus be at unity). The measurement was made a short time after bias application.

The leakage current is determined by both the bulk material resistivity and the Schottky barrier height. It is presently not clear how the latter might be affected by bulk displacement damage. For an ideal Schottky contact, the barrier height depends only on the difference in metal Fermi level and the conduction or valence band energy of the semiconductor. In principle, changes in effective doping can therefore affect the barrier height, as the doping determines the band levels relative to the Fermi level. In practice, however, Fermi level pinning at the interface can modify this behaviour substantially.

An increasing resistivity from displacement damage has previously been seen [42, Section 2.3.2], attributed to a change in the compensation mechanism. A resistivity increase was also found to correlate with an increased concentration of a mid-gap deep trap at an energy of 0.77 eV above the valance band [39].

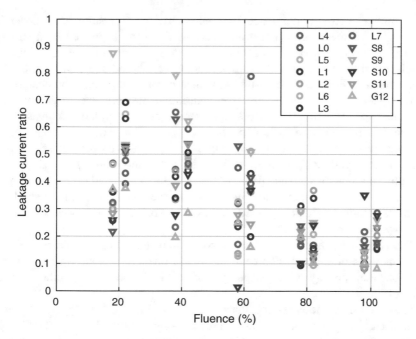

Fig. 8 Evolution of the leakage current, relative to its pre-irradiation value, for 8 large and 4 small pixels (L,S), and the guard ring (G), as function of fluence. Bias −200 V, temperature −20 °C. The data points from different crystals are slightly offset to the left and right

3.5.6 Polarization Time Scale

As described in the first section, the leakage current generally increases with time after application of bias voltage, one symptom of the polarization effect. The time scale of current increase was found to be less than an hour for a sensor exposed to the maximum fluence and measured at −300 V and −23 °C, whereas it was stable over multiple days before irradiation.

If indeed the density of deep acceptors does increase with fluence, and consequently the space charge density and electric field near the anode also increase, aggravated polarization with fluence is to be expected.

3.6   Conclusions on Radiation Hardness of CdTe

The available data indicates that the spectral resolution of CdTe sensors is affected only little up to a proton fluence of ∼10^{10} cm^{-2}, in particular if the detector is sufficiently cooled in operation. Although the electron drift length reduction is already substantial for low fluences, this does not become apparent in the spectra for sensors of only a few millimetres thickness.

A further increase in fluence to around $10^{11}\,cm^{-2}$ results in considerable degradation of the spectral performance. This might be explained by a strongly decreasing electron lifetime, reducing by more than one order of magnitude, in which case higher energy photon lines become very broad and develop a double-peak structure (Fig. 6). Spectral tailing is then dominated by electron charge loss. Also low-energy line widths increase noticeably, possibly due to an increased capacitance of the irradiated sensors, and average signal amplitudes decrease.

The polarization time scale appears to become a lot shorter at large fluence. Although polarization can be removed completely by a brief time of zero applied bias, too fast polarization might still render CdTe unsuitable for high spectral resolution applications at such particle fluences due to the dead time from frequent bias resets. Further insights about the origins of the observed spectral degradation might be obtained by more detailed studies of the polarization process as function of bias voltage and temperature.

The general mechanism for the observed parameter changes could conceivably be an increase of acceptor-like deep traps with fluence. It is known they exist in CdTe and their density is influenced by irradiation with protons and other particles [45, 46].

Sensor cooling increases the fluence up to which good energy resolution is maintained.[13] An open point is how further cooling to well below $-20\,°C$ might benefit the performance of strongly irradiated material.

Other detector materials than CdTe might be a better choice in challenging radiation environments. They may not have the good initial energy resolution, but if the radiation degradation is more favourable, better average performance might be provided over the mission duration. A potential argument for CZT under such condition is that the absence of a delicate Schottky electrode might make thermal annealing an option to periodically re-establish good performance (the literature on this subject is, however, not conclusive, see, for example, [47]). In addition, bias polarization might remain absent or less severe in detectors with near ohmic contacts.

4 Summary

Substantial progress has been made in understanding CdTe, both from a manufacturing point of view, from laboratory studies, and by its behaviour in operating X-ray detecting instruments. Fine details, for example, about native defects and those from radiation-induced displacement damage, are not nearly as well investigated as for silicon or germanium, but are known empirically well enough to predict most practical aspects of detector operation.

---

[13]To illustrate the opposite, spectral measurements at room temperature showed severe degradation of CdTe sensors already after a proton fluence of only $3.6 \times 10^{10}\,cm^{-2}$ [43].

Only a few examples of space-based instruments using high-resolution CdTe sensors with a Schottky electrode exist, none so far with long-term operation experience. A larger number of instruments have chosen CZT, however, with comparatively lower energy resolution and higher low-energy limit. Only NuSTAR achieves a resolution in the range of 1 keV at low photon energy, albeit with small pixels and a comparatively complicated read-out to obtain depth of interaction information.

As far as the radiation environment is not too severe, for example, in near-Earth orbit, Schottky-CdTe sensors allow a relatively simple, robust X-ray detection, also achieving resolutions of 1 keV, while pixels can have larger areas and simple peak sensing front-end electronics are sufficient.

References

1. Owens, A. (2012). *Compound semiconductor radiation detectors*. Boca Raton: CRC Press.
2. National Institute of Standards and Technology. (1995). *Tables of X-Ray Mass Attenuation Coefficients and Mass Energy-Absorption Coefficients*. NISTIR 5632. Available online at https://dx.doi.org/10.18434/T4D01F
3. Kosyachenko, L. A., et al. (2011). Band Gap of CdTe and $Cd_{0.9}Zn_{0.1}Te$ crystals. *Semiconductors, 45*, 1273.
4. Del Sordo, S., et al. (2009). Progress in the development of CdTe and CdZnTe semiconductor radiation detectors for astrophysical and medical applications. *Sensors, 9*, 3491.
5. Biswas, K., & Du, M.-H. (2012). What causes high resistivity in CdTe. *New Journal of Physics, 14*, 063020.
6. Krasikov, D., et al. (2013). Why shallow defect levels alone do not cause high resistivity in CdTe. *Semiconductor Science and Technology, 28*, 125019.
7. Iwańczyk, J., Szymczyk, W. M., & Traczyk, M. (1979). X-ray fluorescence escape peaks in gamma-ray spectra detected by CdTe detectors. *Nuclear Instruments and Methods, 165*, 289.
8. Richter, M., & Siffert, P. (1992). High resolution gamma ray spectroscopy with CdTe detector systems. *Nuclear Instruments and Methods A, 322*, 529.
9. Cola, A., & Farella, I. (2009). The polarization mechanism in CdTe Schottky detectors. *Applied Physics Letters, 94*, 102113.
10. Malm, H. L., & Martini, M. (1974). Polarization phenomena in CdTe nuclear radiation detectors. *IEEE Transactions on Nuclear Science, 21*, 322.
11. Grimm, O. (2020). *Signal Generation in CdTe X-ray Sensors*. Report ETHZ-IPP-2020-01 (8 May 2020). http://arxiv.org/abs/2005.09264
12. Bolotnikov, A. E., et al. (2002). Properties of Pt Schottky type contacts on high-resistivity CdZnTe detectors. *Nuclear Instruments and Methods A, 482*, 395.
13. Pavesi, M., et al. (2017). Electric field reconstruction and transport parameter evaluation in CZT X-ray detectors. *IEEE Transactions on Nuclear Science, 64*, 2706.
14. Auricchio, N., et al. (2010). Charge transport properties in CZT detectors grown by the vertical Bridgman technique. In *IEEE Nuclear Science Symposium & Medical Imaging Conference, Knoxville* (Vol. 3947).
15. Lebrun, F., et al. (2003). ISGRI: The INTEGRAL soft gamma-ray imager. *Astronomy & Astrophysics, 411*, L141.
16. Barthelmy, S. D., et al. (2005). The Burst Alert Telescope (BAT) on the Swift MIDEX Mission. *Space Science Reviews, 120*, 143.
17. Vadawale, S. V., et al. (2014). Hard X-ray continuum from lunar surface: Results from High Energy X-ray spectrometer (HEX) onboard Chandrayaan-1. *Advances in Space Research, 54*, 2041.

18. Harrison, F. A., et al. (2013). The Nuclear Spectroscopic Telescope Array (NuSTAR) high-energy X-ray mission. *The Astrophysical Journal, 770* , 103.
19. Bhalerao, V., et al. (2017). The cadmium zinc telluride imager on AstroSat. *Journal of Astrophysics & Astronomy, 38*, 31.
20. Nakazawa, K., et al. (2018). Hard X-ray imager onboard Hitomi (ASTRO-H). *Journal of Astronomical Telescopes, Instruments, and Systems, 4*, 021410.
21. Furukawa, K., et al. (2019). Development of 60 μm pitch CdTe double-sided strip detectors for the FOXSI-3 sounding rocket experiment. *Nuclear Instruments and Methods A, 924*, 321.
22. Østgaard, N., et al. (2019). The Modular X- and Gamma-Ray Sensor (MXGS) of the ASIM payload on the International Space Station. *Space Science Reviews, 215*, 23.
23. Pavlinsky, M., et al. (2018). ART-XC/SRG overview. In *Proceedings of the SPIE 10699, Space Telescopes and Instrumentation 2018: Ultraviolet to Gamma Ray*, 106991Y.
24. Krucker, S., et al. (2013). The spectrometer/telescope for imaging X-rays on board the ESA Solar Orbiter spacecraft. *Nuclear Instruments and Methods A, 732* , 295.
25. European Space Agency (2009). *Solar Orbiter: Exploring the Sun-Heliosphere Connection.* Assessment Study Report, ESA/SRE-2009-5.
26. Grimm, O., et al. (2012). *The front-end electronics of the Spectrometer Telescope for Imaging X-Rays (STIX) on the ESA Solar Orbiter satellite*, 2012 JINST 7 C12015.
27. Meuris, A., et al. (2012). Caliste-SO X-ray micro-camera for the STIX instrument on-board Solar Orbiter space mission. *Nuclear Instruments and Methods A, 695*, 288.
28. Grimm, O., et al. (2015). *Performance and Qualification of CdTe Pixel Detectors for the Spectrometer/Telescope for Imaging X-rays*, 2015 JINST 10 C02011.
29. Šik, O., Škvarenina, L., & Caha, O., et al. (2018). Determining the sub-surface damage of CdTe single crystals after lapping. *Journal of Materials Science: Materials in Electronics, 29*, 9652.
30. Grimm, O. (2020). Spectral signature of near-surface damage in CdTe X-ray detectors. *Nuclear Instruments and Methods A, 953*, 163104.
31. Tanabashi, M., et al. (Particle Data Group) (2018). *Review of Particle Physics: Physical Review D, 98*, 030001.
32. Sørensen, J. (2012). *Solar Orbiter Environmental Specification.* TEC-EES-03-034/JS (Issue 3.1, 13 June 2012).
33. Jun, I., et al. (2003). Proton Nonionizing Energy Loss (NIEL) for device applications. *IEEE Transactions on Nuclear Science, 50* (1924).
34. Srour, J. R., Marshall, C. J., & Marshall, P. W. (2003). Review of displacement damage effects in silicon devices. *IEEE Transactions on Nuclear Science, 50*, 653 (2003).
35. Eisen, Y., et al. (2002). Radiation damage of Schottky CdTe detectors irradiated by 200 MeV protons. *Nuclear Instruments and Methods A, 491*, 176.
36. Limousin, O., et al. (2015). ASTRO-H CdTe detectors proton irradiation at PIF. *Nuclear Instruments and Methods A, 787*, 328.
37. Ahoranta, J., et al. (2009). Radiation hardness studies of CdTe and HgI$_2$ for the SIXS particle detector on-board the BepiColombo spacecraft. *Nuclear Instruments and Methods A, 605*, 344.
38. Fraboni, B., Cavallini, A., & Castaldini, A. (2007). Charge collecting properties of proton irradiated CdTe detectors. *IEEE Transactions on Nuclear Science, 54*, 807.
39. Fraboni, B., et al. (2005). Recovery of radiation damage in CdTe detectors. *IEEE Transactions on Nuclear Science, 52*, 3085.
40. Wong, A., Harrison, F. A. , & Varnel, L. S. (1996). Effects of proton-induced radiation damage on cadmium zinc telluride pixel detectors. *Proceedings of the SPIE 2806, Gamma-Ray and Cosmic-Ray Detectors, Techniques, and Missions.*
41. Bartlett, L. M., et al. (1996). Radiation damage and activation of CdZnTe by intermediate energy neutrons. *Proceedings of the SPIE 2859, Hard X-Ray/Gamma-Ray and Neutron Optics, Sensors, and Applications*
42. Claeys, C., & Simoen, E. (2002). *Radiation effects in advanced semiconductor materials and devices.* Berlin: Springer.
43. Grimm, O., et al. (2020). Changes in detection characteristics of CdTe X-ray sensors by proton irradiation. *Nuclear Instruments and Methods A, 972*, 164116.

44. Amekura, H., Kishimoto, N., & Saito, T. (1995). Photoconductivity evolution due to carrier trapping by defects in 17 MeV proton irradiated silicon. *Journal of Applied Physics, 77*, 4984.
45. Fraboni, B., Cavallini, A., & Dusi, W. (2004). Damage induced by ionizing radiation in CdTe and CdZnTe detectors. *IEEE Transactions on Nuclear Science, 51*, 1209.
46. Cavallini, A., et al. (2002). Radiation effects on II–VI compound-based detectors. *Nuclear Instruments and Methods A, 476*, 770.
47. Fraboni, B., et al. (2006). Time and thermal recovery of irradiated CdZnTe detectors. *Semiconductor Science and Technology, 21*, 1034.

X-Ray Detectors in Medical Imaging

Witold Skrzynski

1 Introduction

Various imaging modalities are used in medical imaging. In most of them, some kind of energy is emitted toward the patient. The energy interacts within the patient's tissues. The interaction leads to the creation of a signal that can be measured with detectors, usually placed outside the patient. If the measured signal is used to create an image, there is a need to map it to the location within the patient. To get a meaningful image, the measured characteristics of the signal have to differ depending on the type of the tissue or its state. Different modalities of medical imaging can show various characteristics of the tissue because they are based on various physical interactions. In ultrasonography, a mechanical wave with ultrasonic frequency is reflected by the tissues, which differ by acoustic impedance. In magnetic resonance imaging, an electromagnetic wave with radiofrequency gets absorbed by the tissues placed in a magnetic field and then is reemitted from the tissues. The intensity of reemitted wave depends on proton density and relaxation times. The wave's frequency depends on local magnetic field induction, allowing to separate signals from different locations.

Medical X-ray imaging is based on a measurement of the X-ray beam attenuation. The attenuation is different for different tissues and may be additionally modified by the presence of contrast agents (e.g., intravenously administered iodine) if needed. There are several X-ray modalities widely used in medicine. Depending on their application, they can utilize different energies of radiation (e.g., lower energy in mammography than in radiography), or a different method of image creation (radiography vs computed tomography), but the basic scheme is always

W. Skrzynski (✉)
Medical Physics Department, Maria Sklodowska-Curie National Research Institute of Oncology, Warsaw, Poland
e-mail: witold.skrzynski@pib-nio.pl

© Springer Nature Switzerland AG 2022
K. Iniewski (ed.), *Advanced X-ray Detector Technologies*,
https://doi.org/10.1007/978-3-030-64279-2_6

kept: the patient is placed between a radiation source and radiation detector. The information from the detector is used to create an image.

2 Radiography

In the early days of X-ray imaging, glass plates with photographic emulsion were used as detectors, later replaced with films. The film is coated with a thin layer of photographic emulsion containing silver halide, which is not a very efficient X-ray absorber. The film performs much better in detecting visible light than X-rays. This is why the film has been quickly accompanied with an intensifying screen. The intensifying screen is a sheet of fluorescent material, converting the energy of each absorbed X-ray photon to many visible light photons. When the screen is in direct contact with the film, the film gets exposed by visible light. A combined screen-film detector has a much higher sensitivity for X-ray detection than the film alone. The sensitivity depends on the material used in the screen. As an example, screens with rare earth elements more effectively absorb and convert X-ray radiation compared to calcium tungstate screens ($CaWO_4$). With screen-film detectors, it is possible to obtain a radiographic image with a reasonably low patient dose. In medicine, the film is almost always used with intensifying screens, placed permanently in a light-tight radiographic cassette. In most cases, two screens are placed on both sides of the film, which has two layers of emulsion on both sides (Fig. 1a). The light emitted from the intensifying screen can penetrate the film base, reaching the second layer of emulsion and causing image blur. Compared with the film-only detector, the use of the screen-film decreases patient dose but at the cost of a lower spatial resolution of the image. Similarly, a film with larger silver halide grains is more sensitive but has a lower spatial resolution. In mammography, a good spatial resolution is essential for good visualization of microcalcifications, and only one intensifying screen and one layer of emulsion are usually used. In dental radiography, small objects are imaged, and a good spatial resolution is important, while radiation dose is not of much concern, and the film can be used alone [35].

Radiographic films require photochemical processing to convert latent images to visible images. It is a time-consuming process; it is not environmentally friendly and has to be done in the darkroom. Various alternatives have been proposed over the years. Some of them gained initial interest but then were practically forgotten due to low performance, high cost, or impracticality. As an example, the xeroradiographic detector consists of a plate covered with an amorphous selenium photoconductor. Before use, the surface of the plate is uniformly positively charged. During exposure, electrons are excited in the photoconductor, causing local loss of positive charge. After exposure, a negatively charged toner is applied. The toner is attracted to the residual positive charge on the plate, and then it can be moved to paper or plastic sheet, creating an image [39]. Interestingly, films that do not require photochemical processing and can be used in daylight conditions also exist. Radiochromic films become blackened directly by ionizing radiation, thanks to

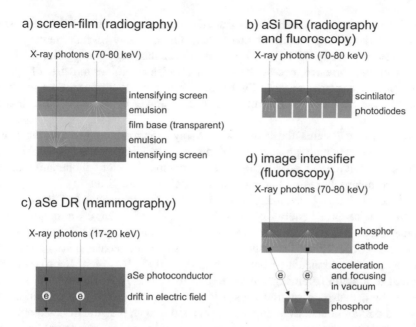

Fig. 1 Several detectors used in planar medical X-ray imaging (typical photon energies given for reference): (**a**) screen-film detector (radiography), (**b**) aSi detector (digital radiography), (**c**) aSe detector (digital mammography), (**d**) image intensifier (fluoroscopy).

radiation-induced polymerization of diacetylene (dark as polymer, colorless as a monomer). Radiochromic films are not sensitive enough to be used for imaging, but they may be used as radiation detectors to monitor skin doses in interventional radiology [9].

Modern radiography has largely moved away from film detectors toward digital imaging. Computed radiography (CR) imaging plates and digital radiography (DR) flat panel detectors do not need photochemical processing to get an image. They also have a much wider dynamic range then screen-film detector, e.g., 10^4:1 instead of 10:1 [8], and offer possibilities for digital processing, enhancement, and storage of the images.

In the CR systems, imaging plates are used as a radiation detector. The plate is placed in a radiographic cassette and emits visible light when absorbing X-ray radiation, which makes it similar to the intensifying screen in screen-film systems. At the same time, a sort of latent image is created on the plate, which makes it similar to film. During exposure, some electrons are trapped. During readout in the CR scanner, the imaging plate is illuminated with red laser, causing trapped electrons to relax to their ground state. The relaxation is associated with emission of blue light (photoluminescence), which is measured with a photomultiplier and used as an information to create an image. The light signal is proportional to the number of trapped electrons, which is in turn proportional to the received dose of radiation. Proper choice of storage phosphor materials for CR plates is essential for good performance of the system. The efficiency of radiation-to-light conversion of the material is important, but other characteristics may be important as well. Some materials (e.g., CsBr:Eu2+) perform better than others (e.g., BaFX:Eu2$^+$) because of their oriented, needlelike structure, which resembles a matrix of optical fibers. Such a structure reduces light scatter in the readout phase, thus resulting in less image blur even if the material layer is made thicker to achieve higher sensitivity [8].

In screen-film systems, the energy of the X-ray radiation is converted to the image with a time-consuming photochemical process. In CR systems, photochemical processing is not needed, but the plate is still a passive detector – it operates without any active means but does not provide a direct readout. Scanning of CR plates takes time; an image is available a few minutes after exposure. In DR flat panel detectors, the absorbed energy is converted to an electric charge within a detector, and the image is practically instantly displayed on the computer screen. In a vast majority of clinically used radiographic DR systems, the conversion is done with two steps, with intermediate conversion to light (Fig. 1b). First, X-ray radiation captured in a scintillation layer is converted to visible light. Second, the light is detected by a matrix of photodiodes. Electric charge is stored in a capacitor and then read out, amplified, digitized, and converted into an image. DR detectors with such indirect conversion are often referred to as aSi (amorphous silicon), although their amorphous silicon elements are not directly used in the detection of X-rays. This is done with the scintillation layer, which is made, e.g., of Gd_2O_2S or CsI. The materials differ in structure similarly as those used for CR imaging plates – while Gd_2O_2S is granular, CsI has a needlelike structure. Direct-conversion DR systems also exist and are often used in mammography. The energy of the absorbed X-ray radiation is converted directly to electrical charge within the amorphous selenium (a-Se) photoconductor (Fig. 1c). Electrons excited to the conduction band move along electrical field lines, without spreading to the sides, and are collected by the capacitors, similarly as in indirect DR detectors [32].

Not only detectors have evolved since the beginning of medical X-ray imaging history. Modern X-ray tubes are very different from those used in the first experiments: they have a higher vacuum, hot filament, line focus, proper shielding, higher filtration, usually rotating target, sometimes flat electron emitter, and a liquid bearing [6]. Tubes were constructed specifically for fluoroscopic/angiographic applications (long exposures, pulsed beam), mammography (lower voltage, smaller focal spot), and computed tomography (large g-force during rotation around a patient) [34]. Let the measure of improvement in the construction of X-ray radiation sources and detectors be a reduction in the average skin dose in pelvic radiographs: since 1896, it was reduced by a factor of about 400(!) [17].

3   Fluoroscopy

The screen-film detector integrates signals during whole exposure, which makes it suitable to capture an image of a stationary object. If the object moves, its image is blurred. Movements of the patient during imaging can be minimized with short exposure time and with immobilization. Patients can be asked to hold a breath, but it is not possible to stop internal movements such as heartbeat or peristalsis. In some situations, it is essential to visualize movements within a patient. Besides the movement of the patient's tissues, the flow of the contrast medium may be observed, movement of surgical instruments, or movement of implants, which are being placed in the patient. That is why a radiation detector is needed, which can provide a live view with a good time resolution.

In the early days of radiology, the live image was observed directly on the fluorescent screen. This resulted in large radiation exposure not only for the patient but also for the operator, who often kept the screen in hand. Since the image was not very bright, fluoroscopy could be only used in darkened rooms, which was inconvenient. In the 1950s, a new image detector for fluoroscopy was introduced, namely, X-ray image intensifier. The operation of the image intensifier consists of several steps. X-ray photons interact with a scintillation layer, converting each absorbed X-ray photon into many visible light photons (as with a simple fluorescent screen). The light strikes photocathode, knocking out electrons, which are then accelerated with electric potential and then fall again on another (smaller) fluorescent screen (Fig. 1d). As an effect, there are many more photons on the smaller screen, resulting in a much brighter (intensified) image with a smaller dose to a patient. The bright image can be observed in normal lighting conditions, recorded with a video camera, etc. Currently, image intensifiers are being replaced with flat panel (DR) detectors with a fast readout [2].

4   Tomographic Imaging

In radiography, three-dimensional structures are visualized in a two-dimensional planar image. The third dimension is lost in the imaging process. Images of objects located at different depths overlap, making some anatomical structures invisible. From the early days of medical X-ray imaging, researchers have tried to image selected planes within the patient [41]. Many experiments involved movement of the radiation source (tube) and/or detector (film) during the exposure. If the tube and detector are synchronously translated in opposite directions during exposure, only the objects located at one plane will be projected all the time at the same positions on the detector, resulting in a sharp image. For other objects, their projected image on the detector will move, resulting in blur. This idea was used in classical tomography (also called laminography) to obtain a sharp image of anatomical structures located at the focal plane (Fig. 2). Typically, only the image of one plane was obtained in

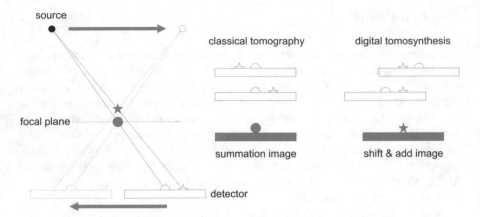

Fig. 2 Image acquisition in classical tomography (with film detector) and digital tomosynthesis (with flat panel DR detector)

one exposure. Another film and other exposure had to be used to image structures located on another plane. It was not only time-consuming but also associated with a multiplication of patient exposure. This limitation can be overcome quite easily if a different detector is used. If the film is replaced with a flat panel DR detector, tens of images at different angles can be acquired during the movement of the tube-detector system. If the images are simply summed up together, the same image as in classical tomography is obtained. If the images are shifted before summation, a different plane is sharply imaged. Thanks to an image detector with good time resolution and a very simple shift-and-add algorithm, images of several planes can be obtained from a single exposure. This solution, called digital tomosynthesis, is used clinically [22], although usually in a slightly different scheme (X-ray tube is moving over a segment of a circle, while the detector remains stationary).

Digital tomosynthesis is currently used mainly in breast imaging. Cross-sectional images of the head and body are usually obtained with X-ray computed tomography (CT). During the CT examination, tube and radiation detectors rotate around a patient, and a cross-sectional image is reconstructed with mathematical algorithms. Despite similarities between digital tomosynthesis and computed tomography, the geometry of the system is fundamentally different. In digital tomosynthesis, the image plane is perpendicular to the plane, in which the tube moves. In computed tomography, the image lays in the tube rotation plane.

In current multislice CT systems, there may be several rows of detectors (e.g., 64 rows with ca. 1000 detectors in each row). Detectors may perform two to three revolutions around a patient per second, registering the signal several hundred times during each rotation. Detectors for modern CT systems need not only to have good sensitivity and high dynamic range but also to be very fast and capable of working in large g-force conditions. For many years, xenon detectors have been used in CT, due to their good time characteristics (fast decay of signal). However, the low density of xenon – even pressurized – does not make it a very efficient detector for X-rays

[10]. Currently, CT detectors are usually made of scintillators and photodiodes [34]. The scintillators are chosen to have a high light output but also a low afterglow. The performance of the detectors, and thus also image quality, can be improved not only by choice of a better scintillator or more efficient photodiode but also by moving the analog-to-digital electronics as close as possible to the detectors, to minimize electronic noise [20].

5 Spectral Imaging

Attenuation coefficients for X-rays depend on the energy of photons. This dependence is different for different materials and tissues. This fact is used in bone densitometry (DXA, dual-energy X-ray absorptiometry), in which bone mineral density is calculated by comparing attenuation of radiation for two different energies [7, 23]. The typical radiographic image provides information about the attenuation of the radiation by the tissues, but it does not bring direct information about the density of the tissue. Identical attenuation can be observed e.g. for the thicker bone of lower density, as well as for thinner bone of higher density. In bone densitometry, quantitative information on bone density is calculated from absorption data measured for two different energies of radiation. Similar approach can be used to obtain other information. In contrast-enhanced spectral mammography, two exposures are made for two different beam energies (different tube voltage and different filtration). Both exposures are made after administration of an intravenous contrast agent, but detector data obtained for two energies may be processed to create virtual non-contrast image [29]. Possibility of using two energies to determine electron density and effective atomic number of the examined tissues has already been proposed for the first commercially available computed tomography system [31]. Currently, spectral CT allows for virtual monochromatic imaging, creation of virtual non-contrast images, better quantification of iodine content, and differentiation of renal stones based on their atomic composition [11, 15]. In some spectral computed tomography systems, the examination has to be performed twice to obtain data for two energies. Since the datasets are not obtained at the same time, patient movement can be an issue. Other spectral CT systems are capable of truly simultaneous imaging for two different energies, thanks to duplication of the tube-detector system or fast alternating tube voltage switching [13].

All the spectral imaging methods mentioned so far are based on the use of two radiation beams with two different energies. Another approach is also possible, based on a single beam, and provided that detector separately measures signals in two (or more) energy ranges. In one design of a spectral CT scanner, a layered detector is used, with two layers sensitive to two different energy ranges [13]. Another possibility is the use of photon-counting detectors (PCD), e.g., cadmium telluride semiconductors [21, 37]. The signal from most radiation detectors used in medical X-ray imaging is simply proportional to the total absorbed energy of radiation. In PCD, each detected X-ray photon generates a separately measured electrical pulse. The height of each pulse can be compared with the threshold, or several different thresholds, to assign it to one of the energy bins. For each pixel, PCD provides information on the number of pulses separately for each energy bin. This allows for truly simultaneous acquisition of separate images for several energy ranges during one exposure. If several contrast agents are administered to a patient before an examination, with different radiation absorption characteristics (e.g., iodine and gadolinium), the virtual reconstruction of images with individual contrast agents is possible. Besides energy-resolving capabilities, photon-counting detectors have no electronic noise, which allows better imaging [3, 20].

6 Imaging at the Higher Energy of X-Ray Photons

In diagnostic X-ray imaging, the energy of X-ray radiation is optimized to achieve good visibility of anatomical structures with patient dose as low as possible. Depending on the application, tube voltage may be in the order of 25–35 kilovolts in mammography, several dozen kV in radiography and fluoroscopy, and up to 120–140 kV in computed tomography. This electric potential is used to accelerate electrons within the X-ray tube, and it sets a limit for the maximum energy of emitted photons (e.g., 140 keV in CT). Most of the photons reaching detectors have lower energy. However, imaging with photons of higher energy is also used in medicine.

In nuclear medicine imaging, a radioisotope is administered to the patient. After its uptake within the tissues, γ-radiation produced in the radioisotope's decay is detected from the outside of the patient. Tc-99 m, which is the most widely used radioisotope for imaging, emits monochromatic γ quanta with an energy of 140 keV. In positron emission tomography (PET) imaging, a radioisotope is administered to the patient, which emits β^+ particles (positrons). Emitted positron annihilates with an electron, resulting in simultaneous emission of two 511 keV photons in opposed directions. Radiation detectors are distributed around the patient (Fig. 3a). If two of them coincidentally detect two photons, it is assumed that the photons originate from the same annihilation event. It is also assumed that the annihilation event has occurred on the line connecting the two detectors (LOR, line of response). This information is a base for image creation. Scintillators, which are used in PET as detectors, need to have a good time response with a very fast decay of light pulse to allow coincidence detection. They also need to have a good detection efficiency for 511 keV photons, which is usually associated with high density and high atomic number. Materials used in radiography (e.g., CsI, the density of 4.5 g/cm^3) or computed tomography (e.g., CWO, the light decay time of 14.5 μs) are not well suited for that task, compared to, e.g., LSO or LYSO scintillators (density 7–7.5 g/cm^3, decay time ca. 40 ns) [26]. With a fast scintillator and a fast light detector (e.g., silicon photomultiplier), it is even possible to estimate time between detection of the two coincidence photons and to approximately determine the position of annihilation along the LOR. Inclusion of the additional data in image reconstruction

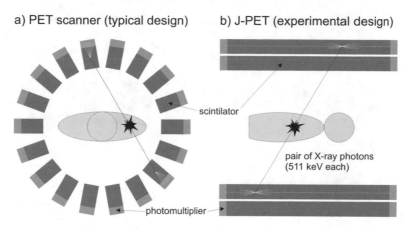

Fig. 3 Signal detection in positron emission tomography (PET) systems: (**a**) typical PET scanner design, (**b**) J-PET (experimental design)

improves image quality in the so-called time-of-flight (TOF) PET scanners [40]. Potentially, detectors with a much worse absorption efficiency could be also used in PET scanners. Compared to crystals commonly used in PET scanners, plastic scintillators have a low density (approximately 1 g/cm³). Additionally, 511 keV photons usually do not leave all of their energy in the plastic scintillator during an interaction. However, the light pulse generated at each interaction is extremely short (1–2 ns), which allows us to obtain information on the place of interaction within the detector. Low-density plastic scintillator, used in a J-PET prototype in a non-typical geometrical setup (Fig. 3b), allows obtaining results comparable as for typical PET scanners [18, 24].

In oncology, radiotherapy with high-energy photon beams is used to treat patients with cancers. The maximum energy of photons in the therapeutic radiation beam is much higher in diagnostic radiology, with a typical value of 6 MeV. Although the beam is not optimized for imaging, it can be used for that task. The images obtained with the therapeutic beam (so-called portal images) can be used to verify the realization of the therapy, to check the alignment of the realized radiation field with a planned field and to verify the proper positioning of the patient. Originally, portal imaging was performed with films, and then various electronic portal imaging devices (EPIDs) were introduced. In one design, a fluorescent screen and a vidicon camera were used. The camera could not be placed directly on the beamline, because it would be hit with high-energy X-rays. Therefore, the light from the fluorescent screen was deflected with a mirror (Fig. 4a). In other designs, a matrix of liquid-filled ion chambers was used as the detector. An ion chamber filled with air, or even with a pressurized gas, would have a low detection efficiency for high-energy photons. Currently EPIDs are usually constructed similarly to radiographic flat panel DR detectors, with a scintillator layer and an array of photodiodes. The main difference between the DR detector and EPID is the presence of an additional metal layer (e.g., 1 mm of copper) in front of the scintillation layer (Fig. 4b).

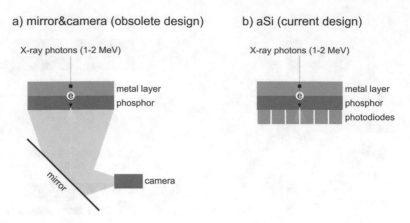

Fig. 4 Electronic portal imaging detectors (EPIDs) used in radiotherapy with high-energy X-ray photons (typical photon energies given for reference): (**a**) mirror and camera system (obsolete design), (**b**) aSi system (current design)

The metal layer is responsible for a generation of recoil electrons, which interact in the scintillator [1, 25, 42]. Additionally, detector elements (pixels) are usually significantly bigger in EPIDs than in radiography detectors (e.g., 400–800 μm vs 100 μm). Although bigger pixels lower detector's spatial resolution, such design allows each pixel to gather a higher signal in a shorter time.

7   Scattered Radiation

In traditional medical X-ray imaging, scattered photons add noise to an attenuation-based image. This is why in radiography anti-scatter grids are generally used, eliminating a large portion of the scattered photons. At the same time, an anti-scatter grid reduces the signal at the detector. That usually needs to be compensated with a higher patient dose compared to a no-grid setup. Other scatter-reduction methods include air gap, meaning the increased distance between patient and detector, and narrow beam geometry. If a narrow beam is used (pencil or slit), the trajectory of scattered photons will fall outside the beam and detector area. Slit-beam geometry is natural for CT scanners but may be also used in other imaging modalities, such as mammography [19].

In nuclear medicine imaging, monochromatic photons are used (radioisotope-emitted gamma rays). Detectors in gamma cameras can discriminate energy of each absorbed γ-ray quantum. A signal from scattered quanta, which has lower energy, can be eliminated with energy windowing, and only the monochromatic peak is used for image creation. That approach cannot be used in radiography. Even in the case of photon-counting detectors, the scattered photons cannot be separated over the wide spectrum of photons emitted from the X-ray tube. It could be possible if

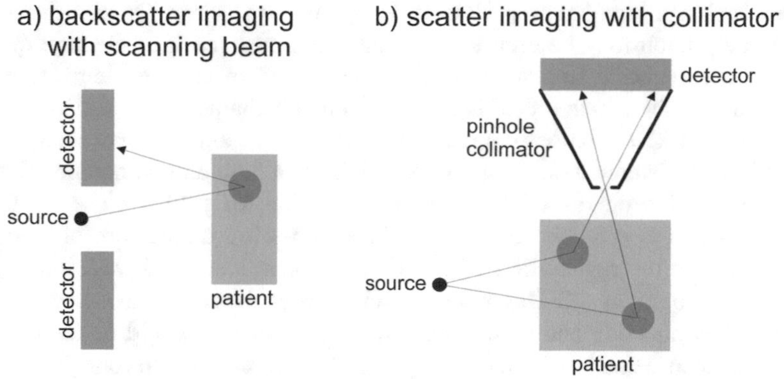

Fig. 5 Examples of scatter-based schemes for X-ray imaging: (**a**) backscatter imaging with scanning beam, (**b**) scatter imaging with collimator

a monochromatic beam was used, e.g., created with a conventional X-ray source combined with a mosaic crystal [12].

Photons, which are scattered in directions other than toward the detector, obviously do not have an impact on the quality of transmission image. While they are usually only of radiological protection concern, they could also be used for imaging. X-ray backscatter systems are used at many airports for passenger screening. Passenger, in a standing position, is scanned with a pencil radiation beam (Fig. 5a). Backscatter radiation is detected, allowing to image items hidden under clothes [16] with a low dose [36]. A similar scheme has also been considered for the imaging of humans [38]. A scanning beam has also been used to limit the volume, in which the radiation is scattered. If a large-area beam was used, the whole irradiated volume would be the source of scattered radiation, and it would be hard to distinguish between signals emitted by different organs. A similar situation can be observed in nuclear medicine imaging, where the whole patient is the source of radiation emitted by the radiopharmaceutical product present in his tissues. In nuclear medicine, this is solved by the use of slit collimators or pinhole collimators. This approach could be adapted to scatter-based imaging in lung radiotherapy (Fig. 5b). Lung tumor has a much higher density than the surrounding lung tissue and is a major source of radiation scatter. A slit X-ray kilovoltage beam may be used for imaging [43], as well as the megavoltage therapeutic beam itself [30]. Experiments show that scatter-based images may provide better tumor contrast as compared to transmission images obtained with EPID [30].

8   Forced Fluorescence

In the description of image formation in X-ray imaging, a photon that interacts through the photoelectric effect is referred to as removed from the beam. This is true because such a photon transfers all of its energy to the electron. On the other side, a new X-ray photon may be emitted from the atom as a result of electron ejection (forced fluorescence). Human tissues are composed mainly of elements with a low atomic number, so the resulting monochromatic characteristic radiation has an energy of a few keV and does not reach the detector. The situation can change, if the patient is administered a contrast agent containing a high atomic number element, such as gold. The energy of characteristic radiation for gold is close to 70 keV. Theoretically, it is possible to create an image showing the capture of a contrast agent based on the registration of its forced fluorescence. In X-ray fluorescence computed tomography (XFCT), many detectors would be placed around the patient, and energy windowing and collimators with parallel holes would be used [5, 14]. Such fluorescence can be also caused by different beams, e.g., protons [4].

9 Phase-Contrast Imaging

In the above description, the X-ray beam was treated as a flux of particles (photons). At the same time, it can be described as a wave, with phase and trajectory changing while traveling through matter. The refraction of X-rays is much lower than the refraction of visible light, but it is observable. Possibility to obtain medical images dependent on refraction and phase change has been shown in experiments (phase-contrast imaging). One possibility is to pass the radiation coming from the patient through two grates differing in spatial frequency [27]. As a result, we obtain a stronger or weaker signal not in the shadow of the entire structure but in places where the wave refracted on the patient's tissues. The image may appear similar to an attenuation-based X-ray image with an edge detection filter applied. In recent years, it has been shown that attenuation-based images and phase-contrast images of the mammary gland can be obtained simultaneously, during a several second exposure with a dose similar to regular mammography [33]. It has also been shown that phase-based mammographic images may be obtained at higher beam energies than traditionally used in mammography, with simultaneous improvement in lesion detectability and a patient dose saving compared to traditional attenuation-based imaging [28].

10 Summary

Various types of passive and active radiation detectors are used in medical imaging. In some of the active detectors, the energy of X-ray radiation is converted directly to electric charge and then to a digital image. In many systems, an intermediate conversion of the signal carriers takes place, e.g., to visible light. Different detector materials and designs can be chosen depending on the energy of radiation. The perfect detector could be characterized with high detection efficiency, high spatial, time, and energy resolution. Usually, there is an interplay between the parameters, and depending on the application, one may want to, e.g., increase detector's sensitivity at the cost of its spatial resolution. While X-ray imaging in medicine is well-established, the availability of new detectors opens new possibilities. On the other hand, experimental imaging modalities create new possibilities for use of the existing radiation sources and detectors.

Bibliography

1. Baily, N. A., Horn, R. A., & Kampp, T. D. (1980). Fluoroscopic visualization of megavoltage therapeutic x ray beams. *International Journal of Radiation Oncology, Biology, Physics, 6*, 935–939. https://doi.org/10.1016/0360-3016(80)90341-7.
2. Balter, S. (2019). Fluoroscopic technology from 1895 to 2019 drivers: Physics and physiology. *Medical Physics International, 7*, 111–140.
3. Barber, W. C, Nygard, E., Iwanczyk, J. S., et al. (2009). *Characterization of a novel photon counting detector for clinical CT: Count rate, energy resolution, and noise performance.* In: Proceedings of SPIE.
4. Bazalova-Carter, M., Ahmad, M., Matsuura, T., et al. (2015). Proton-induced x-ray fluorescence CT imaging. *Medical Physics, 42*, 900–907. https://doi.org/10.1118/1.4906169.
5. Bazalova, M., Kuang, Y., Pratx, G., & Xing, L. (2012). Investigation of X-ray fluorescence computed tomography (XFCT) and K-edge imaging. *IEEE Transactions on Medical Imaging, 31*, 1620–1627. https://doi.org/10.1109/TMI.2012.2201165.
6. Behling, R. (2018). X-ray tubes development- IOMP history of medical physics. *Medical Physics International, 6*, 8–55.
7. Cameron, J. R., & Sorenson, J. (1963). Measurement of bone mineral in vivo: An improved method. *Science, 80*(142), 230–232. https://doi.org/10.1126/science.142.3589.230.
8. Cowen, A. R., Davies, A. G., & Kengyelics, S. M. (2007). Advances in computed radiography systems and their physical imaging characteristics. *Clinical Radiology, 62*, 1132–1141. https://doi.org/10.1016/j.crad.2007.07.009.
9. delle Canne, S., Carosi, A., Bufacchi, A., et al. (2006). Use of GAFCHROMIC XR type R films for skin-dose measurements in interventional radiology. *Physica Medica, 22*, 105–110.
10. Fuchs, T., Kachelrie, M., & Kalender, W. A. (2000). Direct comparison of a xenon and a solid-state CT detector system: Measurements under working conditions. *IEEE Transactions on Medical Imaging, 19*, 941–948. https://doi.org/10.1109/42.887841.
11. Grajo, J. R., Patino, M., Prochowski, A., & Sahani, D. V. (2016). Dual energy CT in practice. *Applied Radiology, 45*, 61–62.
12. Green, F. H., Veale, M. C., Wilson, M. D., et al. (2016). Scatter free imaging for the improvement of breast cancer detection in mammography. *Physics in Medicine and Biology, 61*, 7246–7262. https://doi.org/10.1088/0031-9155/61/20/7246.
13. Johnson, T. R. C. (2012). Dual-energy CT: general principles. *AJR American Journal of Roentgenology, 199*, 3–8. https://doi.org/10.2214/AJR.12.9116.
14. Jones, B. L., & Cho, S. H. (2011). The feasibility of polychromatic cone-beam x-ray fluorescence computed tomography (XFCT) imaging of gold nanoparticle-loaded objects: A Monte Carlo study. *Physics in Medicine and Biology, 56*, 3719–3730. https://doi.org/10.1088/0031-9155/56/12/017.
15. Karçaaltıncaba, M., & Aktaş, A. (2011). Dual-energy CT revisited with multidetector CT: Review of principles and clinical applications. *Diagnostic and Interventional Radiology, 17*, 181–194. https://doi.org/10.4261/1305-3825.DIR.3860-10.0.
16. Kaufman, L., & Carlson, J. W. (2011). An evaluation of airport x-ray backscatter units based on image characteristics. *Journal of Transportation Security, 4*, 73–94. https://doi.org/10.1007/s12198-010-0059-7.

17. Kemerink, G. J., Kütterer, G., Kicken, P. J., et al. (2019). The skin dose of pelvic radiographs since 1896. *Insights Into Imaging, 10*(39). https://doi.org/10.1186/s13244-019-0710-1.
18. Kowalski, P., Wiślicki, W., Shopa, R. Y., et al. (2018). Estimating the NEMA characteristics of the J-PET tomograph using the GATE package. *Physics in Medicine and Biology, 63*. https://doi.org/10.1088/1361-6560/aad29b.
19. Lai, C. J., Shaw, C. C., Geiser, W., et al. (2008). Comparison of slot scanning digital mammography system with full-field digital mammography system. *Medical Physics, 35*, 2339–2346. https://doi.org/10.1118/1.2919768.
20. Lell, M. M., Wildberger, J. E., Alkadhi, H., et al. (2015). Evolution in computed tomography: The battle for speed and dose. *Investigative Radiology, 50*, 629–644. https://doi.org/10.1097/RLI.0000000000000172.
21. Leng, S., Bruesewitz, M., Tao, S., et al. (2019). Photon-counting detector CT: System design and clinical applications of an emerging technology. *Radiographics, 39*, 729–743. https://doi.org/10.1148/rg.2019180115.
22. Machida, H., Yuhara, T., Tamura, M., et al. (2016). Whole-body clinical applications of digital tomosynthesis. *Radiographics, 36*, 735–750. https://doi.org/10.1148/rg.2016150184.
23. Mazess, R. B., Peppler, W. W., & Gibbons, M. (1984). Total body composition by dual-photon (153Gd) absorptiometry. *The American Journal of Clinical Nutrition, 40*, 834–839. https://doi.org/10.1093/ajcn/40.4.834.
24. Moskal, P., Rundel, O., Alfs, D., et al. (2016). Time resolution of the plastic scintillator strips with matrix photomultiplier readout for J-PET tomograph. *Physics in Medicine and Biology, 61*, 2025–2047. https://doi.org/10.1088/0031-9155/61/5/2025.
25. Munro, P. (1995). Portal imaging technology: Past, present, and future. *Seminars in Radiation Oncology, 5*, 115–133. https://doi.org/10.1016/S1053-4296(95)80005-0.
26. Nassalski, A., Kapusta, M., Batsch, T., et al. (2007). Comparative study of scintillators for PET/CT detectors. *IEEE Transactions on Nuclear Science, 54*, 3–10. https://doi.org/10.1109/TNS.2006.890013.
27. Olivo, A., & Speller, R. (2007). A coded-aperture technique allowing x-ray phase contrast imaging with conventional sources. *Applied Physics Letters, 91*, 111–114. https://doi.org/10.1063/1.2772193.
28. Omoumi, F. H., Ghani, M. U., Wong, M. D., et al. (2020). The potential of utilizing mid-energy X-rays for in-line phase sensitive breast cancer imaging. *Biomedical Spectroscopy and Imaging*, 1–14. https://doi.org/10.3233/BSI-200204.
29. Patel, B. K., Lobbes, M. B. I., & Lewin, J. (2018). Contrast enhanced spectral mammography: A review. *Seminars in Ultrasound, CT, and MR, 39*, 70–79. https://doi.org/10.1053/j.sult.2017.08.005.
30. Redler, G., Jones, K. C., Templeton, A., et al. (2018). Compton scatter imaging: A promising modality for image guidance in lung stereotactic body radiation therapy. *Medical Physics, 45*, 1233–1240. https://doi.org/10.1002/mp.12755.
31. Rutherford, R., Pullan, B. R., & Isherwood, I. (1976). Measurement of effective atomic number and electron density using an EMI scanner. *Neuroradiology, 11*, 15–21. https://doi.org/10.1007/BF00327253.
32. Samei, E., & Flynn, M. J. (2003). An experimental comparison of detector performance for direct and indirect digital radiography systems. *Medical Physics, 30*, 608–622. https://doi.org/10.1118/1.1561285.
33. Scherer, K., Willer, K., Gromann, L., et al. (2015). Toward clinically compatible phase-contrast mammography. *PLoS One, 10*, 6–12. https://doi.org/10.1371/journal.pone.0130776.
34. Shefer, E., Altman, A., Behling, R., et al. (2013). State of the art of CT detectors and sources: A literature review. *Current Radiology Reports, 1*, 76–91. https://doi.org/10.1007/s40134-012-0006-4.
35. Sprawls, P. (2018). Film-screen radiography receptor development – A historic perspective. *Medical Physics International, 6*, 56–81.
36. Stepusin, E. J., Maynard, M. R., O'Reilly, S. E., et al. (2017). Organ doses to airline passengers screened by X-ray backscatter imaging systems. *Radiation Research, 187*, 229–240. https://doi.org/10.1667/RR4516.1.

37. Taguchi, K., & Iwanczyk, J. S. (2013). Vision 20/20: Single photon counting x-ray detectors in medical imaging. *Medical Physics, 40*, 100901. https://doi.org/10.1118/1.4820371.
38. Towe, B. C., & Jacobs, A. M. (1981). X-ray backscatter imaging. *IEEE Transactions on Biomedical Engineering, BME-28*, 646–654. https://doi.org/10.1109/TBME.1981.324755.
39. Udoye, C. I., & Jafarzadeh, H. (2010). Xeroradiography: Stagnated after a promising beginning? A historical review. *European Journal of Dentistry, 04*, 095–099. https://doi.org/10.1055/s-0039-1697816.
40. Vandenberghe, S., Mikhaylova, E., D'Hoe, E., et al. (2016). Recent developments in time-of-flight PET. *EJNMMI Physics, 3*. https://doi.org/10.1186/s40658-016-0138-3.
41. Webb, S. (1992). Historical experiments predating commercially available computed tomography. *The British Journal of Radiology, 65*, 835–837. https://doi.org/10.1259/0007-1285-65-777-835.
42. Woźniak, B., Ganowicz, M., Bekman, A., & Maniakowski, Z. (2005). A comparison of the dosimetric properties of the Electronic Portal Imaging Devices (EPIDs) LC250 and aS500. *Reports of Practical Oncology and Radiotherapy, 10*, 249–254. https://doi.org/10.1016/S1507-1367(05)71097-X.
43. Yan, H., Tian, Z., Shao, Y., et al. (2016). A new scheme for real-time high-contrast imaging in lung cancer radiotherapy: A proof-of-concept study. *Physics in Medicine and Biology, 61*, 2372–2388. https://doi.org/10.1088/0031-9155/61/6/2372.

Status of DEXA Instrumentation Using Direct and Indirect Detectors

Jihoon Kang

1 Introduction

Osteoporosis is the most common bone disease in humans, representing a major public health threat. Osteoporosis is defined as "a skeletal disease characterized by compromised bone strength predisposing a person to an increased risk of fracture" [1]. According to the National Osteoporosis Foundation (NOF), approximately 54 million American adults, representing 50% of the US adult population, are affected by osteoporosis (10.2 million) and osteopenia (43.4 million) [2]. One-tenth of women aged 60, one-fifth of women aged 70, two-fifths of women aged 80, and two-thirds of women aged 90 worldwide are estimated to have osteoporosis [3]. Approximately 1 in 3 women over the age of 50 and 1 in 50 men over the age of 5 worldwide experience osteoporotic fractures [4–6]. The prevalence of osteoporosis, osteopenia, and fractures is expected to increase in all countries because the number of older people is increasing.

Bone densitometry is a noninvasive and inexpensive examination for the quantitative measurement of bone mass or density. Numerous methods have been proposed and utilized to measure bone mineral density (BMD) and strength [7–14]. There are still certain technical challenges and performance trade-offs in these diagnostic methods, which include plain radiography, radiographic absorptiometry (RA), single-energy X-ray absorptiometry (SXA), and dual-energy X-ray absorptiometry (DEXA) and quantitative computed tomography (QCT) using X-ray sources; single-photon absorptiometry using gamma sources (SPA based on iodine-125 radionuclide with an energy of 35 keV or americium-241 with an energy of 60 keV) and dual-photon absorptiometry (DPA based on gadolinium-153 radionuclide with

J. Kang (✉)
Chonnam National University, Yeosu, Korea

© Springer Nature Switzerland AG 2022
K. Iniewski (ed.), *Advanced X-ray Detector Technologies*,
https://doi.org/10.1007/978-3-030-64279-2_7

energies of 44 and 100 keV); quantitative ultra-sonometers (QUS) using ultrasound; and quantitative magnetic resonance (QMR) using radio-frequency signals.

DEXA is the most widely used densitometry method for BMD assessment because it gives very precise measurements at clinically relevant skeletal sites. Even though DEXA has inevitable disadvantages including a non-portable bulky system and radiation exposure (albeit in a very low dose), it is the standard test used to diagnose osteoporosis, according to the classification of the World Health Organization (WHO) [15, 16]. Since the first DEXA system was introduced by Hologic in 1987 [17, 18], advances in technology have increased the need to overcome various technical challenges and replace older-generation DEXA systems based on NaI scintillation crystals coupled to photomultiplier tube (PMT).

More than 30 years after its implementation, semiconductor detectors based on cadmium-zinc-telluride (CZT) or cadmium telluride (CdTe) are regarded as one of the most useful DEXA detectors. Currently, CZT or CdTe detectors with direct conversion mechanisms have been widely utilized on a commercial basis [19, 20]. This is due to their outstanding features of excellent energy resolution, high detection efficiency, room temperature operation, and compact four-side tillable scalability. These compound semiconductor detectors have a number of features that may render their prolonged implementation difficult. They are susceptible to irradiation damage, incomplete carrier collection, and a large dark current. Their high price and fragile structure also limits their applicability. As an alternative, an X-ray diagnostic detector combining high-performance scintillation crystals with a compact photosensor has attracted interest for use in DEXA applications. The first commercial DEXA scanner based on this technology was recently introduced [21].

This chapter focuses on DEXA instrumentation, the most common densitometry method currently, to review its principles and use techniques. The main features of commercial systems, with differences in creating the dual-photon beams, scanning the patient body, and employing the photon detectors, are described and compared. We also present recent results obtained from indirect detectors consisting of a Geiger-mode avalanche photodiode (GAPD) coupled to widely utilize lutetium-yttrium oxyorthosilicate (LYSO) and newly developed gadolinium aluminum gallium garnet (GAGG) for potential application in DEXA systems.

2   Basic Principles and Techniques for DEXA Instrumentation

Dual-energy X-ray absorptiometry exploits the difference in the transmission and attenuation of tissue and bone at low and high energies. These X-rays transfer their energy to the body as they pass through it. The energy level should be sufficient to transmit the matter within the imaging area and still be efficiently detected by the DEXA detector.

In the basic physics of the interaction of radiation with matter, the transmission of a monoenergetic photon beam having the same energy through matter is described using the following exponential equation:

$$I = I_0 e^{-\mu_l x} \tag{1}$$

where I is the beam intensity transmitted through the matter, I_0 is the initial beam intensity recorded with no matter present, μ_l [cm^{-1}] is the linear attenuation coefficient of the matter at the photon energy of interest, and x [cm] is the thickness of the matter.

It should be noted that the negative sign signifies that the beam intensity decreases with increasing matter thickness. The linear attenuation coefficient is found to increase linearly with matter density ρ and can be obtained from:

$$\mu_l \left(\text{cm}^{-1}\right) = \mu_m \left(\text{cm}^2/\text{g}\right) \times \rho \left(\text{g/cm}^3\right) \tag{2}$$

where μ_m is the mass attenuation coefficient of the matter and depends upon the effective atomic number Z_{eff} of the matter and incident photon energy E.

The exponential part of Eq. (1) can be changed as in Eq. (3) below:

$$\begin{aligned}
\mu_l &\left(\text{cm}^{-1}\right) \times x \,(\text{cm}) \\
&= \mu_m \left(\text{cm}^2/\text{g}\right) \times \rho \left(\text{g/cm}^3\right) \times x \,(\text{cm}) \\
&= \mu_m \left(\text{cm}^2/\text{g}\right) \times d_m \left(\text{g/cm}^2\right)
\end{aligned} \tag{3}$$

where d_m is the area density (g/cm^2).

Thus, from Eqs. (1, 2, and 3), the general form expressed in terms of BMD is given by:

$$I = I_0 e^{-\mu_m d_m} \tag{4}$$

A fundamental assumption in calculating bone density values is that the body is composed of a two-compartment model of bone mineral and soft tissue (Fig. 1). Bone mineral includes phosphorus and calcium molecules having relatively higher effective atomic numbers and physical density, whereas soft tissue is a mixture of muscle, fat, skin, and water. For a monoenergetic photon beam, the mass attenuation coefficient for a mixture of elements can be obtained from the values for its component elements with bone (B) and soft tissue (T) as in Eq. (5) below:

$$\mu_m \,(\text{mix}) \; d_m \,(\text{mix}) = \mu_m(B) \, d_m(B) + \mu_m(T) \, d_m(T) \tag{5}$$

For polyenergetic photon beams, realistic X-ray photon beams with the different X-ray energies, the mass attenuation coefficient (μ_m) for a mixture of elements can be obtained from the values for its component elements with bone (B) and soft tissue

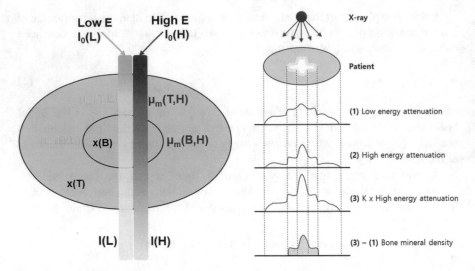

Fig. 1 The fundamental principle of DEXA. Bone density values can be calculated by the monoexponential attenuation process that transmits or attenuates X-rays through the body using low and high energy

(T) for each high energy (H) and low energy (L) as in the following equations:

$$\mu_m\,(\text{mix},\,L)\,d_m\,(\text{mix}) = \mu_m\,(B,\,L)\,d_m(B) + \mu_m\,(T,\,L)\,d_m(T) \qquad (6)$$

$$\mu_m\,(\text{mix},\,H)\,d_m\,(\text{mix}) = \mu_m\,(B,\,H)\,d_m(B) + \mu_m\,(T,\,H)\,d_m(T) \qquad (7)$$

Thus, from Eqs. (1, 6, and 7), the mathematical equations become:

$$I(L) = I_0(L)e^{-(\mu_m(B,L)d_m(B)+\mu_m(T,L)d_m(T))}$$

$$\frac{I(L)}{I_0(L)} = e^{-(\mu_m(B,L)d_m(B)+\mu_m(T,L)d_m(T))} \qquad (8)$$

$$\ln\left(\frac{I(L)}{I_0(L)}\right) = -\,(\mu_m\,(B,\,L)\,d_m(B) + \mu_m\,(T,\,L)\,d_m(T))$$

$$I(H) = I_0(H)e^{-(\mu_m(B,H)d_m(B)+\mu_m(T,H)d_m(T))}$$

$$\frac{I(H)}{I_0(H)} = e^{-(\mu_m(B,H)d_m(B)+\mu_m(T,H)d_m(T))} \qquad (9)$$

$$\ln\left(\frac{I(H)}{I_0(H)}\right) = -\,(\mu_m\,(B,\,H)\,d_m(B) + \mu_m\,(T,\,H)\,d_m(T))$$

Considering k factor, Eq. (9) becomes:

$$\ln\left(\frac{I(H)}{I_0(H)}\right)\left(\frac{\mu_m(T,L)}{\mu_m(T,H)}\right) = -\left(\mu_m(B,H)\,d_m(B) + \mu_m(T,H)\,d_m(T)\right)\left(\frac{\mu_m(T,L)}{\mu_m(T,H)}\right)$$

$$\ln\left(\frac{I(H)}{I_0(H)}\right)(k) = -\mu_m(B,H)\,d_m(B)(k) - \mu_m(T,L)\,d_m(T)$$

$$(10)$$

These equations are solved for $d_m(B)$ for the area density of bone.

$$\ln\left(\frac{I(L)}{I_0(L)}\right) - \ln\left(\frac{I(H)}{I_0(H)}\right)(k) = -\mu_m(B,L)\,d_m(B) - \mu_m(T,L)\,d_m(T)$$

$$+\mu_m(B,H)\,d_m(B)(k) + \mu_m(T,L)\,d_m(T)$$

$$\ln\left(\frac{I(L)}{I_0(L)}\right) - \ln\left(\frac{I(H)}{I_0(H)}\right)(k) = -\mu_m(B,L)\,d_m(B) + \mu_m(B,H)\,d_m(B)(k)$$

$$\ln\left(\frac{I(L)}{I_0(L)}\right) - \ln\left(\frac{I(H)}{I_0(H)}\right)(k) = -\left(\mu_m(B,L) - (k)\mu_m(B,H)\right)d_m(B)$$

$$d_m(B) = -\frac{\ln\left(\frac{I(L)}{I_0(L)}\right) - \ln\left(\frac{I(H)}{I_0(H)}\right)(k)}{\left(\mu_m(B,L) - (k)\mu_m(B,H)\right)}$$

$$(11)$$

The k factor can be acquired from the patient scan by measuring the transmitted intensity of the beam at points at which there is no bone. Once the k factor is determined, the equation can be solved to calculate the area bone density, $d_m(B)$.

BMD results are generally scored by two measures, the T-score and the Z-score, for interpretation. A clear understanding of T-score and Z-score is essential in clinical practice. Both scores are derived by comparison to the mean value of a reference population on a standard deviation of patients of the same gender. Negative scores indicate lower bone density, and positive scores indicate higher.

The recommended reference group for T-score is a young adult group of healthy subjects aged 20–35 years with peak bone mass. The Z-score is derived from an age-matched reference population. Following WHO criteria, a T-score greater than −1 standard deviation (SD) is classified as normal bone mineral density. T-score ranging from −1 to −2.5, from −2.5 to −3.5, and −3.5 or less are used for the diagnosis of osteopenia, osteoporosis, and severe osteoporosis, respectively. Each SD decrease in the T-score is associated with an average increase in fracture risk of 1.5- to 3-fold in adults.

A Z-score helps diagnose secondary osteoporosis. It is most useful for premenopausal women, adult men under age 50, and children. A Z-score of less than −2 standard deviation (SD) is classified as a BMD risk below the expected range for the patient's age, and a careful inspection for coexisting illnesses or treatments that may contribute to osteoporosis is needed.

3 Different Features of Commercial Systems

Significant differences in the technologies used by different manufacturers are present in all DEXA systems on the market. Differences are seen in the irradiating X-ray energy levels, the filtering X-ray energy ranges, the creation of dual-photon beams, the types of photon detectors employed, the calibration methods used, the use of bone edge detection algorithms, how the region of interest ranges are calculated, and the reference databases used. For all of these reasons, it is difficult to make quantitative comparisons of BMD values from different DEXA systems unless a cross calibration study has been done. Three features, how to create dual-photon beams, acquire DEXA images, and convert incident photons to electrical signals, are described in the following section.

3.1 Generating Low- and High-Energy Beams: Voltage Switching Versus K-Edge Filtering

In most DEXA systems, the X-ray tubes used are standard tungsten anode tubes with focal spot sizes on the order of 0.5–1 mm^2. However, there are differences in how X-ray beams of two peak energies are produced to create dual-energy images.

One method is the voltage switching system utilized by Hologic (Marlborough, MA, USA). The X-ray tube potential is switched rapidly from low (~70 kVp) to high (~140 kVp), alternating at 60 Hz with an internal rotating disc of additional beam-filtering materials. Low- and high-energy beams are alternated with very short pulses (~8.3 ms) during DEXA image acquisition. An aluminum filter is used to harden the low- and high-energy beams. Additional copper or brass is used to remove the low-energy part of the spectrum in high-energy beams and minimize the overlap between low and high X-ray spectra in dual-energy beams. The tube voltage switches, beam-filtering materials, and photon detectors are all electronically and mechanically synchronized to sequentially acquire low- and high-energy information during DEXA scans.

The other method is the K-edge filtering system employed by Norland (Fort Atkinson, WI, USA) and GE Lunar (Madison, WI, USA). The X-ray tube unit is operated in a steady direct current mode and the potential X-ray source is constant. One of the several rare earth filters is placed between the X-ray tube and the patient bed for the generation of low- and high-energy beams. These filters selectively absorb X-ray photons around the K-shell energy of the filter material due to the K-edge of the atomic structure of the element. With an X-ray voltage of 100 kVp, the samarium (Z = 62) filter module was used to separate high-energy from low-energy photons by Norland Corporation. An X-ray tube with 80 kVp combined with a cerium (Z = 58) filter module by GE Lunar was used to split a single split X-ray spectrum into two components of low-and high-energy photons.

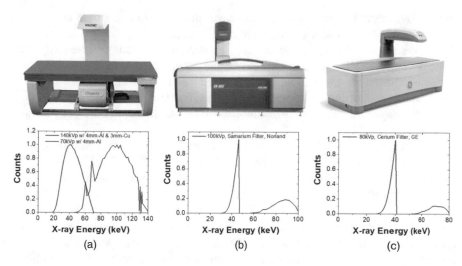

Fig. 2 Commercial DEXA systems with different approaches to creating dual-energy beams (top) and their representative energy spectra (bottom). (**a**) Discovery, Hologic, (**b**) XR-800, Norland, (**c**) Prodigy, GE Lunar

Figure 2 shows the representative DEXA systems and their simulated X-ray energy spectra to create dual-photon beams for each method. These representative energy spectra were generated using the tungsten anode spectral model using interpolating polynomials (TASMIP) algorithm [22]. In summary, each method generates different dual-energy beams and has inherent strengths and weaknesses. An additional discussion of these methods is presented in [23].

3.2 Acquiring DEXA Images: Pencil Beams, Fan Beams, and Cone Beams

All components in a typical DEXA system, including the X-ray unit, source collimator, filtering module, patient table, detector collimator, and DEXA detector have a fixed geometry on a C-shape gantry. These X-ray beam geometry and image acquisition methods determine the scan speed and image quality.

A pencil-beam system was employed in the first and original DEXA system. They use a highly collimated or narrowed X-ray beam (2–3 mm) in conjunction with a single detector that moves in a rectilinear fashion in tandem across the measurement region. Single detector elements receive the single thin X-ray beam passing through the patient at a time and have to "raster" scan over the region of interest (ROI). This pencil-beam system produces the most geometrically correct information, with little or no magnification of the area being scanned. Although the pencil beam is considered the gold standard for precision, the scan time is increased due to travel time for the single beam to cover the body. Imaging time is typically 20 min for whole-body scans, whereas hip and spine scans done using a pencil-beam DEXA system take up to 3–5 min to complete. Because of the long whole-body scan time, fan-beam DEXA systems were developed to speed up the testing process, and the time required for a whole-body scan was reduced to as few as 3 min. The pencil beam is an excellent choice for practices scanning fewer than 15–20 patients per day. The X-ray exposure dose is very small (average natural radiation dose per year) for the patient.

Examples of excellent pencil-beam DEXA systems are QDR® 1000 and QDR® 2000 from Hologic; DPX® Plus, DPX®-L, DPX-IQ™, DPX®-SF, DPX®-A, DPX-MD™, DPX-MD+™, and DPX-NT™ from GE Lunar; and XR-36™, XR-46™, Excell™, and Excell™plus from Norland.

Fan-beam systems were introduced in the early 2000s as an alternative to pencil beams. They use a much broader or fan-shaped beam, which is generated by a slit-type collimator. The array detector has multiple linear elements and collects ten or more pixels at a time. This enables the entire scan line to be quantified instantly and can shorten the scan time. Thus, fan-beam systems are much faster than pencil-beam systems for equivalent imaging properties. Images of the spine or hip are typically acquired in less than 30 s, compared to the 3–5 min required for pencil-beam systems. Whole-body scans take ~3 min to complete, allowing greater patient throughput. Also, fan-beam systems produce better-resolution images due to the considerably higher-energy photon intensities and greater photon flux, compared to older pencil-beam machines. The trade-off for the increased scan speed and the improved image resolution of fan-beam DEXA systems is higher radiation exposure, which is 60 times greater in some scan cases. An additional significant downside is the magnification and/or distortion of the image caused by fan-shaped geometry. The degree of magnification depends upon the distance of the source to the object; the closer the body part (bone or tissue) is to the source, the greater the magnification. These magnification issues result in a variation of up to 37% in BMC assessments. In summary, the fan-beam scanner offers shorter scan times allowing greater patient throughput, better resolution, and slightly higher radiation doses. Examples of fan-array DEXA scanners are QDR® 4500 A, QDR® 4500 C, QDR® 4500 W, QDR® 4500 SL, Delphi™, and Discovery™ from Hologic and Expert®, Prodigy™, DPX Bravo™, and DPX Duo™ from GE Lunar.

The most recent advancement has been the introduction of cone-beam systems. Cone-beam geometry uses a two-dimensional (2D) detector array to take snapshot style images of the entire scan field in a single exposure. There is no movement of the DEXA components, including the X-ray unit and detector module. Thus, the image acquisition time can be reduced to ~1 s for heel, hand, spine, and hip scans [24]. The first commercial cone-beam DEXA system (DMS Lexxos) was introduced by Diagnostic Medical Systems (Perols, France), and a clinical evaluation was undertaken by G. Blake et al. in 2005 [25]. Although the cone-beam system could offer the benefit of short scan times, the technical challenges caused by the Compton scatter effect needed to be solved for DEXA application. The dominant attenuation process in soft tissue is Compton scatter for X-ray energies of less than

100 keV used in DEXA scanners. For pencil-beam and fan-beam systems, most scattered radiation, the ratio of the number of detected scattered photons to the number of non-scattered photons is quite low because of the limited solid angle and acceptable detector area. Thus, X-ray scatter events are rarely contained in the DEXA images and accurate image acquisition is possible without complicated correction algorithms. However, there is a significant effect on image quality caused by the Compton scattering in cone-beam systems. An effective and acceptable correction should be applied over the image field of view to eliminate all the variable effects of scattered radiation on the measurements. Another limitation is the magnification effect (similar to those of the fan-beam systems described previously), whereby the size of the detected area of an ROI depends upon the position of the object between the X-ray tube and the detector. Magnification and its associated errors occur in 2D directions, compared to the fan-beam system with magnification in only the fan length dimension.

In summary, there are different methods to project the three-dimensional (3D) human body onto a 2D image (Table 1). With the pencil-beam system, all events come from the face of the detector, and the projection data are acquired with linear sampling intervals in the X and Y direction. Fan-beam systems project the image under a certain angle in a direction parallel to the fan width, and the 2D image is acquired by a scanning motion perpendicular to the fan direction. Cone-beam systems simultaneously acquire data for all projection angles in both image directions, and there is no movement of the detector arm during image acquisition. Indeed, the degree of scattering is the lowest for pencil beams, medium for fan beams, and largest for cone beams, where instantaneous dose rates of around 200 μSv/h at a distance of 1 m from the patient have been reported [24].

3.3 Photon Detectors: Direct Versus Indirect Mechanisms

DEXA detectors must meet several property standards as follows:

(1) Good energy resolution: The detector has to distinguish low- and high-energy from incident dual-energy X-ray.
(2) High stopping power: All X-ray photons must be detected and converted to electric signals. It mainly is dependent upon density and the effective atomic number of the detector material.
(3) High count rate: Detectors must be sufficient to handle incident photons. A shorter time constant of output pulse is preferred in high counting rate applications. Thus, the unipolar output having one polarity may be suitable to avoid the problems of pulse pileup and counting loss. It is generally true that all the factors providing high count rate capabilities in front-end and back-end electronics degrade the signal-to-noise ratio and energy resolution. Special circuitry, including pole-zero cancellation, or baseline restoration are required

Table 1 Different X-ray beam geometries and image acquisition methods

	Pencil-beam (First-generation DEXA)	Fan-beam (Current most DEXA)	Cone-beam (Next-generation DEXA)
Beam shape and geometry			
Detector movement	Both X- and Y-direction	One X- or Y-direction	No
Scan time (whole-body/peripheral)	20 min/5 min	3 min/30 s	1 ~ 3 s
Compton scatter effect	Low	High	Very High
Image quality	Low	Good	Good

to avoid baseline shift problems, allowing good pulse-height determination at high counting rates.

(4) Short- and long-term stability: it is very important that the response must be uniform during DEXA scans. Although detector performance remains very stable over time, the gain of the detector and electronics invariably changes as the system ages. A method to "tune" the detector is, therefore, necessary to ensure consistent performance over time. A typical quality assurance program might involve daily measurements, weekly checks, and monthly evaluations. All examinations should be undertaken using the same conditions as those used for routine clinical studies.

There are currently two main types of DEXA detectors, direct and indirect. The main difference between the two types is the conversion process for incident photons.

For indirect detectors, a scintillation crystal converts the incident radiation to visible light photons that are subsequently detected by the photosensors in the form of electronic charges. These detectors are being used to detect the X-rays indirectly, and the total charge produced in the photosensors is proportional to the visible light emitted from a scintillator and the energy deposited by the incident radiation in the scintillator. The original DEXA scanner was the DPX system (Bravo and Duo) made by GE Lunar Corporation. It employed an indirect detector, which was composed of a single NaI(Tl) crystal coupled to a PMT. This indirect detector enabled energy-selective radiation counting, and the output signal was proportional to the detected energy, providing affordable DEXA image acquisition. Light yield of NaI is approximately 38 photons/keV, and 2280 scintillation light is generated by interacting with incident radiation of 60 keV. Considering a 70% geometrical light collection efficiency at the scintillator and a 25% quantum efficiency at PMT, the entered light photons and converted initial photoelectrons in the photosensors are ~1600 photons and 400 photoelectrons, respectively. The Poisson statistical variation of this number is 20 (square root of 400), or the percentage uncertainty is 5% for 60 keV energy. This system continues to work well and achieve high precision in patient scanning. Despite its superior stopping power and detection efficiency of the scintillation detector, a number of technical problems may render its eternal implementation difficult. These include bulky size, multi-row array scalability, poor energy resolution, and long-term stability. Also, there are clinical reasons to pursue an improved DEXA detector design, including reduced scan time and improved image resolution.

The easiest way to solve the technical hurdles and achieve the clinical requirements was to develop and employ a direct detector for a next-generation DEXA scanner. Currently, semiconductor detectors based on CZT or CdTe have been widely utilized on a commercial basis. Compared to the NaI(Tl)-PMT detector, the compactness of CZT allows a scalable design in 2D arrays. Thus, it is possible to make multi-row detectors for fan-beam scanners and reduce scan time. For direct detectors, a semiconductor can effectively absorb the incoming radiation and directly convert the X-ray energy into electrical signals large enough to permit the integration and counting of individual photons, which are read-out by a preamplifier and back-end electronics. Thus, it produces a signal proportional to the amount of absorbed radiation energy. Recent advances in flip-chip bonding and electronic packaging technology have facilitated direct X-ray detector assembly. Several interconnection processes have explored different capabilities to yield a proper bond using multiple flip-chip bump materials including gold, indium, and lead-tin solders. The bumps create both electrical and mechanical connections to other components. This allows the integrated circuits (ICs) to have shorter and denser routed connections to the substrate and provide lower impedance. Several direct detectors have been fabricated by bump-bonding CZT or CdTe detectors to a dedicated custom application-specific integrated circuit (ASIC) read-out for various medical applications. Good energy resolution is one of the attractive features of improved image resolution. The conversion efficiency of X-ray photons to an electrical signal in CdTe and CZT can be much higher than that in scintillation detectors. Also, the created charges do not spread and do not escape to neighboring pixels in direct detectors, but the generated scintillation light is dispersed within the pixel volume and disappears by escape/absorption/extinction in the scintillation detectors. Thus, the overall SNR in the direct detection approach can be significantly better than that achievable by indirect detectors. The bandgap of these detectors is approximately 1.55 eV, and one ionization, creating an electron-hole pair, is produced per ~4.4 eV of radiation energy absorbed. Therefore, a 60 keV photon results in 13,600 electron-hole pairs. The Poisson statistical variation of this number is 117 (square root of 13,600), or percentage uncertainty is 0.8% for 60 keV energy. Considering a Fano factor of ~0.15, the statistical variation on 13,600 is actually closer to $0.15 \times 117 = 17$, and the theoretical energy resolution is 0.1%. In the current state of technology, it would be difficult to achieve this energy resolution in realistic systems where electron/charge trapping phenomena, current leakage, electronic noise, and temperature effects may influence the results. Experimental measurements revealed that CZT detectors could provide better energy resolution in a typical range of 2–5%, compared to ~20% by NaI-PMT detectors [26].

As mentioned previously, semiconductor detectors such as CZT or CdTe have been utilized widely on a commercial basis. Recently, new indirect detector consisting of fast scintillator and GAPD photosensor has been studied actively for use in DEXA system. Figure 3 shows the X-ray energy spectra which were acquired from indirect detector (LYSO-GAPD) and direct detector (CZT) used in a commercial DEXA (DEXXUM-T, Osteosys, Republic of Korea). This comparison study of different DEXA detectors was performed under the same experimental condition. Even though the low- and high-energy peaks were well identified in both detectors, the overall performance including energy resolution and peak-to-valley ratio of CZT detector was relatively better compared to that seen in the indirect detector. Also, it was reported that the energy resolutions of direct and indirect detector were 18% and 34% for gamma energy spectra using Am-241, respectively [27].

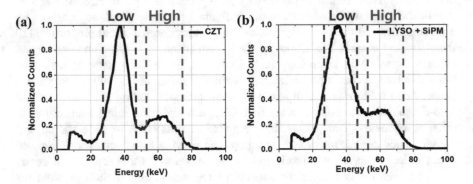

Fig. 3 A performance comparison of direct (CZT) and indirect (LYSO-GAPD) detector for DEXA application

4 LYSO-GAPD and GAGG-GAPD Detector for DEXA Potential Application

4.1 GAPD Photosensor

Geiger-mode avalanche photodiodes (GAPDs), also called solid-state photomultiplier (SSPM), silicon photomultiplier (SiPM), multi-pixel photon counters (MPPC), and micro-pixel avalanche photodiodes (MAPDs), have been utilized as next-generation photosensor in radiation detection systems, and many research groups have also demonstrated the usefulness of GAPDs in different medical applications. Currently, various types of devices with different pixel sizes are commercially available from several vendors (SensL, Hamamatsu, RMD, Zecotek, FBK, etc.).

GAPDs consist of a densely packed matrix with many microcells (~1000 to ~10,000) ranging from 5×5 to 100×100 μm^2 in size, and each microcell operates independently in a Geiger mode as an on/off switch for the photons. The amplitude of the output pulse is proportional to the total number of fired microcells, reflecting the number of absorbed photons. GAPDs provide several key properties [28, 29]. Compared to PMT, GAPDs have compactness and insensitivity to high magnetic fields. In contrast to APDs (Avalanche photodiodes), GAPDs have a high gain (~10^6) and a low excess noise factor (~1.1), allowing operation with a simple preamplifier. Their fast response time (< 1 ns) allows improved timing performance. Further advantages over PMT and APDs are the low operating voltage (< 100 V) and high uniformity (< 20%) among the pixels. In addition, the fabrication costs can be reduced substantially because it can be manufactured using a standard metal oxide semiconductor (MOS) production process.

4.2 Scintillation Crystals

A wide range of inorganic scintillator materials is used to convert radiation (X-ray and gamma-rays) to scintillation light for radiation detectors and medical applications. Even though the individual atoms and molecules of these substances do not scintillate, the characteristics of their crystal structure based on electronic band structure scintillate for the incoming radiation. The properties of ideal crystals are the following:

(1) High stopping power dependent upon density and effective atomic number, which determines the system sensitivity
(2) High light output dependent upon luminosity and light yield, which determines the energy resolution
(3) Fast response time dependent upon decay time constants, which determines the time performance, including count rate and dead time loss

At present, lutetium-based oxide scintillation crystals like cerium-doped lutetium oxyorthosilicate (LSO), lutetium-yttrium oxyorthosilicate (LYSO), and lutetium gadolinium oxyorthosilicate (LGSO) are most commonly used in conventional radiation detection systems as they offer the best trade-offs of the aforementioned factors.

Recently, a Ce-doped gadolinium aluminum gallium garnet (GAGG) single-crystal scintillator was developed, which has a relatively high density (6.63 g/cm^3), high light yield (46 photons/keV), and short decay time (~88 ns). Furthermore, its non-hydroscopic characteristics enable the fabrication of scintillator blocks suited for radiation imaging modalities. Compared to LSO/LYSO/LSGO scintillators, GAGG does not contain lutetium atoms and does not emit natural radioactivity. Thus, the background counts from [176]Lu is eliminated. GAGG has been actively studied for the development of radiation detectors and reported elsewhere [30, 31].

4.3   Approaches Used at Chonnam National University

As mentioned previously, semiconductor detectors such as CZT or CdTe, with excellent energy resolution, have been utilized widely on a commercial basis. As an alternative, X-ray diagnostic detectors, combining high-performance scintillation crystals with compact photosensors, have attracted interest for use in DEXA applications. The first commercial DEXA scanner was recently introduced [32–34].

A research team from Chonnam National University (CNU), composed of the Research Center for Healthcare & Biomedical Engineering and the Medical Imaging System & Equipment Lab, has actively studied GAPDs as radiation detector photosensors since 2007. The characterization of two types of DEXA scanners based on LYSO-GAPD and GAGG-GAPD detectors was conducted. The intrinsic performance of the DEXA detector module was characterized under a variety of conditions. Initial imaging studies were performed to demonstrate the feasibility of X-ray imaging with these prototypes [35, 36].

4.3.1   DEXA Detector Configurations

LYSO and GAGG scintillation crystals were coupled to GAPDs with optical grease. All surfaces of the crystal were polished and wrapped with Teflon except for the surface facing the photosensor. A GAPD (MicroFJ-SMPTA-30035; SensL, Cork, Ireland) based on silicon photomultiplier (SiPM) technology was used for the scintillation light read-out. It had an active area of 3.07×3.07 mm^2 and 5676 microcells of 35×35 μm^2.

The physical and scintillation properties of the scintillation crystals and CZT have been summarized in Table 2. LYSO and GAGG had a good stopping power for DEXA energy in the range of 20–100 keV. The estimated detection efficiency of >99.9% for 80 keV was adequate even with 2-mm-thick LYSO and 3-mm-thick

Table 2 Summary of the various properties for CdTe/CZT, LYSO-GAPD, and GAGG-GAPD detector for DEXA application

	CZT, CdTe	LYSO-GAPD	GAGG-GAPD
Energy resolution at 100 keV	<5%	<20%	<25%
Physical hardness	Fragile	Stiffness	Stiffness
Hazardous substances	Cadmium	No	No
Temperature dependency	High	Low	Low
Decay time	~5 μs	40 ns	88 ns
Max. count rate	1 Mcps	30 Mcps	20 Mcps
Density	5.85 g/cm^3	7.35 g/cm^3	6.63 g/cm^3
Thickness at 80 keV (stopping power)	>2.5 mm(>99%) >5 mm(>99.9%)	>1.5 mm(>99%) >2 mm(>99.9%)	>2 mm(>99%) >3 mm(>99.9%)
Supply voltage	300 V ~ 1000 V	<50 V	<50 V
Mass production	Difficult	Possible	Possible

GAGG, whereas CZT should increase the thickness to 5 mm. Therefore, DEXA detectors had pixel dimensions of $3 \times 3 \times 2$ mm^3 and $3 \times 3 \times 3$ mm^3 for LYSO-GAPD and GAGG-GAPD, respectively.

4.3.2 Experimental Setup

The detector and X-ray tube were positioned at opposite sides. The collimator formed a pencil-shaped beam of irradiation followed by a Ce filter. The source-object distance (SOD) was 186 mm and the source-detector distance (SDD) was 584 mm.

The X-ray source consisted of an integrated monoblock system (XRB80; Spellman, Germany) with a high-voltage power supply (HVPS), filament supply, X-ray tube, and oil encapsulant. The X-ray tube, in a sealed tank, had a fixed tungsten anode and a focal spot size of 0.5 mm. The emitted X-ray beam angle was about 75° with a fan-shaped geometry. The maximum tube voltage and tube current were 80 kVp and 1.25 mA, respectively. The X-ray source was operated in a continuous mode with an intrinsic beam filter, including an amorphous of 3.3 mm, oil of 8 mm, beryllium of 0.8 mm, and glass 1-mm thick [37].

A lead collimator with a 1-mm hole diameter and a 3-mm thickness was installed on the X-ray tube. In addition, a K-edge cerium filter module was used to separate high-energy from low-energy photons by selectively absorbing X-ray photons around the K-shell energy of the filter material [38, 39]. It could generate effective energy peaks of 30 keV and 60 keV. The K-edge filter module consisted of six cerium sheet components 0.1-mm-thick and a 50 mm × 50 mm surface area (Alfa Aesar, Ward Hill, MA, USA). It had different filter thicknesses ranging from 0.1 to 0.6 mm and allowed for X-ray beam intensity adjustment according to the patient thickness.

The output signal from the DEXA detector was fed into custom-made current feedback charge-sensitive preamplifiers using high-gain operational amplifiers. The amplified signal was digitized and recorded by the custom-made data acquisition (DAQ) system based on a field-programmable gate array (FPGA) or oscilloscope with a bandwidth of 1 GHz, a sampling rate of 10 GSPS, and channel memory of 10 Mpts. The digitized output was post-processed and converted into energy and counts. The other signal was fed into four comparators, including LL (low window at low energy), LH (high window at low energy), HL (low window at high energy), and HH (high window at high energy), to generate effective digital count signals if the analog signal voltage was above the specific threshold levels of each discriminator. Each comparator offered a short propagation delay of ~3.5 ns. It had different threshold voltages adjusted by an external LDO (low dropout) circuit. The GAPD was operated at a bias voltage of 30.0 V, exceeding the breakdown voltage by 2.5 V and corresponding to a gain of approximately 10^6. The detector module and read-out electronics were operated at room temperature (25 °C) without additional cooling equipment.

4.3.3 Characterization of the Intrinsic Performance of the DEXA Detectors

Figure 4 shows the measured energy spectra, which were acquired using Am-241 (60 keV) and Co-57 (122 keV) for 10 min for each LYSO-GAPD and GAGG-GAPD detector module. The measured photopeak positions of Am-241 and Co-57 were 1100 mV and 2200 mV and 850 mV and 1750 mV for the LYSO-GAPD and GAGG-GAPD detector modules, respectively. The energy resolutions of Am-241 and Co-57 were 26.8% and 17.2% and 38.5% and 23.8% for the LYSO-GAPD and GAGG-GAPD detector modules, respectively. It should be noted that the energy resolution value (17.2% and 23.8%) at 122 keV was considerably improved compared to that seen in previous studies (27% and 25%) [40, 41].

Figure 5a shows the estimated energy spectra as a function of tube voltage from 40 to 80 kVp to provide reference X-ray energy spectra. The simulation was performed using a tungsten anode spectral model using a TASMIP algorithm. As shown in Fig. 5b and d, the measured spectra were similar in shape to the simulated ones. The K-edge peak filtered by the tungsten anode target could be observed in the energy spectrum at an X-ray tube voltage of 80 kVp in LYSO-GAPD detector. These results demonstrated that the spectral energy performance obtained using LYSO-GAPD was noticeably better than the one obtained using GAGG-GAPD. The estimated beam peak energy increased from 28.4 to 36 keV with increasing tube voltage from 40 to 80 kVp. The measured peak positions of the LYSO-GAPD and GAGG-GAPD detectors increased from 482 to 776 mV and from 378 to 595 mV, respectively. The measured energy linearity of the LYSO-GAPD and GAGG-GAPD detectors was similar to that estimated in the simulation studies.

Regarding the effect of tube current on count rate performance, the measured spectra were similar in shape to the simulated ones at a tube current of <0.05 mA. However, a tail in the spectrum above the maximum tube voltage was observed due to the pulse pileup effect under extremely high incident photon conditions. Thus, the resulting data were fit with paralyzable dead time models used for most radiation

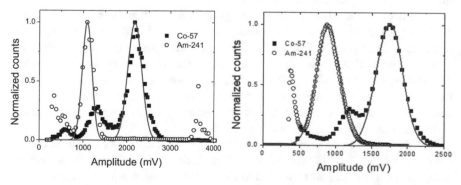

Fig. 4 Acquired energy spectra of Am-241 (o) and Co-57 (■) for LYSO-GAPD (left) and GAGG-GAPD (right) detector module

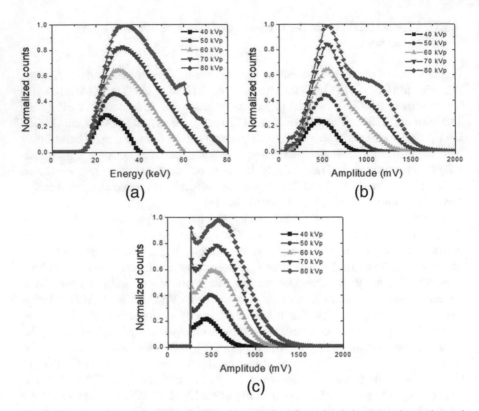

Fig. 5 X-ray spectra as a function of tube voltage obtained from simulation (**a**) and experimental measurement from LYSO-GAPD (**b**) and GAGG-GAPD (**c**)

detectors exhibiting pileup effects at high counting rates. The peak of the count rate curve was observed at ~0.4 mA tube current. There was a leveling-off when the tube current was increased. The maximum count rate was 2 Mcps, mainly due to the dead time of the DAQ system used in this study.

The effect of the cerium filter on energy spectra was examined for six different thicknesses ranging from 0.1 to 0.6 mm, with steps of 0.1 mm. The X-ray tube was operated at 80 kVp and 0.1 mA, and filtered dual-energy spectra were acquired for 1 min. Each spectrum was normalized with respect to its maximum count value. The separated energy window (the difference between the lower and upper level) for each low and high energy was set to 30% and centered around the photopeak positions. The events falling within the selected energy window were accepted and binned into each low- and high-energy bands.

The dual-energy peaks were located at around 32 and 65 keV, for both LYSO-GAPD and GAGG-GAPD modules. Each peak could be isolated with a Ce filter thickness of >0.3 mm in GAGG-GAPD, whereas a thickness of >0.1 mm was feasible for LYSO-GAPD. The peak-to-valley ratio as a function of the K-edge cerium filter thickness increased from 2.8 to 14.2 at low energy and from 1.7 to 3.8 at a high energy in the LYSO-GAPD detector.

4.3.4 Imaging Performance and Long-Term Stability

The imaging capability and dynamic range were evaluated using BMD phantoms with different thicknesses of aluminum and acryl to simulate bone and soft tissue. The encapsulated spine phantom (Fig. 6a) consisted of an aluminum bar block with different thicknesses ranging from 4 to 12 mm with a step of 2 mm in a block of solid acrylic with a thickness of 120 mm. The aluminum step-wedge phantom (Fig. 6b) with 15 steps in increments of 0.4 mm had a surface area of 10 mm × 50 mm for each step and different thicknesses ranging from 1.2 to 6.8 mm. A K-edge Ce filter was inserted in front of the collimator, and the X-ray tube was operated at 80 kVp and 0.1 mA. A pencil beam oriented parallel to the short axis of the phantom was irradiated, and a transverse pattern across the phantom was scanned. The X-ray source and detector were moved toward the short axis in 1-mm steps. The data were acquired for 0.05 s for each interval. The counted photons were separated into low- and high-energy bands. These dual-energy images were converted into density images using a conventional dual-energy subtraction algorithm, as described in Fig. 1.

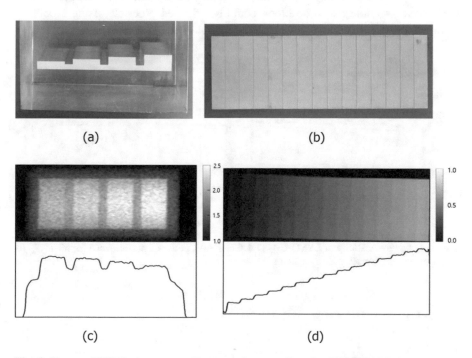

(a)

(b)

(c)

(d)

Fig. 6 Picture of DEXA phantoms and image performance by using LYSO-GAPD detector

Figure 6c and 6d presents the DEXA images and line profiles, acquired from BMD phantoms. Each step was clearly resolved for both DEXA detectors. The image intensity increased linearly from 2.07 to 2.41 and from 1.48 to 1.79 for LYSO-GAPD and GAGG-GAPD, respectively. These preliminary image performances were well-characterized by identifying the imaging capability and dynamic range.

The precision of a diagnostic test such as DEXA is an indication of the reproducibility of replicate measurements. Imprecise measurements can be caused by random error, gain variations, and electronic noise for a given X-ray source, radiation detector, or amplifier circuit. Therefore, the assessment of long-term stability is a very important parameter to monitor the temporal changes in a scanner. A setup identical to that used for spectral measurements was employed at room temperature (25 °C) without additional cooling modules or temperature stabilizing equipment. X-ray energy spectra and DEXA phantom images were acquired to examine the reproducibility of the results (Fig. 7). These experimental tests were repeated for 50 days. The slight variation in the count rate and image pixel value could be partly attributed to temperature fluctuations of the X-ray unit. The average count rates for low- and high-energy were 163.3 ± 3.0 kcps and 124.2 ± 3.6 kcps, respectively. The variation in image pixel values was considerably small ($< 3\%$). These results indicate that the LYSO-GAPD detector could provide consistent count rates and phantom images over time. The long-term stability results showed the possibility of using LYSO-GAPD to develop DEXA with stable performance over time, although a longer-term stability study is needed. If the temperature of the detector module and tube block could be adjusted precisely, using a cooling chamber, the stability of the detector module would be improved.

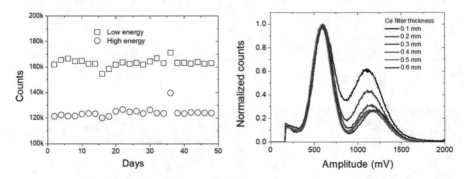

Fig. 7 Long-term stability of LYSO-GAPD detector. (**a**) Count rate performance for 50 days, (**b**) representative dual-energy X-ray spectra

5 Summary

The chapter introduced DEXA instrumentation that has seen an exponential increase in use in clinical practice for bone densitometry as the number of older people has increased. DEXA technology has numerous strengths compared to other diagnostic methods, including its availability, short scan times, minimal radiation exposure, and excellent precision. In elderly adults, DEXA BMD is also a sufficiently robust predictor of osteoporotic fractures and can be used to define the disease.

The fundamental principle of DEXA is the measurement of differences in the transmission and attenuation of tissue and bone for low and high energies. There are different approaches to project the 3D human body onto the 2D image: creating low- and high-energy beams using voltage switching and K-edge filtering systems, acquiring DEXA images using pencil beams, fan beams, and cone beams and converting incident X-ray photons to electrical signals using indirect and direct detectors.

CZT has been considered a representative detector for DEXA implementation due to its good stopping power, excellent energy resolution, and well-established techniques. The recent investigations of alternative LYSO-GAPD and GAGG-GAPD detectors in DEXA application were described. Simulation studies and experimental measurements confirmed that it was feasible to develop a DEXA detector based on the conventionally used LYSO and newly developed GAGG scintillation crystals coupled to a GAPD photosensor. New detectors provided several potential merits including a relatively high count rate owing to its fast decay time and small crystal volume due to its good stopping power. In addition, they could be potentially affordable and cost-effective because of the availability of suppliers for the components, enabling mass production.

References

1. National Institutes of Health. (2001). NIH consensus development panel on osteoporosis prevention, diagnosis, and therapy. Highlights of the conference. *Southern Medical Journal, 94*, 569–573.
2. https://www.nof.org/about-us/nof-background/
3. Kanis, J. A. (2007). *Assessment of osteoporosis at the primary health care level*. WHO Collaborating Centre for Metabolic Bone Diseases. WHO Collaborating Centre for Metabolic Bone Diseases.
4. Melton, L. J., III, Atkinson, E. J., O'Connor, M. K., O'Fallon, W. M., & Riggs, B. L. (1998). Bone density and fracture risk in men. *Journal of Bone and Mineral Research, 13*, 1915–1923.
5. Melton, L. J., III, Chrischilles, E. A., Cooper, C., Lane, A. W., & Riggs, B. L. (2005). How many women have osteoporosis? *Journal of Bone and Mineral Research, 20*, 886–892.
6. Kanis, J. A., Johnell, O., Oden, A., Sernbo, I., Redlund-Johnell, I., Dawson, A., Laet, D., & Jonsson, B. (2000). Long-term risk of osteoporotic fracture in Malmö. *Osteoporos International, 11*, 669–674.
7. Fogelman, I., & Blake, G. M. (2000). Different approaches to bone densitometry. *Journal of Nuclear Medicine, 41*, 2015–2025.

8. Mazess, R. B. (1984). Advances in single-and dual-photon absorptiometry. *Osteoporosis, 1,* 57–63.
9. Mazess, R. B., & Barden, H. S. (1987). Single-and dual-photon absorptiometry for bone measurement in osteoporosis. Osteoporosis update. 73–80.
10. Mazess, R. B., Barden, H., Vetter, J., & Ettinger, M. (1989). Advances in noninvasive bone measurement. *Annals of Biomedical Engineering, 17,* 177–181.
11. Mazess, R., Collick, B., Trempe, J., Barden, H., & Hanson, J. (1989). Performance evaluation of a dual-energy X-ray bone densitometer. *Calcified Tissue International, 44,* 228–232.
12. Sartoris, D. J., & Resnick, D. (1989). Dual-energy radiographic absorptiometry for bone densitometry: Current status and perspective. *AJR. American Journal of Roentgenology, 152,* 241–246.
13. Johnson, P. C., & Jhingran, S. G. (1986). Precision of dual photon absorptiometry measurements. *Journal of Nuclear Medicine, 27,* 1362–1365.
14. Wahner, H. W., Dunn, W. L., Brown, M. L., Morin, R. L., & Riggs, B. L. (1988). Comparison of dual-energy x-ray absorptiometry and dual photon absorptiometry for bone mineral measurements of the lumbar spine. *Mayo Clinic Proceedings, 63,* 1075–1084.
15. Cosman, F., de Beur, S. J., LeBoff, M. S., Lewiecki, E. M., Tanner, B., Randall, S., & Lindsay, R. (2014). Clinician's guide to prevention and treatment of osteoporosis. *Osteoporosis International, 25,* 2359–2381.
16. World Health Organization. (1994). *Assessment of fracture risk and its application to screening for postmenopausal osteoporosis: Report of a WHO study group [meeting held in Rome from 22 to 25 June 1992].* WHO.
17. FDA 510k K871553, October 10, 1987.
18. Kelly, T. L., Slovik, D. M., & Neer, R. M. (1989). Calibration and standardization of bone mineral densitometers. *Journal of Bone and Mineral Research, 4,* 663–669.
19. Wear, J., Buchholz, M., Payne, R. K., Gorsuch, D., Bisek, J., Ergun, D. L., Grosholz, J., & Falk, R. (2000). CZT detector for dual-energy x-ray absorptiometry (DEXA). *SPIE, 4142,* 175–188.
20. Takahashi, T., & Watanabe, S. (2001). Recent progress in CdTe and CdZnTe detectors. *IEEE Transactions on Nuclear Science, 48,* 950–959.
21. https://www.gehealthcare.com/products/bone-and-metabolic-health/prodigy
22. Blake, G. M., & Fogelman, I. (1997). Technical principles of dual energy x-ray absorptiometry. *Journal of Nuclear Medicine, 27,* 210–228.
23. Wahner, H., Steiger, P. and Von Stetten, E. 1994. Instruments and measurement techniques in Wahner HW, Fogelman I (Eds): The evaluation of osteoporosis: Dual energy X-ray absorptiometry in clinical practice. London: Martin Dunitz
24. Boudousq, V., Kotzki, P. O., Dinten, J. M., Barrau, C., Robert-Coutant, C., Thomas, E., & Goulart, D. M. (2003). Total dose incurred by patients and staff from BMD measurement using a new 2D digital bone densitometer. *Osteoporosis International, 14,* 263–269.
25. Blake, G. M., Knapp, K. M., & Fogelman, I. (2005). Dual X-ray absorptiometry – Clinical evaluation of a new cone-beam system. *Calcified Tissue International, 76,* 113–120.
26. Knoll, G. F. (2000). *Radiation detection and measurement.*
27. Park, C., Song, H., Joung, J., Kim, Y., Kim, K. B., & Chung, Y. H. (2020). Feasibility study of SiPM based scintillation detector for dual-energy X-ray absorptiometry. *Nuclear Engineering and Technology, 52,* 2346–2352.
28. Kang, J., Choi, Y., Hong, K. J., Jung, J. H., Hu, W., Huh, Y. S., Lim, H., & Kim, B. T. (2010). A feasibility study of photosensor charge signal transmission to preamplifier using long cable for development of hybrid PET-MRI. *Medical Physics, 37,* 5655–5664.
29. Kang, J., Choi, Y., Hong, K. J., Hu, W., Jung, J. H., Huh, Y., & Kim, B. T. (2011). A small animal PET based on GAPDs and charge signal transmission approach for hybrid PET-MR imaging. *Journal of Instrumentation, 6,* P08012.
30. Yeom, J. Y., Yamamoto, S., Derenzo, S. E., Spanoudaki, V. C., Kamada, K., Endo, T., & Levin, C. S. (2013). First performance results of Ce:GAGG scintillation crystals with silicon photomultipliers. *IEEE Transactions on Nuclear Science, 60,* 988–992.

31. Yamamoto, S., Yeom, J. Y., Kamada, K., Endo, T., & Levin, C. S. (2013). Development of an ultrahigh resolution block detector based on 0.4 mm pixel Ce: GAGG scintillators and a silicon photomultiplier array. *IEEE Transactions on Nuclear Science, 60*, 4582–4587.
32. https://www.hologic.com/
33. https://www.gehealthcare.com/
34. http://www.swissray.com/SRI/
35. Yang, J., Min, B. J., & Kang, J. (2020). A feasibility study of LYSO-GAPD detector for DEXA applications. *Journal of Instrumentation, 15*, P05017.
36. Yang, J., Min, B. J., & Kang, J. (2019). A feasibility study of GAGG-GAPD detector for development of DEXA. *Journal of Magnetics, 24*, 739–743.
37. https://www.spellmanhv.com/ko/high-voltage-power-supplies/XRB80N100
38. Crabtree, N. J., Leonard, M. B., & Zemel, B. S. (2007). Dual-energy X-ray absorptiometry. *Bone Densitometry in Growing Patients*, 41–57.
39. Boone, J. M., & Seibert, J. A. (1997). An accurate method for computer-generating tungsten anode x-ray spectra from 30 to 140 kV. *Medical Physics, 24*, 1661–1670.
40. Habte, F., Olcott, P. D., Levin, C. S., & Foudray, A. M. (2007). Prototype parallel readout system for position sensitive PMT based gamma ray imaging systems. *IEEE Transactions on Nuclear Science, 54*, 60–65.
41. Olcott, P. D., Talcott, J. A., Levin, C. S., Habte, F., & Foudray, A. M. K. (2005). Compact readout electronics for position sensitive photomultiplier tubes. *IEEE Transactions on Nuclear Science, 52*, 21–27.

Medical Photon-Counting CT – Status and Clinical Application Review

Thomas Flohr, Martin Petersilka, Andre Henning, Stefan Ulzheimer, and Bernhard Schmidt

1 Introduction

Computed tomography (CT) is the backbone of radiological diagnosis. The application spectrum of CT has been continuously expanded by technological advances, among them the introduction of spiral CT [1], the rapid progress of multi-detector row CT [2, 3], and new system concepts such as wide detector CT [4] or dual-source CT [5] with their specific clinical benefits.

Today, CT is a mature modality in its saturation phase. Yet, there are still clinical limitations for current CT technology, mainly caused by insufficient spatial resolution for some applications such as CT angiography (CTA) of the coronary arteries and limited potential to further reduce the radiation dose to the patient. Furthermore, CT is in general very sensitive, but not very specific – the morphological information obtained with a CT scan is often not sufficient to guide clinical decisions without further workup of the patient. Dual-energy CT has gained momentum as a technique to enhance the clinical value of CT by adding functional information to morphology. The reviews on clinical applications of dual-energy CT may be found in [6–12]. Dual-energy CT data can currently be acquired with dual-source CT systems [5, 13], CT systems with fast kV switching [14], or dual-layer detector CT systems [15]. However, each of these solutions has inherent limitations.

Photon-counting detectors are a new technology with the potential to overcome these drawbacks, by providing CT data at very high spatial resolution, without electronic noise and with inherent spectral information. Photon-counting detectors and their potential benefits have already been evaluated in experimental CT benchtop systems more than 10 years ago (e.g., [16]). The detectors used in these

T. Flohr (✉) · M. Petersilka · A. Henning · S. Ulzheimer · B. Schmidt
Siemens Healthcare GmbH, Computed Tomography, Forchheim, Germany
e-mail: thomas.flohr@siemens-healthineers.com

© Springer Nature Switzerland AG 2022
K. Iniewski (ed.), *Advanced X-ray Detector Technologies*,
https://doi.org/10.1007/978-3-030-64279-2_8

early systems, however, did not tolerate the high x-ray flux rates required in medical CT, and stable image quality without ring artifacts caused by signal drift and other processes could not be achieved. The significant progress in detector material synthesis and detector electronics design has meanwhile helped to overcome these problems, and preclinical whole-body photon-counting CT prototypes could be installed for preclinical testing in human subjects. Still there is a way to go before photon-counting detectors can be broadly released in the commercial CT systems.

This review article gives an overview of the basic principles of medical photon-counting detector CT and of the clinical experience gained so far in preclinical installations. Other reviews of photon-counting detector CT are available in [17–20].

2 Principles of Medical Photon-Counting CT

2.1 Properties of Solid-State Scintillation Detectors

Solid-state scintillation detectors are used in all current medical CT scanners. They consist of individual detector cells with a side length of 0.8–1 mm, made of a scintillator with a photodiode attached to its backside (see Fig. 1). The absorbed X-rays produce visible light in the scintillator which is detected by the photodiode and converted into an electrical current. Both the intensity of the scintillation light and the amplitude of the induced current pulse are proportional to the energy E of the absorbed X-ray photon. All current pulses registered during the measurement time of one reading (projection) are integrated. X-ray photons with lower energy E, which carry most of the low-contrast information, contribute less to the integrated detector signal than the X-ray photons with higher energy. This energy weighting reduces the contrast-to-noise ratio (CNR) in the CT images. This is a particular challenge in CT scans after the administration of iodinated contrast agent – the majority of all medical CT scans. The X-ray absorption of iodine is highest at lower energies closely above its K-edge at 33 keV; these energies are down-weighted in the signal of a scintillation detector.

The low-level analogue electric signal of the photodiodes is distorted by electronic noise which becomes larger than the quantum noise (Poisson noise) of the X-ray photons at low X-ray flux. It causes a disproportional increase of image noise and an instability of the low CT-numbers (e.g., in CT scans of the lungs) if the flux is further reduced and sets a limit to radiation dose reduction in medical CT.

The individual detectors are separated by optically in-transparent layers with a width of about 0.1 mm to prevent optical crosstalk. X-ray photons absorbed in the separation layers do not contribute to the measured signal even though they have passed through the patient – they are wasted radiation dose. Current medical CT detectors with a size of about 0.8×0.8 mm^2 to 1×1 mm^2 have a geometric dose efficiency of about 80–90%. Significantly reducing the size of the scintillators

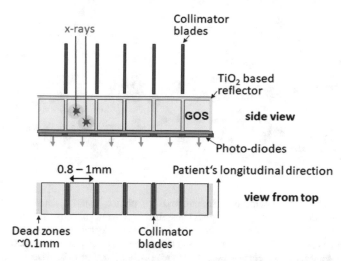

Fig. 1 Schematic drawing of an energy-integrating scintillator detector. Side view and view from the top. Individual detector elements made of a scintillator such as gadolinium-oxide or gadolinium-oxysulfide (GOS) absorb the X-rays (red arrows) and convert their energy into visible light (orange circles). The light is registered by photodiodes and converted into an electrical current. The detector elements are separated by optically in-transparent layers (e.g., based on TiO_2) to prevent optical crosstalk – these layers are "dead zones." Collimator blades above the separation layers suppress scattered radiation

to increase spatial resolution beyond today's performance levels while keeping the width of the separation layers constant is problematic because of reduced geometric efficiency.

2.2 Properties of Photon-Counting Detectors

Photon-counting detectors are made of semiconductors such as cadmium telluride (CdTe) or cadmium zinc telluride (CZT). High voltage (800–1000 V) is applied between the cathode on top of the detector and pixelated anode electrodes at the bottom (see Fig. 2, left). The absorbed X-rays produce electron-hole pairs which are separated in the strong electric field. The electrons drift to the anodes and induce short current pulses (10^{-9} s). A pulse-shaping circuit transforms them to voltage pulses with a full width at half maximum (FWHM) of 10–15 nanoseconds; their pulse-height is proportional to the energy E of the absorbed X-ray photons. The pulses are counted if they exceed a threshold T_0 (see Fig. 2 right). In a photon-counting detector for medical CT, T_0 is about 25 keV.

Photon-counting detectors have several advantages compared to solid-state scintillation detectors. The individual detectors are defined by the strong electric field between common cathode and pixelated anodes (Fig. 2, left); there are no

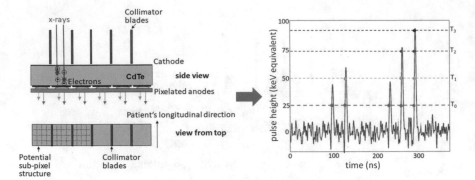

Fig. 2 Left: Schematic drawing of a direct-converting photon-counting detector. Side view and view from the top. The X-rays (red arrows) absorbed in a semiconductor such as CdTe or CZT produce electron-hole pairs that are separated in a strong electric field between cathode and pixelated anodes and induce fast signal pulses at the anodes. The individual detectors are formed by the pixelated anodes and the electric field; there are no separation layers between them. Collimator blades are needed to suppress scattered radiation. Right: The pulses are counted as soon as they exceed a threshold T_0 (dashed blue line, "counting" is indicated by a blue dot). T_0 has a typical energy of 25 keV. Three additional thresholds at higher energies (T_1 at 50 keV, T_2 at 75 keV, T_3 at 90 keV) are also indicated

additional separation layers. The geometrical dose efficiency of a photon-counting detector is only reduced by anti-scatter collimator blades or grids. Each "macro-pixel" confined by the collimator blades may be divided into smaller sub-pixels which are read out separately to increase the spatial resolution (indicated in Fig. 2, left. The pixelated anodes must then be structured correspondingly, which is not shown in order to not overload the drawing).

Low-amplitude baseline noise is well below the threshold T_0 (see Fig. 2, right) and does not trigger counts – even at low X-ray flux, only the statistical Poisson noise of the X-rays is present in the signal. CT scans at very low radiation dose or CT scans of obese patients show therefore less image noise, less streak artifacts, and more stable CT numbers than the scans with a scintillation detector, and radiation dose reduction beyond today's limits seems possible.

The detector responsivity in the X-ray energy range from 30 keV to 100 keV is approximately constant. All X-ray photons registered during the time of a projection (reading) contribute equally to the measurement signal regardless of their energy E, as soon as E exceeds T_0. There is no down-weighting of lower energy X-rays as in solid-state scintillation detectors. Photon-counting detectors can therefore provide CT images with potentially improved CNR, in particular in CT scans after the administration of iodinated contrast agent.

Several counters operating at different threshold energies can be introduced for energy discrimination (see Fig. 2, right). Physically, the different thresholds are realized by different voltages which are fed into pulse-height comparator circuits. In the example of Fig. 4, four different energy thresholds (T_0, T_1, T_2, and T_3) are realized. The photon-counting detector simultaneously provides four signals S_0, S_1,

S_2, and S_3 with different lower energy thresholds T_0, T_1, T_2, and T_3. Subtracting the detector signals with adjacent lower energy thresholds results in "energy bin" data. Energy bin $b_0 = S_1 - S_0$ as an example contains all X-ray photons detected in the energy range between T_1 and T_0.

Simultaneous readout of CT data in different energy bins opens the potential of spectrally resolved measurements and material differentiation in any CT scan. Established dual-energy applications – mainly based on decomposition into two base materials such as iodine and water or iodine and calcium – are routinely feasible, and virtual monoenergetic images (VMIs), iodine maps, and virtual non-contrast images (with the iodine removed) or virtual non-calcium images can be computed whenever needed for the diagnosis. Data acquisition with more than two thresholds enables multi-material decomposition under certain preconditions. Unfortunately, the availability of N energy thresholds does not imply potential differentiation of N base materials. Compton scattering and the photoelectric effect are the only two relevant interaction mechanisms in the X-ray energy range accessible to CT (35 keV–140 keV). Their energy dependence is similar for all elements without K-edge in this range – this applies to all materials naturally occurring in the human body including iodinated contrast agent. As soon as the two base materials, e.g., water and iodine, have been chosen, the energy-dependent attenuation of any other material can be described by a linear combination of the two base materials. It is therefore not possible to differentiate this other material from a mixture of the two base materials. The differentiation of the two base materials requires two measurements at different energies – dividing the energy range into more than two energy bins will not provide relevant new information. The situation changes if a material with K-edge in the energy range accessible to CT, such as gadolinium with a K-edge at 55 keV, is added to the two base materials. For a K-edge material, the energy dependence of X-ray attenuation is different, and CT measurements at three or more energies can be used for three-material decomposition (two base materials plus the K-edge material). Three- or more-material decomposition with CT data in three or more energy bins will be limited to clinical scenarios in which two contrast agents (e.g., iodine and gadolinium or iodine and bismuth) are applied and need to be separated, or other heavy elements are introduced into the human body (e.g., tungsten or gold nanoparticles).

In addition to potential material decomposition, the CNR of the images can be further improved by optimized weighting of the different energy bins. Instead of just adding the bin data for the reconstruction of an image, higher weights can be assigned to the low-energy bin data, in particular in CT scans using iodinated contrast agent.

2.3   Challenges for Photon-Counting Detectors in Medical CT

Compared to established dual-energy CT acquisition techniques, photon-counting detectors are often assumed to provide better energy separation and less spectral overlap. However, unavoidable physical effects reduce the energy separation of CdTe- or CZT-based photon-counting detectors. The current pulses produced by the X-rays absorbed close to the pixel borders are split between adjacent detectors ("charge sharing"). This leads to the registration of a high-energy X-ray photon as several lower-energy events. Cd and Te have K-edges at 26.7 and 31.8 keV, respectively. Absorbed X-rays kick out the K-electrons of the detector material and lose the K-edge energy. The empty K-shells are immediately refilled, and the characteristic X-rays are released which are counted in the same detector pixel or "escape" to neighboring pixels and are counted there ("K-escape"). In summary, high-energy X-ray photons are erroneously counted at lower energies, and spectral separation and spatial resolution (in the case of K-escape) are reduced. Charge sharing and K-escape are illustrated in Fig. 3 (top). If the detector is read out in several energy bins, the low-energy bins will contain wrong high-energy information ("high-energy tails"). Increasing the size of the detector pixels reduces the high-energy tails, because boundary effects such as charge sharing and K-escape contribute less to the total detector signal (see Fig. 3, bottom). For a realistic detector model with detector sizes as in Fig. 3, including charge sharing, fluorescence, and K-escape, the spectral separation with two energy bins is equivalent to that of a dual-kV technique with optimized prefiltration [21].

Even the larger detector pixels would further improve spectral separation. Unfortunately, there is an effect that limits the maximum size of the detector pixels: the finite width of the voltage pulses after pulse-shaping (FWHM ≥ 10 ns). Medical CTs are operated at high X-ray flux rates up to 10^9 counts per s and mm^2 – if the detector pixels are too large, too many X-ray photons hit them too closely in time to be registered separately. Overlapping pulses are then counted as one hit only ("pulse pileup"). Pulse pileup leads to nonlinear detector count rates and finally to detector saturation. Even though the signal can be linearized before the onset of saturation, significant quantum losses, increased image noise, and reduced energy discrimination cannot be avoided. Finding the optimum size of the detector cells to balance pulse pileup, charge sharing, and K-escape is one of the most challenging tasks in designing a photon-counting detector.

Yet another challenge for photon-counting detectors is count-rate drift at higher X-ray flux rates. Nonhomogeneously distributed crystal defects in the sensor material cause the trapping of electrons and holes – the resulting space charges modify the electric field distribution differently in the individual detector pixels. Depending on the "irradiation history" of the pixels, the characteristics of the signal pulses are changed, and this deviation from calibration may cause severe ring artifacts in the images. While count-rate drift was one of the major problems for the use of photon-counting detectors in medical CT, it could meanwhile be reduced to clinically tolerable values by refined material synthesis.

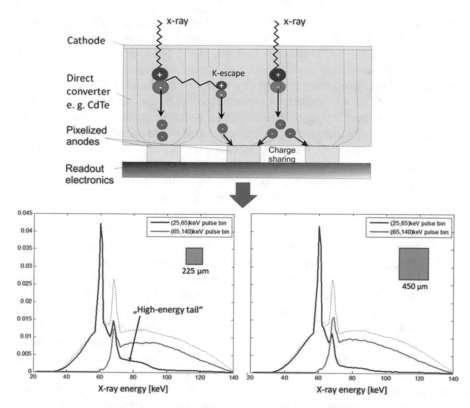

Fig. 3 Top: Schematic illustration of charge sharing at pixel boundaries and K-escape, which result in multiple counting of X-ray photons at wrong lower energies. Bottom: Charge sharing and K-escape lead to a characteristic high-energy tail of low-energy bins. The computer simulation of the X-ray spectra recorded in the two energy bins of a realistic photon-counting detector for an incident 140 kV spectrum (bin$_1$: 25–65 keV, blue line; bin$_2$: 65–140 keV, green line). Left: Pixel size 0.225×0.225 mm^2. Right: Pixel size 0.45×0.45 mm^2. Increasing the pixel size reduces the high-energy tail

3 Preclinical Evaluation of Photon-Counting CT

Photon-counting detectors are a promising new technology for future medical CT systems. Currently, the preclinical prototypes are used to evaluate the potential and limitations of photon-counting CT in clinical practice. We will focus on these preclinical installations and leave out other more experimental solutions, benchtop systems, and photon-counting micro CT systems.

A preclinical single-source CT system with photon-counting detector based on CZT (Philips Healthcare, Haifa, Israel) provides an in-plane field of view of 168 mm and a z-coverage of 2.5 mm, with a rotation time of 1 s [22]. The size of the detector pixels is 0.5×0.5 mm^2. The photon-counting detector has five energy thresholds. The system has so far been used for scans of phantoms and animals. Improved assessment of lung structures due to higher resolution [22] was demonstrated as well as improved visualization of the in-stent lumen and in-stent restenosis in the coronary stents [23]. Differentiation between blood and iodine in a bovine brain was shown by computing iodine maps and virtual noncontrast images [30]. The separation of multiple contrast agents by means of multi-material decomposition and K-edge imaging was evaluated in several studies [24–29], with various potential clinical applications. The separation of iodine, gadolinium, and calcium might be helpful in CTAs of the aorta for the assessment of aortic endoleak dynamics and distinction from intra-aneurysmatic calcifications in a single scan [26]. Virtual CT colonoscopy might benefit from the differentiation of gadolinium-tagged polyps and iodine-tagged fecal material [27]. The sequential administration of several contrast agents might enable the imaging of multiple uptake phases in an organ with a single scan [28, 29].

A preclinical hybrid dual-source CT scanner is equipped with a conventional scintillation detector and a CdTe photon-counting detector (Siemens Healthcare GmbH, Forchheim, Germany). The photon-counting detector consists of sub-pixels with a size of 0.225×0.225 mm^2. The detector provides two energy thresholds per sub-pixel. The 2×2 sub-pixels can be binned to a "sharp" pixel or "UHR" pixel with a pixel size of 0.45×0.45 mm^2; the 4×4 sub-pixels can be binned to a "macro" pixel with a size of 0.9×0.9 mm^2 comparable to today's medical CT systems. By assigning alternating low-energy and high-energy thresholds to adjacent detector sub-pixels in a "chess pattern mode," the detector provides four energy thresholds in "macro" pixels. The in-plane field of view of the photon-counting detector is 275 mm and the z-coverage is 8–16 mm, depending on the readout mode. A completion scan with the energy-integrating subsystem can be used to extend the photon-counting field of view to 500 mm. The shortest rotation time of the system is 0.5 s. The X-ray tubes can be operated at voltages up to 140 kV, with a tube current up to 550 mA. The preclinical hybrid dual-source CT scanner was described and evaluated in [31–33].

Yet another preclinical single-source CT scanner (Siemens Healthcare GmbH, Forchheim, Germany) is equipped with a CdTe photon-counting detector with a similar pixel geometry as the hybrid dual-source CT. Its field of view is 500 mm; its z-coverage is 57.6 mm (144×0.4 mm) in standard resolution mode ("macro" pixels) and 24 mm (120×0.2 mm) in "UHR" mode. The shortest rotation time of the system is 0.3 s.

The imaging performance of the preclinical hybrid dual-source CT was evaluated by means of phantom and cadaver scans [34, 35], demonstrating the clinical image quality at clinically realistic levels of X-ray photon flux with negligible effect of pulse pileup [34]. In contrast-enhanced abdominal scans of several human volunteers, photon-counting detector images showed similar qualitative and quantitative scores as conventional CT images for image quality, image noise, and artifacts, while additionally providing spectral information for material decomposition [36].

The predicted CNR improvements with photon-counting detectors, expected as a result of the missing down-weighting of low-energy X-ray photons, were confirmed both for contrast-enhanced scans (CNR of iodinated contrast agent versus soft tissue) and noncontrast scans (CNR of different tissues). Improved iodine CNR by photon-counting CT was demonstrated in a study using four anthropomorphic phantoms simulating four patient sizes [35]. Mean increases in iodine CNR of 11%, 23%, 31%, and 38% relative to the scintillation detector system at 80, 100, 120, and 140 kV, respectively, were shown. A similar overall improvement of the iodine CNR by about 25% was already found in [34]. The improvements in iodine CNR can potentially be translated into reduced radiation dose or reduced amount of contrast agent. The improvement of soft tissue contrasts was demonstrated in a brain CT study with 21 human volunteers [37]. The higher reader scores for the differentiation of gray and white brain matter for photon-counting CT images compared to conventional CT images were attributed to both higher soft tissue contrasts (10.3 ± 1.9 HU versus 8.9 ± 1.8 HU) and lower image noise for photon-counting CT.

The impact of the missing electronic noise on image quality was assessed for various clinical applications at low radiation dose. Less streaking artifacts and more homogeneous image noise in shoulder images acquired with the photon-counting detector of the preclinical hybrid dual-source CT as compared to its scintillation detector were demonstrated [38]. Symons et al. [39] evaluated the performance of the photon-counting CT prototype for potential low-dose lung cancer screening. Scanning a lung phantom at low radiation dose, the authors found better Hounsfield unit stability for lung, ground-glass, and emphysema-equivalent foams in combination with better reproducibility. The stability of Hounsfield units is an important prerequisite for quantitative CT. Additionally, photon-counting CT showed less noise and higher CNR. The better performance of photon-counting CT at very low radiation dose, attributed to the effective elimination of electronic noise and better weighting of low-energy X-ray photons, might enable further reduced radiation dose in CT lung cancer screening. In a study with 30 human subjects undergoing dose-reduced chest CT imaging [40], photon-counting CT demonstrated higher diagnostic quality with significantly better image quality scores for the lung, soft tissue, and bone and fewer beam-hardening artifacts, lower image noise, and higher CNR for lung nodule detection (see Fig. 4a, b).

The improved quality of the coronary artery calcium (CAC) scoring at low radiation dose was shown in a combined phantom, ex vivo and in vivo study [41]. The agreement between standard-dose (average $CTDI_{vol}$ = 5.4 mGy) and low-dose (average $CTDI_{vol}$ = 1.6 mGy) CAC scores in ten volunteers was significantly better for photon-counting CT than for the conventional CT, with better low-dose CAC score reproducibility. This finding was attributed to the absence of electronic noise in combination with improved calcium-soft tissue contrasts due to the missing down-weighting of the low-energy X-ray photons. The authors concluded that photon-counting CT technology may play a role in further reducing the radiation dose of CAC scoring.

The improvements in spatial resolution with the preclinical hybrid dual-source CT enabled by the smaller pixels of its photon-counting detector in "sharp" mode and in "UHR" mode were evaluated in several phantom and patient studies. 150 μm in-plane spatial resolution, with a cutoff spatial frequency of the modulation transfer

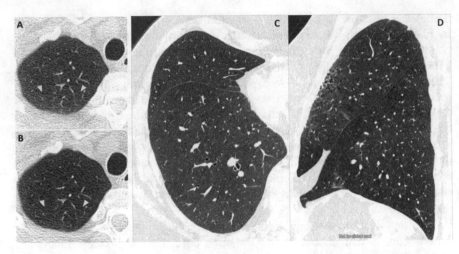

Fig. 4 Lung imaging with photon-counting CT. (**a–b**): Low-dose imaging. Low-dose lung scan acquired with a preclinical hybrid dual-source CT prototype with photon-counting detector. (**a**) scintillation detector image. (**b**) photon-counting detector image, demonstrating less image noise (arrowheads) because of the absence of electronic noise. Courtesy of R Symons, NIH, Bethesda, USA. (**c–d**): High-resolution imaging. A lung scan of a 74-year-old woman with breast cancer and signs of fibrosis after radiation therapy, acquired with a preclinical single-source CT prototype with photon-counting detector. Data acquisition: "UHR mode," 120 × 0.2 mm collimation, 0.3 s rotation time, CTDI$_{vol}$ = 3.89 mGy, DLP = 126 mGycm. Image reconstruction: sharp convolution kernel, 1024 × 1024 image matrix, 0.4 mm slice width. An excellent visualization of fibrosis and fine details such as fissures. Courtesy of Dr. J. Ferda, Pilsen, Czech Republic

function (MTF) at 32.4 lp/cm, and minimum slice widths down to 0.41 mm were demonstrated, and a better spatial resolution was confirmed in the clinical images of the lung, shoulder, and temporal bone (Fig. 5a, b) [42]. At matched in-plane spatial resolution, the photon-counting images had less image noise than the conventional CT images because of the better modulation transfer function (MTF) of the measurement system. Significant improvements of coronary stent lumen visibility in "sharp" and "UHR" mode were found [43], as well as superior qualitative and quantitative image characteristics for the coronary stent imaging when using a dedicated sharp convolution kernel [44]. Figure 5d and e shows a coronary stent scanned with the photon-counting CT prototype in the "macro" and in the "sharp" acquisition mode.

In a small study with eight human volunteers undergoing scans of the brain, the thorax, and the left kidney, Pourmorteza et al. [45] observed improved spatial resolution and less image noise with the "UHR" mode compared with standard-resolution photon-counting CT in "macro" mode. A substantially better delineation of the temporal bone anatomy scanned with the "UHR" mode compared with the ultra-high-resolution mode of a commercial energy-integrating-detector CT scanner was demonstrated [46].

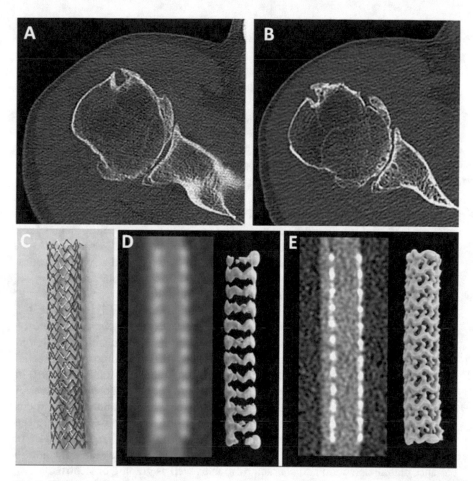

Fig. 5 High-resolution imaging with photon-counting CT. All images are acquired with the preclinical hybrid dual-source CT prototype. (**a** and **b**): An example of a shoulder scan. (**a**) Scintillation detector image. (**b**) Photon-counting detector image in the "sharp" mode, demonstrating higher spatial resolution and significantly improved visualization of the bony structures. (**c, d,** and **e**): The coronary stent (**c**) scanned in "macro" mode (**d**), corresponding to the resolution level of today's medical CT systems, and in "sharp" mode (**e**). Courtesy of Clinical Innovation Center, Mayo Clinic Rochester, MN, USA

Lung imaging is another clinical application that might benefit from the increased spatial resolution. Superior visualization of higher-order bronchi and third-/fourth-order bronchial walls at preserved lung nodule conspicuity compared with clinical reference images could be demonstrated in 22 adult patients referred for clinically indicated high-resolution chest CT [47]. The authors combined photon-counting CT in "sharp" acquisition mode with image reconstruction at 1024 × 1024 matrix size using a dedicated sharp convolution kernel. According to the authors, photon-counting CT is beneficial for high-resolution imaging of airway diseases and potentially for other pathologies, such as fibrosis, honeycombing, and emphysema. An example of high-resolution photon-counting chest CT is shown in Fig. 4c and d.

Increased resolution comes at the expense of increased image noise if the radiation dose is kept constant. Increased radiation dose to the patient to compensate for the higher noise may not be acceptable and may not even be available in all clinical situations. Spatial resolution does not only depend on the detector pixel size but also on the focal spot size of the X-ray tube which needs to be correspondingly small. The smaller the focal spot is, the less X-ray tube power is usually available, which may limit the clinical applicability of high-resolution photon-counting CT. Nonlinear data and image denoising techniques will therefore play a key role in harnessing the high-resolution potential of photon-counting detectors (see, e.g., [48–50]).

A key benefit of photon-counting CT is the spectrally resolved data acquisition in any scan. Spectral information can readily be added to the anatomical images for a better visualization of structures by means of virtual monoenergetic images (VMIs), for material classification and quantitation, and to obtain quantitative information about local perfusion by means of iodine maps and virtual noncontrast images.

The spectral performance of the preclinical hybrid dual-source prototype with photon-counting detector was evaluated in phantom studies [51], and the CT number accuracy in VMIs and iodine quantification accuracy were found to be comparable to dual-source dual-energy CT. According to the authors, photon-counting CT offers additional advantages, such as perfect temporal and spatial alignment to avoid motion artifacts, high spatial resolution, and improved CNR. In an anthropomorphic head phantom containing tubes filled with aqueous solutions of iodine (0.1–50 mg/ml), an excellent agreement between actual iodine concentrations and iodine concentrations measured in the iodine maps was observed [52]. The authors assessed the use of iodine maps and VMIs in the head and neck CTA in 16 asymptomatic volunteers and proposed VMIs at different keV as a method to enhance plaque detection and characterization as well as grading of stenosis.

The routine availability of VMIs with photon-counting CT may pave the way to further standardization of the CT protocols. VMIs are based on a two-material decomposition into water and iodine; they show the correct attenuation values of iodine at the selected keV. VMIs at standardized keV levels tailored to the clinical question (e.g., 50–70 keV for contrast-enhanced examinations of parenchymal organs, 40–50 keV for CTAs) may serve as a primary output of any CT scan regardless of the acquisition protocol (see Fig. 6). Follow-up scans will then be easily comparable, because their image impression and quantitative CT numbers depend on the standardized keV level only and no longer on the acquisition protocol. Prerequisite for the routine use of VMIs is the availability of refined processing techniques (see, e.g., [53]) to enhance CNR and image quality. Going one step further, the acquisition protocol may be standardized as well. Some authors [54] already recommend a standardized acquisition protocol with 140 kV X-ray tube voltage for contrast-enhanced abdominal CT examinations in all patient sizes, with standardized VMI reconstruction at 50 keV. According to the authors, optimal or near-optimal iodine CNR for all patient sizes is obtained with this protocol.

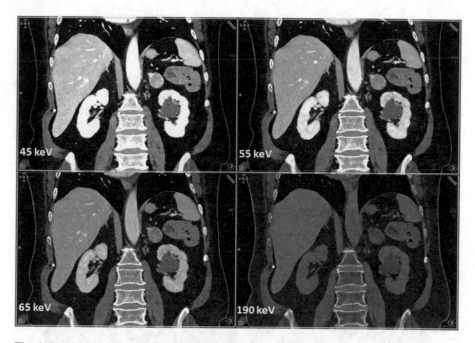

Fig. 6 The use of VMIs as standardized output of photon-counting CT. Abdominal images of a 67-year-old woman with adrenal adenoma and parapelvic renal cyst, acquired with a preclinical single-source CT prototype with photon-counting detector. Data acquisition: "UHR" mode, 120 × 0.2 mm collimation, 0.3 s rotation time, $CTDI_{vol}$ = 10.2 mGy, DLP = 450 mGycm. Image reconstruction: VMIs at various keV levels, 0.4 mm slice width. Note the decreasing contrast of iodine and calcium with increasing keV. The 45 keV images may serve as the standard output of CTAs; the 55 keV or 65 keV images are optimal for the examinations of the parenchymal organs; and the 190 keV images may substitute virtual noncontrast images with the iodine removed. Courtesy of Dr. J. Ferda, Pilsen, Czech Republic

Several authors assessed the performance of spectral photon-counting CT for detection and characterization of kidney stones, another established dual-energy CT application [55–57]. They found a comparable overall performance of the state-of-the-art dual-energy CT in differentiating stone composition, while photon-counting CT was better able to help characterize small renal stones [57]. Because of its higher spatial resolution, photon-counting CT can provide both, a high-resolution image of the stone structure and a material map image of the stone composition (see Fig. 7).

If the photon-counting detector is operated with more than two energy bins, multi-material decomposition is possible if K-edge elements are present. In a canine model of myocardial infarction, Symons et al. [58] determined the feasibility of the dual-contrast agent imaging of the heart to simultaneously assess both first-pass and late enhancement of the myocardium with the preclinical hybrid dual-source CT prototype. In a canine model of a myocardial infarction, gadolinium was injected 10 min prior to CT, while iodinated contrast agent was given immediately before the CT scan. The authors concluded that combined first-pass iodine and late gadolinium

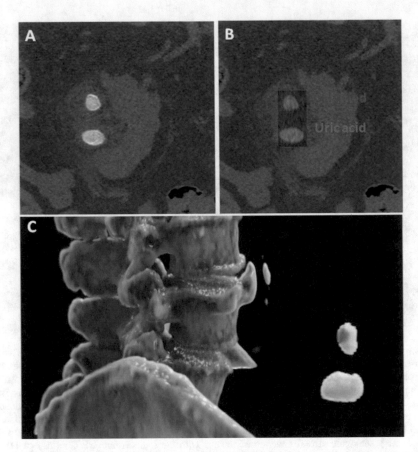

Fig. 7 The combination of high-resolution morphological and functional imaging with photon-counting CT. The abdominal CT scan of a volunteer acquired with the preclinical hybrid dual-source CT prototype in "sharp" mode. The high-resolution image at 0.25 mm nominal slice thickness shows the internal structure of two kidney stones (**a**), while a material map reveals their composition (**b**). A volume rendered image puts both stones into their anatomical context (**c**, note the different color schemes for uric acid and calcium). Courtesy of S. Leng, Mayo Clinic Rochester, MN, USA

maps allowed the quantitative separation of blood pool, infarct scar, and remote myocardium. The same authors also investigated the feasibility of simultaneous material decomposition of three contrast agents (bismuth, iodine, and gadolinium) in vivo in a canine model [59]. They observed tissue enhancement at multiple phases in a single CT acquisition, opening the potential to replace multiphase CT scans by a single CT acquisition with multiple contrast agents (see Fig. 8).

In the clinical practice, the use of multi-material maps may be hampered by the unavoidable increase of image noise in a multi-material decomposition. Similar to the ultra-high-resolution scanning, nonlinear data and image denoising techniques

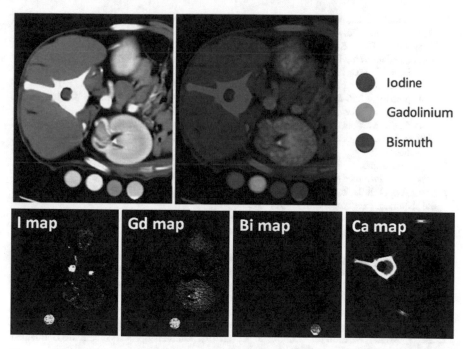

Fig. 8 Simultaneous imaging of three different contrast agents (iodine, gadolinium, and bismuth) by multi-material decomposition in a dog model. Scan data were acquired with the preclinical hybrid dual-source CT prototype. Bismuth was administered more than 1 day prior to scanning, followed by intravenous administration of gadolinium-based contrast agent 3–5 min before the scan and iodine-based contrast agent immediately before the scan to visualize the different phases of renal enhancement in a single CT acquisition. Scan data were read out in four energy bins (25–50 keV, 50–75 keV, 75–90 keV, and 90–140 keV). Top left: Standard CT image combining the data of the four energy bins. The three contrast agents cannot be differentiated. Top right: Grayscale image with overlay of the colored material maps. Bottom: The individual material maps. Note the noise amplification by multi-material decomposition. Courtesy of R Symons, NIH, Bethesda, USA

will play a key role in fully exploiting the potential of multi-material decomposition (see, e.g., [60]).

In this review article, we have outlined the basic principles of medical photon-counting CT and its potential clinical applications. Once remaining challenges of this technology have been mastered, photon-counting CT has the potential to bring clinical CT to a new level of performance.

References

1. Kalender, W., Seissler, W., Klotz, E., & Vock, P. (1990). Spiral volumetric CT with single-breath-hold technique, continuous transport and continuous scanner rotation. *Radiology, 176*, 181–183.

2. McCollough, C. H., & Zink, F. E. (1999). Performance evaluation of a multi-slice CT system. *Medical Physics, 26*, 2223–2230.
3. Klingenbeck-Regn, K., Schaller, S., Flohr, T., Ohnesorge, B., Kopp, A. F., & Baum, U. (1999). Subsecond multi-slice computed tomography: Basics and applications. *European Journal of Radiology, 31*, 110–124.
4. Mori, S., Obata, T., Nakajima, N., Ichihara, N., & Endo, M. (2005). Volumetric perfusion CT using prototype 256-detector row CT scanner: Preliminary study with healthy porcine model. *AJNR. American Journal of Neuroradiology, 26*(10), 2536–2541.
5. Flohr, T. G., McCollough, C. H., Bruder, H., Petersilka, M., Gruber, K., Süß, C., et al. (2006). First performance evaluation of a dual-source CT (DSCT) system. *European Radiology, 16*, 256–268.
6. Lu, G. M., Zhao, Y., Zhang, L. J., & Schoepf, U. J. (2012). Dual-energy CT of the lung. *AJR. American Journal of Roentgenology, 199*(5 Suppl), S40–S53.
7. Marin, D., Boll, D. T., Mileto, A., & Nelson, R. C. (2014). State of the art: Dual-energy CT of the abdomen. *Radiology, 271*(2), 327–342.
8. Odisio, E. G., Truong, M. T., Duran, C., de Groot, P. M., & Godoy, M. C. (2018). Role of dual-energy computed tomography in thoracic oncology. *Radiologic Clinics of North America, 56*(4), 535–548.
9. Albrecht, M. H., De Cecco, C. N., Schoepf, U. J., et al. (2018). Dual-energy CT of the heart current and future status. *European Journal of Radiology, 105*, 110–118.
10. De Santis, D., Eid, M., De Cecco, C. N., et al. (2018). Dual-energy computed tomography in cardiothoracic vascular imaging. *Radiologic Clinics of North America, 56*(4), 521–534.
11. Rajiah, P., Sundaram, M., & Subhas, N. (2019). Dual-energy CT in musculoskeletal imaging: What is the role beyond gout? *AJR. American Journal of Roentgenology, 213*(3), 493–505.
12. Siegel, M. J., & Ramirez-Giraldo, J. C. (2019). Dual-energy CT in children: Imaging algorithms and clinical applications. *Radiology, 291*(2), 286–297.
13. Johnson, T. R. C., Krauß, B., Sedlmair, M., et al. (2007). Material differentiation by dual-energy CT: Initial experience. *Eur Radiol 2007, 17*(6), 1510–1517.
14. Zhang, D., Li, X., & Liu, B. (2011). Objective characterization of GE discovery CT750 HD scanner: Gemstone spectral imaging mode. *Medical Physics, 38*(3), 1178–1188.
15. Rassouli, N., Etesami, M., Dhanantwari, A., & Rajiah, P. (2017). Detector-based spectral CT with a novel dual-layer technology: Principles and applications. *Insights Into Imaging, 8*(6), 589–598.
16. Feuerlein, S., Roessl, E., Proksa, R., et al. (2008). Multienergy photon-counting K-edge imaging: Potential for improved luminal depiction in vascular imaging. *Radiology, 249*(3), 1010–1016.
17. Taguchi, K., & Iwanczyk, J. S. (2013). Vision 20/20: Single photon-counting x-ray detectors in medical imaging. *Medical Physics, 40*(10), 100901.
18. Taguchi, K. (2017). Energy-sensitive photon-counting detector-based X-ray computed tomography. *Radiological Physics and Technology, 10*(1), 8–22.
19. Willemink, M. J., Persson, M., Pourmorteza, A., Pelc, N. J., & Fleischmann, D. (2018). Photon-counting CT: Technical principles and clinical prospects. *Radiology, 289*(2), 293–312.
20. Leng, S., Bruesewitz, M., Tao, S., et al. (2019). Photon-counting detector CT: System design and clinical applications of an emerging technology. *Radiographics, 39*(3), 729–743.
21. Kappler, S., Niederlöhner, D., Stierstorfer, K., & Flohr, T. (2010). Contrast-enhancement, image noise and dual-energy simulations for quantum-counting clinical CT. *Proceedings of the SPIE Medical Imaging Conference, 7622*, 76223H.
22. Kopp, F. A., Daerr, H., Si-Mohamed, S., et al. (2018). Evaluation of a pre-clinical photon-counting CT prototype for pulmonary imaging. *Scientific Reports, 8*(1), 17386.
23. Bratke, G., Hickethier, T., Bar-Ness, D., et al. (2019 Sep 12). Spectral photon-counting computed tomography for coronary stent imaging: Evaluation of the potential clinical impact for the delineation of in-stent restenosis. *Investigative Radiology*. https://doi.org/10.1097/RLI.0000000000000610. [Epub ahead of print].

24. Riederer, I., Bar-Ness, D., & Kimm, M. A. (2019). Liquid embolics agents in spectral x-ray photon-counting computed tomography using tantalum K-edge imaging. *Scientific Reports, 9*, 5268.
25. Cormode, D. P., Si-Mohamed, S., Bar-Ness, D., et al. (2017). Multicolor spectral photon-counting computed tomography: In vivo dual contrast imaging with a high count rate scanner. *Scientific Reports, 7*(1), 4784.
26. Dangelmaier, J., Bar-Ness, D., Daerr, H., et al. (2018). Experimental feasibility of spectral photon-counting computed tomography with two contrast agents for the detection of endoleaks following endovascular aortic repair. *European Radiology, 28*(8), 3318–3325.
27. Muenzel, D., Bar-Ness, D., Roessl, E., et al. (2017). Spectral photon-counting CT: Initial experience with dual-contrast agent K-edge Colonography. *Radiology, 283*(3), 723–728.
28. Si-Mohamed, S., Bar-Ness, D., Sigovan, M., et al. (2018). Multicolour imaging with spectral photon-counting CT: A phantom study. *Eur Rad Experimental, 2*, 34.
29. Si-Mohamed, S., Tatard-Leitman, V., Laugerette, A., et al. (2019). Spectral photon-counting computed tomography (SPCCT): In-vivo single-acquisition multi-phase liver imaging with a dual contrast agent protocol. *Scientific Reports, 9*(1), 8458.
30. Riederer, I., Si-Mohamed, S., Ehn, S., et al. (2019). Differentiation between blood and iodine in a bovine brain – Initial experience with spectral photon-counting computed tomography (SPCCT). *PLoS One, 14*(2), e0212679.
31. Kappler, S., Hannemann, T., Kraft, E., et al. First results from a hybrid prototype CT scanner for exploring benefits of quantum-counting in clinical CT. In *Medical imaging 2012: Physics of medical imaging* (p. 83130X). San Diego: International Society for Optics and Photonics.
32. Kappler, S., Henning, A., Krauss, B., et al. Multi-energy performance of a research prototype CT scanner with small-pixel counting detector. In *Medical imaging 2013: Physics of medical imaging* (p. 86680O). Lake Buena Vista: International Society for Optics and Photonics.
33. Kappler, S., Henning, A., Kreisler, B., et al. Photon-counting CT at elevated x-ray tube currents: contrast stability, image noise and multi-energy performance. In *Medical imaging 2014: Physics of medical imaging* (p. 90331C). San Diego: International Society for Optics and Photonics.
34. Yu, Z., Leng, S., Jorgensen, S. M., et al. (2016). Evaluation of conventional imaging performance in a research CT system with a photon-counting detector array. *Physics in Medicine and Biology, 61*, 1572–1595.
35. Gutjahr, R., Halaweish, A. F., Yu, Z., et al. (2016). Human imaging with photon-counting-based computed tomography at clinical dose levels: Contrast-to-noise ratio and cadaver studies. *Investigative Radiology, 51*(7), 421–429.
36. Pourmorteza, A., Symons, R., Sandfort, V., et al. (2016). Abdominal imaging with contrast-enhanced photon-counting CT: First human experience. *Radiology, 279*(1), 239–245.
37. Pourmorteza, A., Symons, R., Reich, D. S., Bagheri, M., Cork, T. E., Kappler, S., Ulzheimer, S., & Bluemke, D. A. (2017). Photon-counting CT of the brain: In vivo human results and image-quality assessment. *AJNR. American Journal of Neuroradiology, 38*(12), 2257–2263.
38. Yu, Z., Leng, S., Kappler, S., et al. (2016). Noise performance of low-dose CT_ comparison between an energy integrating detector and a photon-counting detector using a whole-body research photon-counting CT scanner. *Journal of Medical Imaging, 3*(4), 043503.
39. Symons, R., Cork, T., Sahbaee, P., et al. (2017). Low-dose lung cancer screening with photon-counting CT: A feasibility study. *Physics in Medicine and Biology, 62*(1), 202–213.
40. Symons, R., Pourmorteza, A., Sandfort, V., et al. (2017). Feasibility of dose-reduced chest CT with photon-counting detectors: Initial results in humans. *Radiology, 285*(3), 980–989.
41. Symons, R., Sandfort, V., Mallek, M., Ulzheimer, S., & Pourmorteza, A. (2019). Coronary artery calcium scoring with photon-counting CT: First in vivo human experience. *The International Journal of Cardiovascular Imaging, 35*(4), 733–739.

42. Leng, S., Rajendran, K., Gong, H., et al. (2018). 150-μm spatial resolution using photon-counting detector computed tomography technology: Technical performance and first patient images. *Investigative Radiology, 53*(11), 655–662.
43. Symons, R., de Bruecker, Y., Roosen, J., et al. (2018). Quarter-millimeter spectral coronary stent imaging with photon-counting CT: Initial experience. *Journal of Cardiovascular Computed Tomography, 12*, 509–515.
44. von Spiczak, J., Mannil, M., Peters, B., et al. (2018). Photon-counting computed tomography with dedicated sharp convolution kernels – Tapping the potential of a new technology for stent imaging. *Investigative Radiology, 53*(8), 486–494.
45. Pourmorteza, A., Symons, R., Henning, A., Ulzheimer, S., & Bluemke, D. A. (2018). Dose efficiency of quarter-millimeter photon-counting computed tomography: First-in-human results. *Investigative Radiology, 53*(6), 365–372.
46. Zhou, W., Lane, J. I., Carlson, M. L., et al. (2018). Comparison of a photon-counting-detector CT with an energy-integrating-detector CT for temporal bone imaging: A cadaveric study. *AJNR. American Journal of Neuroradiology, 39*(9), 1733–1738.
47. Bartlett, D. J., Koo, W. C., Bartholmai, B. J., et al. (2019). High-resolution chest computed tomography imaging of the lungs: Impact of 1024 matrix reconstruction and photon-counting detector computed tomography. *Investigative Radiology, 54*(3), 129–137.
48. Li, Z., Leng, S., Yu, L., et al. (2017). An effective noise reduction method for multi-energy CT images that exploit spatio-spectral features. *Medical Physics, 44*(5), 1610–1623.
49. Harrison, A. P., Xu, Z., Pourmorteza, A., Bluemke, D. A., & Mollura, D. J. (2017). A multichannel block-matching denoising algorithm for spectral photon-counting CT images. *Medical Physics, 44*(6), 2447–2452.
50. Rajendran, K., Tao, S., Abdurakhimova, D., Leng, S., & McCollough, C. (2018 Mar). Ultra-high resolution photon-counting detector CT reconstruction using spectral prior image constrained compressed-sensing (UHR-SPICCS). *Proceedings of SPIE The International Society for Optical Engineering, 10573*, pii: 1057318. https://doi.org/10.1117/12.2294628.
51. Leng, S., Zhou, W., Yu, Z., et al. (2017). Spectral performance of a whole-body research photon-counting detector CT: quantitative accuracy in derived image sets. *Physics in Medicine and Biology, 62*(17), 7216–7232.
52. Symons, R., Reich, D. S., Bagheri, M., et al. (2018). Photon-counting computed tomography for vascular imaging of the head and neck: First in vivo human results. *Investigative Radiology, 53*(3), 135–142.
53. Grant, K. L., Flohr, T. G., Krauss, B., et al. (2014). Assessment of an advanced image-based technique to calculate virtual monoenergetic computed tomographic images from a dual-energy examination to improve contrast-to-noise ratio in examinations using iodinated contrast media. *Investigative Radiology, 49*(9), 586–592.
54. Zhou, W., Abdurakhimova, D., Bruesewitz, M., et al. (2018 Mar). Impact of photon-counting detector technology on kV selection and diagnostic workflow in CT. *Proceedings of SPIE The International Society for Optical Engineering, 10573*, pii: 105731C. https://doi.org/10.1117/12.2294952.
55. Gutjahr, R., Polster, C., Henning, A., Kappler, S., Leng, S., McCollough, C. H., Sedlmair, M. U., Schmidt, B., Krauss, B., & Flohr, T. G. (2017). Dual-energy CT kidney stone differentiation in photon-counting computed tomography. *Proceedings of SPIE The International Society for Optical Engineering, 10132.*
56. Ferrero, A., Gutjahr, R., Halaweish, A. F., Leng, S., & McCollough, C. H. (2018). Characterization of urinary stone composition by use of whole-body, Photon-counting Detector CT. *Academic Radiology, 25*(10), 1270–1276.

57. Marcus, R. P., Fletcher, J. G., Ferrero, A., et al. (2018). Detection and characterization of renal stones by using photon-counting-based CT. *Radiology, 289*(2), 436–442.
58. Symons, R., Cork, T. E., Lakshmanan, M. N., et al. (2017). Dual-contrast agent photon-counting computed tomography of the heart: Initial experience. *The International Journal of Cardiovascular Imaging, 33*, 1253–1261.
59. Symons, R., Krauss, B., Sahbaee, P., et al. (2017). Photon-counting CT for simultaneous imaging of multiple contrast agents in the abdomen: An in vivo study. *Medical Physics, 44*(10), 5120–5127.
60. Tao, S., Rajendran, K., McCollough, C. H., & Leng, S. (2018). Material decomposition with prior knowledge aware iterative denoising (MD-PKAID). *Physics in Medicine and Biology, 63*(19), 195003.

Linearly Polarized X-ray Fluorescence Computed Tomography with a Photon Counting Detector

Zhijun Chi

1 Introduction

With the rapid development of nanoparticle (NP) technology applied in biomedical researches, especially for cancer diagnosis [1–5], there is an increasing demand for a novel imaging method that can simultaneously determine the spatial distribution and the concentration of NPs uptaken by the target organs in a nondestructive and noninvasive manner. X-ray fluorescence computed tomography (XFCT) provides a promising prospect for this application, since it combines the high sensitivity of X-ray fluorescence analysis (XRF) and the high resolution of computed tomography (CT). The principle of X-ray K_α fluorescence is shown in Fig. 1. When an X-ray beam incidents on a special element (e.g., iodine, gadolinium, or gold) of NPs, a K-shell electron is ejected with maximum probability via photoelectric effect. Then, the free vacancy will be occupied by an L-shell electron, accompanied with the emission of characteristic fluorescence K_α. Although other types of fluorescence emission, e.g., K_β (K-shell de-excitation by M-shell electron), can occur, K_α fluorescence is the most commonly used one in XFCT because of its highest intensity. Using a suitable geometry to collect the K_α fluorescence, a CT scan of an imaging object makes it possible to realize the localization and quantification of NPs simultaneously.

Compared with other advanced imaging modalities, such as K-edge subtraction CT, optical imaging techniques, and nuclear imaging techniques, XFCT has its own advantages. For example, the sensitivity of XFCT has been proved to be more superior than that of K-edge subtraction CT for contrast agents at very low concentration [6, 7]. Moreover, XFCT provides better depth resolution com-

Z. Chi (✉)
Beijing Normal University, Beijing, China
e-mail: chizj18@bnu.edu.cn

© Springer Nature Switzerland AG 2022
K. Iniewski (ed.), *Advanced X-ray Detector Technologies*,
https://doi.org/10.1007/978-3-030-64279-2_9

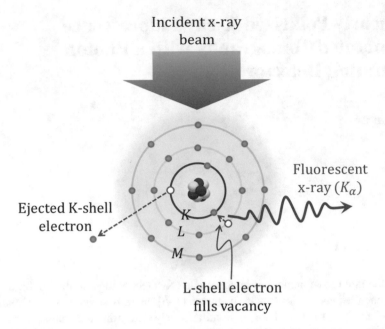

Fig. 1 Principle of X-ray fluorescence (K_α, K-shell electron excitation by primary X-ray and de-excitation by L-shell electron)

pared with optical imaging modalities using near-infrared (NIR) light, such as fluorescence diffuse optical tomography [8], bioluminescence tomography [9], and X-ray luminescence CT [10], and the spatial resolution of XFCT (typical value of 10^1–10^2 μm) is much higher than those NIR-based imaging methods and common nuclear imaging techniques, such as single photon emission computed tomography (SPECT) and positron emission computed tomography (PET). Therefore, XFCT combined with NP technology provides an excellent prospect for functional and molecular imaging.

In terms of XFCT, early experiments are carried out at synchrotron radiation facilities [11–14]. The high-brightness monoenergetic X-rays of synchrotron radiation allow one to map trace element distributions with high sensitivity. However, the translation–rotation data acquisition scheme employed in synchrotron-based XFCT costs a huge amount of time. Although several novel imaging geometries have been proposed to overcome this problem, such as sheet-beam geometry [15], pinhole [15–18], and multi-pinhole [19] collimator geometry, these improvements are effective only for very small imaging samples because of the limited field of view (FOV) of a synchrotron radiation facility, usually at a millimeter level. Besides, the large-scale footprint and expensive cost of a synchrotron facility make it impractical for most biomedical research laboratories. Nowadays, the accessibility of XFCT is promoted by using polychromatic X-ray tubes. However, longer scanning time is required in this technique, even when a fan/cone beam geometry is adopted [20, 21], and the reconstruction precision and sensitivity is also deteriorated due to the strong scattering background.

Alternatively, a Thomson scattering (also called inverse Compton scattering) X-ray source, based on the collision of intense laser and relativistic electrons, provides valuable prospects for XFCT, since it can provide quasi-monochromatic, continuously energy-tunable, polarization-controllable, and high-brightness X-rays. The typical X-ray energy region (10–100 keV) of this type of light source can cover almost all biomedical applications of XFCT, and its quasi-monochromaticity will help to improve the sensitivity of XFCT. The linearly polarized X-rays produced by this type of light source will help to reduce the Compton scattering background in XFCT by properly arranging the X-ray fluorescence detector. Furthermore, the small footprint (room-scale or table-top scale) and moderate cost make it very suitable for laboratory- and hospital-scale imaging application. In this chapter, we will introduce the principle of Compton scattering suppression by using linearly polarized X-rays and the implementation feasibility of linearly polarized XFCT based on this type of light source and a photon counting detector.

2 Linearly Polarized X-ray Generation and Compton Scattering Suppression Theory

2.1 Linearly Polarized X-ray Generation

The physics of Thomson scattering X-ray source can be described using a simplified model where a laser beam (plane wave) causes relativistic electrons to oscillate in its electromagnetic field and emits high-brightness cones of radiation (dipole radiation, linear Thomson scattering) as observed in the laboratory frame $S(x, y, z)$.

For simplicity, we assume that the electron is moving with velocity of $v = \beta c$ (β is the electron velocity normalized by the speed of light in a vacuum c) in the $+z$ direction and the laser with its electric field $\vec{E}_l$ along the x-axis is moving in the $-z$ direction (head-on collision geometry). In the coordinate system $S'(x', y', z')$ co-moving with the electron, the electric dipole moment $\vec{p}'$ of the electron is

$$\vec{p}' = -ex'\vec{\eta}_x, \tag{1}$$

where e is the elementary charge, $\vec{\eta}_x$ is the unit vector in x-direction, and the superscript $'\prime'$ represents that the physical parameter is expressed in the S' coordinate system, similarly hereinafter. According to the classical electromagnetic theory, the radiation electric field $\vec{E}'$ and magnetic field $\vec{B}'$ by this dipole moment are

$$\vec{E}' = \frac{e^{i\left(\vec{k}' \cdot \vec{R}' - \omega' t'\right)}}{4\pi \epsilon_0 c^2 R'} \left(\ddot{\vec{p}}' \times \vec{\eta}_{R'}\right) \times \vec{\eta}_{R'}$$

$$= \frac{e^{i\left(\vec{k}' \cdot \vec{R}' - \omega' t'\right)}}{4\pi \epsilon_0 c^2 R'} \ddot{p}_x'\Big[-\left(1 - \sin^2 \theta' \cos^2 \phi'\right) \vec{\eta}_x$$

$$+ \sin^2 \theta' \sin \phi' \cos \phi' \vec{\eta}_y + \sin \theta' \cos \theta' \cos \phi' \vec{\eta}_z\Big], \tag{2a}$$

$$\vec{B}' = \frac{e^{i\left(\vec{k}' \cdot \vec{R}' - \omega' t'\right)}}{4\pi \epsilon_0 c^3 R'} \left(\ddot{\vec{p}}' \times \vec{\eta}_{R'}\right)$$

$$= \frac{e^{i\left(\vec{k}' \cdot \vec{R}' - \omega' t'\right)}}{4\pi \epsilon_0 c^3 R'} \ddot{p}_x' \left(-\cos \theta' \vec{\eta}_y + \sin \theta' \sin \phi' \vec{\eta}_z\right), \tag{2b}$$

where ϵ_0 is the permittivity in a vacuum, $\vec{k}'$ and ω' are the wave vector and angular frequency of the laser, respectively, $\vec{R}'$ and $\vec{\eta}_{R'}$ are the space vector and its unit vector, respectively, θ' and ϕ' are the polar angle and azimuthal angle, respectively, $\vec{\eta}_y$ and $\vec{\eta}_z$ are the unit vectors in y- and z-directions, respectively, and $\ddot{\vec{p}}'$ is the second-order derivative of $\vec{p}'$ with respect to time t'

$$\ddot{\vec{p}}' = -e\ddot{x}'\vec{\eta}_x = -e\vec{a}' = \frac{e}{m_e}e\vec{E}_l' = \frac{e^2 E_{l,x}'}{m_e}\vec{\eta}_x = \ddot{p}_x'\vec{\eta}_x, \tag{3}$$

where m_e is the rest mass of electron, $\vec{a}'$ is the electron accelerator that is calculated by neglecting the contribution of magnetic field $\vec{B}'$ in the Lorentz force (linear Thomson scattering), and

$$\ddot{p}_x' = \frac{e^2 E_{l,x}'}{m_e}. \tag{4}$$

We assume that the radiation is emitted at $z = z' = 0$ at time $t = t' = 0$, thus the detection distance R' can be expressed using physical parameters in the S coordinate system under Lorentz transformation, i.e.,

$$R' = ct' = \gamma (ct - \beta z) = \gamma (R - \beta R \cos \theta)$$

$$= \gamma R (1 - \beta \cos \theta), \tag{5}$$

where $\gamma = 1/\sqrt{(1 - \beta^2)}$ is the relativistic Lorentz factor. Considering the Lorentz transformation of electromagnetic field

$$\begin{cases} E_x = \gamma \left(E'_x + \beta c B'_y \right), \\ E_y = \gamma \left(E'_y - \beta c B'_x \right), \\ E_z = E'_z, \end{cases} \tag{6}$$

the angle relation between S' and S systems

$$\begin{cases} \cos \theta' = \dfrac{\cos \theta - \beta}{1 - \beta \cos \theta}, \\ \sin \theta' = \dfrac{\sin \theta}{\gamma \left(1 - \beta \cos \theta \right)}, \\ \phi' = \phi, \end{cases} \tag{7}$$

and the fact that Lorentz transformation does not change the phase of an electromagnetic wave

$$\vec{k}' \cdot \vec{R}' - \omega' t' = \vec{k} \cdot \vec{R} - \omega t, \tag{8}$$

the radiation electric field $\vec{E}$ observed in the laboratory frame S can be transformed from Eq. (2):

$$\begin{aligned} \vec{E} = &\frac{2 r_e E_{l,x} e^{i(\vec{k} \cdot \vec{R} - \omega t)}}{\gamma R \left(1 - \beta \cos \theta \right)^3} \\ &\times \left[- \left(1 - \beta \cos \theta - \sin^2 \theta \cos^2 \phi \right) \vec{\eta}_x \right. \\ &\left. + \sin^2 \theta \sin \phi \cos \phi \vec{\eta}_y + (\cos \theta - \beta) \sin \theta \cos \phi \vec{\eta}_z \right], \end{aligned} \tag{9}$$

where $r_e = \frac{e^2}{4\pi \epsilon_0 m_e c^2} = 2.82 \times 10^{-15}$ m is the classical electron radius, and the electric field $E'_{l,x}$ in Eq. (4) is expressed by its counterpart in the S system under Lorentz transformation, i.e., $E'_{l,x} = \gamma \left(E_{l,x} + \beta c B_{l,y} \right) \approx 2\gamma E_{l,x}$. In the relativistic limit, where $\beta \approx 1$, the z-component of $\vec{E}$ will be vanished, i.e., $E_z \approx 0$.

To demonstrate the polarization of the radiation field, we give a typical example where the electron energy is 50 MeV and the detection plane locates at $z = 3.0$ m. The degree of polarization (DOP), defined as

$$\mathrm{DOP} = \frac{|E_x|^2 - |E_y|^2}{|E_x|^2 + |E_y|^2}, \tag{10}$$

is shown in Fig. 2. Note that most X-rays within the cone of $1/\gamma$ (half photons concentrate in this cone) are well polarized, and all X-rays within the cone of $1/2\gamma$ have

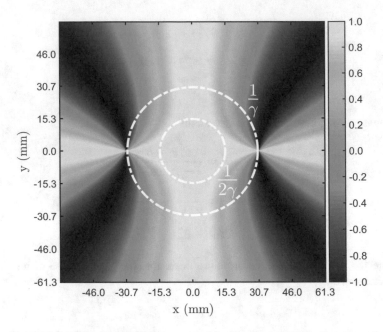

Fig. 2 The typical DOP of the radiation generated by Thomson scattering between relativistic electrons and a laser beam. The parameters are as follows: electron energy 50 MeV and detection plane $z = 3.0$ m. The white dash-dotted circles depicted in the figure are the regions with cone angles of $1/\gamma$ and $1/2\gamma$, respectively

excellent polarization with DOP ≥ 0.9. Therefore, the polarization characteristics of laser photons can be easily transferred to X-rays through a Thomson scattering, and linearly polarized X-rays can be generated straightforwardly in a Thomson scattering light source just by adjusting the laser beam to linear polarization.

2.2 Compton Scattering Suppression

For X-ray with linear polarization, its differential cross section of Compton scattering is described by the Klein–Nishina formula:

$$\frac{\mathrm{d}\sigma_{KN,LP}}{\mathrm{d}\Omega} = \frac{1}{2}r_e^2\varepsilon^2(\varepsilon + \varepsilon^{-1} - 2\sin^2\theta\cos^2\phi), \qquad (11)$$

where ε is the energy ratio between the scattered photon E_f and the incident photon E_i,

$$\varepsilon = \frac{E_f}{E_i} = \frac{1}{1 + \frac{E_i}{m_e c^2}(1 - \cos\theta)}. \qquad (12)$$

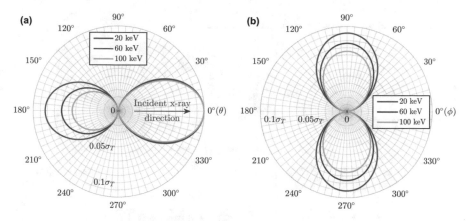

Fig. 3 Differential Compton scattering cross sections at different X-ray energies in angular coordinate: (**a**) $\phi = 0°/180°$ and (**b**) $\theta = 90°/270°$. $\sigma_T = 6.65 \times 10^{-25}$ cm^2 is the total Thomson scattering cross section

The angular dependence of the Compton scattering cross section is shown in Fig. 3. It can be seen that the differential cross section depends on both the polar angle θ and the azimuthal angle ϕ, and it can reach a minimum value at the direction of $\theta = 90°/270°$ and $\phi = 0°/180°$. Based on this fact, if the X-ray fluorescence detector is placed at this direction, the single Compton scattering background in XFCT can be greatly reduced, which is the theoretical basis of scattering suppression based on linearly polarized X-rays.

3 Monte Carlo Simulation

To demonstrate the feasibility of XFCT based on a Thomson scattering X-ray source and a photon counting detector, a full-scale Monte Carlo simulation was performed using the Geant4 toolkit [22] (version 10.05).

3.1 Imaging Geometry, Light Source, Detector, and Phantom

Considering that the FOV of a Thomson scattering light source is very large, usually at the centimeter-scale (see Fig. 2), a fan beam imaging geometry was adopted in the simulation. The layout of XFCT simulation is shown in Fig. 4. X-rays were generated at the interaction point (IP) [23] of the electron bunch and the laser beam and then propagated 12.0 m in the z-axis direction before reaching the sample. The X-ray beam was linearly polarized in the x-axis (horizontal) direction and can be tuned in the y-axis (vertical) direction when polarization-based comparison

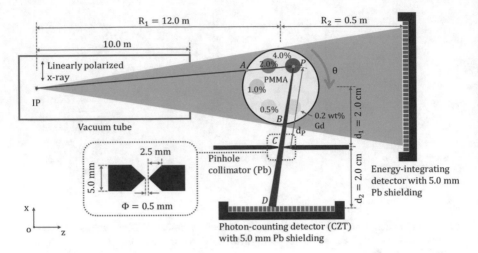

Fig. 4 Schematic for XFCT based on a Thomson scattering X-ray source and a photon counting detector (not to scale)

simulation was carried out. The source spot size used in this simulation was 10.0 μm root mean square (r.m.s.) and the X-ray energy was $E_0 = 60$ keV with an r.m.s. bandwidth of 1.0%. All the source parameters in the simulation, such as the source spot size and bandwidth [23, 24], can be easily achieved based on the existing technology. Since the X-ray energy of a Thomson scattering light source is correlated with the observation angle [25], a large FOV with quasi-monochromatic X-rays can be obtained only by enlarging the imaging distance, that is the reason why we choose such a long source-to-sample distance (12.0 m).

According to the scattering suppression theory in Sect. 2.2, a photon counting detector was placed perpendicular to the x-axis in the $x - z$ plane for the lowest scattering background. The detector was made of cadmium zinc telluride (CZT) and modelled using parameters that are available at present [26]: pixel size of 100 μm and energy range of 20–100 keV. For simplicity, other parameters, e.g., energy resolution, detection efficiency, and statistic noise, were assumed to be ideal. The distance between the photon counting detector and the sample was 4.0 cm. In the middle of them was a lead (Pb) pinhole collimator with diameter of $\Phi = 0.5$ mm and thickness of 5.0 mm. For the image reconstruction in XFCT, the attenuation data of the sample is necessary (see Sect. 3.2). Hence, an ideal energy-integrating detector with pixel size of 100 μm was placed 0.5 m downstream from the sample.

The sample was a cylindrical polymethyl methacrylate (PMMA) of diameter 2.5 cm, inside which there were five contrast agent containers of diameter 5.0 mm. The contrast agents were water solutions loaded with gadolinium (Gd), and the weight fractions of Gd were 0.2, 0.5, 1.0, 2.0, and 4.0 wt%, respectively. For element Gd, its K-edge is located at $E_k = 50.23$ keV and its K_α lines are located at 43.00 keV ($K_{\alpha1}$) and 42.31 keV ($K_{\alpha2}$) [27]. During the simulation, a 5-keV energy interval, ranging from 40 keV to 45 keV, of the photon counting detector was set to collect the K_α fluorescent photons of Gd. The fluorescence yield $\omega_{K\alpha}$ of Gd used in the simulation was 0.75046 (i.e., $\omega_{K\alpha1} = 0.48224$ and $\omega_{K\alpha2} = 0.26822$). For a CT scan, there were 360 projections acquired at a rotation step of $1°$, in each of which the incident X-ray photon number used for simulation was 1.0×10^9. These X-ray photons were uniformly distributed in a fan beam angle θ_c of 2.56 mrad confined by the collimator at the exit of the vacuum tube, corresponding to an FOV of ~3.0 cm at the sample.

3.2   Image Reconstruction

According to the imaging geometry in Fig. 4, the detection process of X-ray fluorescence photons can be divided into three steps:

(i) The incident X-rays with intensity I_0 and energy E_0 will be attenuated by the sample when they travel from point A to point P; then, the X-ray intensity $I(P)$ at P can be written as

$$I(P) = I_0 \exp\left[-\int_A^P \mu(E_0, \mathbf{r})ds\right], \qquad (13a)$$

where μ is the linear attenuation coefficient of the sample and $\mathbf{r}$ denotes a vector in Euclidean space.

(ii) The fluorescent X-ray photons are emitted isotropically if E_0 is higher than E_k ($E_0 = 60$ keV $> E_k$ in our simulation case), and the fluorescence intensity $I_{XRF}(P)$ at P is proportional to $I(P)$ and the local concentration $\rho_{Gd}(P)$ of Gd represented by mass percent:

$$I_{XRF}(P) = I(P)\mu_{Gd,PE}^m(E_0)\rho_{Gd}(P)\omega_{K\alpha}, \qquad (13b)$$

where $\mu_{Gd,PE}^m(E_0)$ is the photoelectric mass absorption coefficient of Gd at E_0.

(iii) The fluorescent X-rays will be collected by the pinhole collimator C after propagating a distance d_p, and then they are detected by the photon counting detector D. Along this path, they are also attenuated by the sample from P to the boundary B. Hence, the fluorescence intensity $I_{det}(P, D)$ coming from P and detected at D can be expressed as

$$I_{det}(P, D) = I_{XRF}(P)\frac{\pi\left(\frac{\Phi}{2}\right)^2}{4\pi d_p^2} \exp\left[-\int_P^B \mu(E_{XRF}, \mathbf{r})ds\right], \qquad (13c)$$

where E_{XRF} is the spectrum-averaged energy of K_α lines of Gd and its value is taken as 42.76 keV in the simulation.

Obviously, the total fluorescence intensity detected at the ith bin D_i ($i = 1, 2, \ldots, M$) of D is a volume integral of $I_{\text{det}}(P, D)$ over the pixel region $V_{P \to D_i}$ in the phantom inside a cone subtended by D_i toward the pinhole:

$$I_{\text{det},i} = \iiint_{V_{P \to D_i}} I_{\text{det}}(P, D) \mathrm{d}V_P. \tag{14}$$

If the attenuation terms $\mu(E_0, \mathbf{r})$ and $\mu(E_{\text{XRF}}, \mathbf{r})$ in Eqs. (13a) and (13c) are known, the relation between the detected fluorescence intensity $\mathbf{I}_{\text{det}}$ of size $M \times 1$ and the unknown Gd concentration ρ_{Gd} of size $N \times 1$, after sample discretization, can be expressed in the form:

$$\mathbf{I}_{\text{det}} = \mathbf{A}\rho_{\text{Gd}}, \tag{15}$$

where $\mathbf{A} = \left[a_{ij}\right]$ is the system matrix of size $M \times N$. Then, the ρ_{Gd} map can be reconstructed using the iterative algorithm called maximum-likelihood expectation maximization (MLEM) [28]:

$$\rho_{\text{Gd},j}^{(k+1)} = \frac{\rho_{\text{Gd},j}^{(k)}}{\sum_{i=1}^{M} a_{ij}} \sum_{i=1}^{M} \frac{a_{ij} I_{\text{det},i}}{\sum_{j'=1}^{N} a_{ij'} \rho_{\text{Gd},j'}^{(k)}} \quad (j = 1, 2, \ldots, N). \tag{16}$$

However, the attenuation of the sample is usually unknown and nonnegligible; hence, an adjoint attenuation CT scan is necessary to calculate the system matrix $\mathbf{A}$, and this can be realized easily since the energy of a Thomson scattering light source is continuously tunable. For determining the attenuation of the fluorescence photons, another attenuation CT scan was simulated at the same imaging geometry, while the photon energy was reduced to 42.76 keV with an r.m.s. bandwidth of 1% and only 180 projections were acquired at the same sampling interval (i.e., 1°). Meanwhile, the incident X-ray photon number used at each projection was unchanged (i.e., 1.0×10^9). The attenuation data of the phantom at both 60 keV and 42.76 keV were reconstructed using the well-known ART-TV iterative algorithm, and 180 projections were used in each attenuation reconstruction.

4 Results and Discussions

To examine the efficiency of the scattering suppression scheme, two typical X-ray polarizations, i.e., horizontal polarization and vertical polarization, were used for comparison in the simulation. The comparison result between the two fluorescence spectra collected by the photon counting detector with an ideal energy resolution is shown in Fig. 5. The spectrum was created by summing all of the pixels of the photon counting detector after a full XFCT scan. Note that the K_α and K_β lines of Gd and the Compton and Thomson scattering backgrounds are all witnessed. The

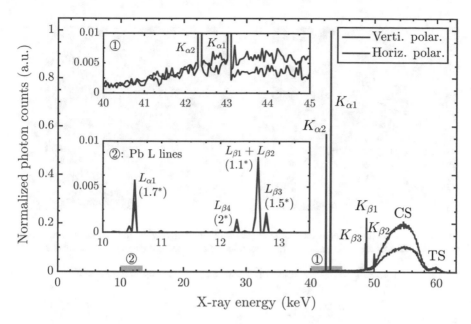

Fig. 5 Comparison between two typical fluorescence spectra detected by the photon counting detector. CS: Compton scattering; TS: Thomson scattering; * The ratio between the vertical polarization spectrum and the horizontal polarization spectrum at the corresponding L lines of Pb

Compton scattering background mainly concentrates in the energy region of 48–59 keV, and hence it can be effectively excluded by the photon counting detector in the signal collection region of 40–45 keV in the simulation. Compared with the vertical polarization case, the Compton scattering background, in the horizontal polarization case, is suppressed by about 1.6 times, verified by calculating the mean value of typical L lines of Pb (see subfigure ② of Fig. 5). Hence, the scattering background in the fluorescence signal region is also reduced at the same level (see subfigure ① of Fig. 5). Different from the theoretical prediction described in Sect. 2.2, the improvement of scattering background is not substantial and this can be attributed to the deviation of scattering angles to the right one. In our simulation geometry, only Compton scattering occurring along the x-axis direction strictly satisfies the right angle described in Sect. 2.2, and the scattering from other parts of the phantom is still significant. By restricting the detected energy range of fluorescent signal and using the X-rays with horizontal linear polarization, the scattering background can be reduced to a negligible level; hence, no scattering correction is needed before the CT reconstruction.

The reconstruction results of the phantom with and without attenuation correction are depicted in Fig. 6. These results are expressed in relative values. It can be seen that the five Gd contrast agents can be identified in the attenuation-corrected XFCT image. In order to quantitatively analyze the reconstruction results,

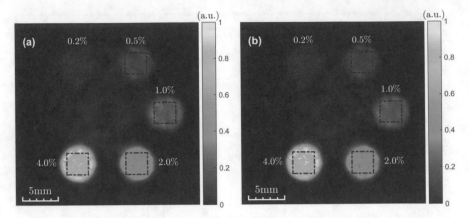

Fig. 6 Gd concentration maps reconstructed (**a**) without and (**b**) with attenuation correction. The red dotted squares are ROIs chosen for the quantitative analysis of the reconstructed results

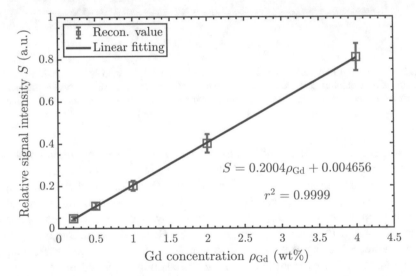

Fig. 7 Relation between the reconstructed result and the actual Gd concentration. The reconstruction value is a mean over all pixels in the ROI and the error bar is the corresponding standard deviation

six regions of interest (ROIs) highlighted by red dotted squares in Fig. 6a and b are chosen. The relation between the reconstructed value S averaged over the ROI pixels in Fig. 6b and the actual Gd concentration ρ_{Gd} is plotted in Fig. 7. Also shown in Fig. 7 is the linear fitting result between S and ρ_{Gd}. Obviously, there is a good linear relation between the reconstructed result and the Gd concentration with $r^2 = 0.9999$. Using this relation for calibration, the relative values in the reconstructed images can be expressed by actual concentration values for quantitatively absolute measurement.

Compared with the reconstruction result in Fig. 6a, the image contrast reconstructed with attenuation correction in Fig. 6b is improved significantly. To quantitatively analyze this contrast improvement, the contrast-to-noise ratio (CNR) is calculated. The CNR is defined as

$$\text{CNR} = \frac{S_{\text{Gd}} - S_{\text{BG}}}{\sigma_{\text{BG}}}, \tag{17}$$

where S_{Gd} and S_{BG} are the means of the ROI pixel values in the Gd and PMMA regions, respectively, and σ_{BG} is the standard deviation in the same ROI region of the PMMA. The calculated CNR results are depicted in Fig. 8. Obviously, the CNR of the reconstructed result with attenuation correction is higher than that of the corresponding result without attenuation correction. Based on the CNR results, we can also evaluate the contrast agent detectability. According to the Rose criterion, an object's CNR must exceed 3–5 in order to be detectable [29]. To determine the actual value, we introduce another evaluation index called limit of detection (LOD) with 95% confidence, which is defined as [30]

$$\text{LOD} = S_{\text{BG}} + 3.29\sigma_{\text{BG}}. \tag{18}$$

This gives a CNR threshold of 3.29 with 95% confidence. Based on this criterion, the four contrast agents with higher Gd concentration in XFCT can be clearly detected (CNR>5, see Fig. 8) and the contrast agent with 0.2 wt% Gd concentration in XFCT can be identified with a confidence of more than 95% (CNR>3.29, also see Fig. 8). Since the attenuation correction can enhance the CNR in XFCT and a higher CNR

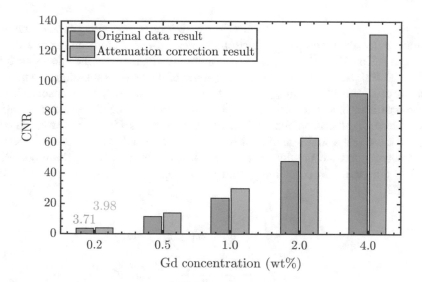

Fig. 8 Comparison of the CNRs in XFCT reconstructed with and without attenuation correction

means a lower LOD, it is necessary to correct the phantom attenuation to improve the LOD in XFCT.

5 Possibility Toward In Vivo Imaging

The simulation results in Sect. 4 have verified the feasibility of linearly polarized XFCT based on a Thomson scattering light source and a photon counting detector. However, the simulation parameters, e.g., the sampling strategy and the photon number, are not optimized. For the purpose of in vivo application of this imaging modality, optimized imaging parameters should be used. Further simulations for this purpose have been carried out, and the details have been described elsewhere[31]. Here, we just give the basic conclusions. Although 180° angular sampling is enough for conventional attenuation CT in parallel or quasi-parallel (small fan angle) imaging geometry, one full cycle (360°) sampling is necessary in XFCT when the attenuation of the phantom cannot be neglected. Using the imaging geometry of Fig. 4, the sampling interval can be reduced to 5° and the photon number can be decreased to 5.0×10^8 per projection without affecting the accurate reconstruction of the phantom. Based on these optimized imaging parameters, we will discuss the typical imaging time, the potential pulse pile-up effect, and the radiation dose using this imaging scheme.

5.1 Imaging Time

For the sampling strategy using 360° at 5° intervals and the optimized photon number 5.0×10^8 per projection, the total incident X-ray photon number of a whole CT scan is 3.6×10^{10}. These photons are uniformly distributed in the fan of $\theta_c = 2.56$ mrad and 100 μm height at 12.5 m downstream of the IP, corresponding to a photon density of 1.1×10^{10} ph/mm^2. By extending this 2D model to 3D, the incident X-ray photon number in a cone of the same polar angle at the same distance can be easily calculated using the same photon density and the value is 9.0×10^{12}. The photons in this cone are just a small fraction of the total photon yield in a Thomson scattering light source, and the fraction F can be calculated by integrating the probability density function $f(\theta, \phi)$ of the spatial distribution of scattered photons over this cone [32]:

$$
\begin{aligned}
F = \int_0^{2\pi} \int_0^{\theta_c/2} \frac{3}{8\pi} \frac{1}{\gamma^2 (1 - \beta \cos\theta)^2} \\
\times \left[1 - \frac{\sin^2\theta \cos^2\phi}{\gamma^2 (1 - \beta \cos\theta)^2} \right] \sin\theta \, d\theta \, d\phi.
\end{aligned} \tag{19}
$$

For a commonly used head-on collision geometry between the electron bunch and laser pulse, γ can be calculated through the relation between the scattered X-ray peak energy E_x and the laser photon energy E_l:

$$E_x = 4\gamma^2 E_l. \tag{20}$$

For an 800-nm laser commonly adopted in a Thomson scattering light source and 60-keV X-rays in our simulation, γ is 98.34 and F is 0.023. Hence, the total photon yield required for a whole XFCT scan is $9.0 \times 10^{12}/F = 4.0 \times 10^{14}$.

For a low-repetition Thomson scattering light source, e.g., the Tsinghua Thomson scattering X-ray source with photon yield of 2.0×10^8 ph/s [33], it would take $\sim$23 days to finish the data acquisition for the whole CT with the same statistics. Obviously, the huge amount of imaging time cannot be acceptable for practical XFCT applications. However, the photon flux of a Thomson scattering light source, based on the high-repetition design technology, can reach 10^{13} ph/s, and a series of laboratory-scaled facilities have been proposed and are under development at present [34, 35]. Using these high flux light sources, the data acquisition time of XFCT can be reduced to the second level for the same parameters in our simulation, which is possible for in vivo imaging.

5.2 Pulse Pile-Up Effect

For a photon counting detector, the pulse pile-up problem will become serious when the photon flux is high, because of which the imaging quality will be degraded. The typical X-ray pulse length of a Thomson scattering light source is very short (ps–fs), far below the time resolution of a photon counting detector. It is, therefore, necessary to analyze the potential pulse pile-up in our imaging geometry. In order to avoid the pulse pile-up problem, the photon number of fluorescent X-rays detected per pixel on the detector cannot exceed 1 in one incident X-ray pulse. Hence, the photon yield Y per pulse should meet the following requirement:

$$\frac{YF}{\pi \left[R_1 \tan(\frac{\theta_c}{2}) \right]^2} \mu^m_{\text{Gd,PE}}(E_0)\rho_{\text{Gd}}(P)$$

$$\times V_{\text{Gd}}\omega_{K\alpha} \frac{\pi \left(\frac{\Phi}{2} \right)^2}{4\pi d_p^2} \frac{1}{N_{\text{pix}}} \leq 1, \tag{21}$$

where V_{Gd} is the phantom volume containing Gd inside the fan beam and $N_{\text{pix}} = 250$ (i.e., $\frac{2.5\,\text{cm}}{100\,\mu\text{m}} \cdot \frac{d_2}{d_1}$) is the number of effective detector pixels collecting fluorescent X-ray photons. For conservative estimate, we neglect the phantom attenuation in Eq. (21) and assume that the phantom is fully filled with contrast agent with Gd concentration of 4.0 wt%. Thus, the value of V_{Gd} is $\sim$49.1 mm^3. The photoelectric mass absorption coefficient $\mu_{Gd,PE}^m(E_0)$ of Gd at E_0 (60 keV) can be obtained from the software XOP (version 2.4)[36] and its value is 11.3128 cm^2/g. The distance d_p can be taken as an approximate value, i.e., $d_p \simeq d_1 = 2.0$ cm, and other parameters in Eq. (21) are the same as before. Substituting these parameters into Eq. (21), we can calculate the photon yield $Y \leq 1.2 \times 10^{11}$ ph/pulse. This value limits the highest photon yield per pulse of a Thomson scattering light source to avoid the pulse pile-up problem using our imaging geometry.

For a Thomson scattering light source with low repetition, the photon yield cannot reach the above limitation at present and the pulse pile-up cannot occur for fluorescent X-ray photons excited by adjacent incident X-ray pulses; the pulse pile-up problem, therefore, can be neglected in our imaging geometry. For a Thomson scattering light source with high repetition, usually in continuous wave (CW) mode, the pulse pile-up for fluorescent X-ray photons excited by adjacent incident X-ray pulses can occur, while this problem cannot occur in one incident X-ray pulse because of its relatively low photon yield per pulse. To estimate this effect, we take the photon flux of 10^{13} ph/s as an example. In this case, the photon flux of fluorescent X-rays on the photon counting detector, based on the above analysis, is ~ 83 ph/(pixel· s) (i.e., $\frac{10^{13}}{1.2 \times 10^{11}}$). However, the count rate for a photon counting detector can reach 0.01–10 Mcps/pixel now [26], and it is far above the photon flux of fluorescent X-rays. Therefore, the pulse pile-up can also be neglected in our imaging geometry for a high-repetition Thomson scattering light source with photon flux of 10^{13} ph/s.

5.3 Radiation Dose

The radiation dose is an important factor that needs to be taken seriously in a practical XFCT application for small animals or human beings. Here, we will estimate the typical radiation dose in our imaging parameters. For a water cylinder phantom with radius of r, when it is irradiated by a parallel monochromatic X-ray beam that is perpendicular to the axis of the phantom and covers the phantom width, the theoretical radiation absorbed dose D is given as [37]

$$D = \frac{2}{\pi r \rho_w}(1 - g)E_0\phi_0 T \mu_w(E_0)/\mu_{w,en}(E_0), \tag{22}$$

where $\rho_w = 1.0 \times 10^3$ kg/m^3 is the density of water, ϕ_0 is the fluence rate, in unit of ph/(m^2·s), of the incident beam in front of the phantom, T is the exposure time, $\mu_w(E_0)$ and $\mu_{w,en}(E_0)$ are the linear attenuation coefficient and the linear energy absorption coefficient of water at X-ray energy E_0, respectively, and the factor g is

$$g = \int_0^{\frac{\pi}{2}} \exp\left[-2r\mu_{w,en}(E_0)\cos\theta\right]\cos\theta d\theta. \tag{23}$$

In our imaging geometry, $r = 1.25$ cm, $E_0 = 60$ keV, $\phi_0 T = 1.1 \times 10^{10}$ ph/mm$^2 \times \left(\frac{12.5m}{11.9875m}\right)^2 = 1.2 \times 10^{16}$ ph/m^2, and the values of $\mu_w(E_0)$ and $\mu_{w,en}(E_0)$ can be obtained from the National Institute of Standards and Technology (NIST) [38] and their values are 20.59 m^{-1} and 3.19 m^{-1}, respectively. Based on these parameters, the factor g can be numerically calculated and the value is 0.9394. Hence, the radiation absorbed dose D is 2.3 Gy, which is less than the generally known value of LD$_{50}$ (50% lethal dose) for mice, i.e., about 7 Gy [39]. This result shows that in vivo applications, especially for small animals, of XFCT using our imaging geometry are possible.

6 Future of This Imaging Modality

The excellent beam quality, small footprint, and moderate cost of Thomson scattering X-ray sources provide a promising prospect for XFCT in laboratories and hospitals. Our simulation results have verified the feasibility of polarized XFCT based on this type of light source and a photon counting detector. Based on the estimation of the imaging time and radiation dose, this imaging modality is possible for in vivo applications, especially for small animals. It is expected that the in vivo imaging requirements of XFCT will push this technology toward the following directions:

- High flux light source: The flux of a Thomson scattering X-ray source is a key matter that determines the imaging time. Based on our simulation results, when the flux of a Thomson scattering light source can reach 10^{13} ph/s or higher, it is possible to finish the data acquisition of XFCT at the second level or less without introducing any pulse pile-up problem. At this time level, the in vivo applications of XFCT can be realized.
- High energy and spatial resolution photon counting detector: In our simulation, only present available parameters of the photon counting detector are used. Based on the result in Fig. 5, we can infer that if the detection interval of the photon counting detector is further decreased, e.g., 1 keV (42.2–43.2 keV), the scattering background can be further reduced and the CNR of the reconstructed image will further improved. In our simulation, the spatial resolution of XFCT can reach to ∼100 μm level using the imaging geometry in Fig. 4. For higher resolution imaging applications, the present spatial resolution of the photon counting detector needs to be improved since it is the most important factor that determines the resolution of XFCT.
- High efficient collimator: The pinhole collimator used in our simulation limits the collection angle of fluorescent X-ray photons. Besides, the pinhole-collimated imaging geometry also makes the scattering suppression nonsignificant, since most Compton scattering photons cannot satisfy the optimized scattering condition described in Sect. 2.2. Therefore, a collimator with high efficiency needs

to be developed. For this application, an X-ray polycapillary lens may play an important role because of its high spatial resolution, large collection angle, and relatively high transmission efficiency.

Acknowledgments This work was supported by the Fundamental Research Funds for the Central Universities (Grant No. 2018NTST05), and the National Natural Science Foundation of China (Grant No. 11905011).

References

1. Qian, X., Peng, X.-H., Ansari, D. O., Yin-Goen, Q., Chen, G. Z., Shin, D. M., et al. (2008). In vivo tumor targeting and spectroscopic detection with surface-enhanced raman nanoparticle tags. *Nature Biotechnology, 26*, 83.
2. Brigger, I., Dubernet, C., & Couvreur, P. (2012). Nanoparticles in cancer therapy and diagnosis. *Advanced Drug Delivery Reviews, 64*, 24–36.
3. Toy, R., Bauer, L., Hoimes, C., Ghaghada, K. B., & Karathanasis, E. (2014). Targeted nanotechnology for cancer imaging. *Advanced Drug Delivery Reviews, 76*, 79–97.
4. Tietze, R., & Alexiou, C. (2017). Improving cancer imaging with magnetic nanoparticles: Where are we now? *Nanomedicine (London), 12*(3), 167–170.
5. Chen, F., Ma, K., Madajewski, B., Zhuang, L., Zhang, L., Rickert, K., et al. (2018). Ultrasmall targeted nanoparticles with engineered antibody fragments for imaging detection of HER2-overexpressing breast cancer. *Nature Communications, 9*, 4141.
6. Bazalova, M., Kuang, Y., Pratx, G., & Xing, L. (2012). Investigation of X-ray fluorescence computed tomography (XFCT) and k-edge imaging. *IEEE Transactions on Medical Imaging, 31*, 1620–1627.
7. Feng, P., Cong, W., Wei, B., & Wang, G. (2013). Analytic comparison between X-ray fluorescence CT and k-edge CT. *IEEE Transactions on Biomedical Engineering, 61*, 975–985.
8. Ntziachristos, V., Ripoll, J., Wang, L. V., & Weissleder, R. (2005). Looking and listening to light: The evolution of whole-body photonic imaging. *Nature Biotechnology, 23*, 313.
9. Wang, G., Cong, W., Durairaj, K., Qian, X., Shen, H., Sinn, P., et al. (2006). In vivo mouse studies with bioluminescence tomography. *Optics Express, 14*, 7801–7809.
10. Lun, M. C., Zhang, W., & Li, C. (2017). Sensitivity study of X-ray luminescence computed tomography. *Applied Optics, 56*, 3010–3019.
11. Boisseau, P. (1986). *Determination of Three Dimensional Trace Element Distributions by the Use of Monochromatic X-ray Microbeams*. Ph.D. thesis, Massachusetts Institute of Technology.
12. Takeda, T., Yuasa, T., Hoshino, A., Akiba, M., Uchida, A., Kazama, M., et al. (1997). Fluorescent X-ray computed tomography to visualize specific material distribution. In *Developments in X-Ray Tomography* (Vol. 3149, pp. 160–172). Bellingham: International Society for Optics and Photonics.
13. Simionovici, A., Chukalina, M., Schroer, C., Drakopoulos, M., Snigirev, A., Snigireva, I., et al. (2000). High-resolution X-ray fluorescence microtomography of homogeneous samples. *IEEE Transactions on Nuclear Science, 47*, 2736–2740.
14. Takeda, T., Wu, J., Huo, Q., Yuasa, T., Hyodo, K., Dilmanian, F. A., et al. (2009). X-ray fluorescent CT imaging of cerebral uptake of stable-iodine perfusion agent iodoamphetamine analog IMP in mice. Journal of Synchrotron Radiation, 16, 57–62.
15. Huo, Q., Sato, H., Yuasa, T., Akatsuka, T., Wu, J., & Lwin, T.-T. (2009). First experimental result with fluorescent X-ray CT based on sheet-beam geometry. *X-ray Spectrometry: An International Journal, 38*, 439–445.
16. Fu, G., Meng, L.-J., Eng, P., Newville, M., Vargas, P., & Riviere, P. L. (2013). Experimental demonstration of novel imaging geometries for X-ray fluorescence computed tomography. *Medical Physics, 40*, 061903.

17. Deng, B., Du, G., Zhou, G., Wang, Y., Ren, Y., Chen, R., et al. (2015). 3d elemental sensitive imaging by full-field XFCT. *Analyst, 140*, 3521–3525.
18. Sasaya, T., Sunaguchi, N., Hyodo, K., Zeniya, T., Takeda, T., Yuasa, T., et al. (2017). Dual-energy fluorescent X-ray computed tomography system with a pinhole design: Use of k-edge discontinuity for scatter correction. *Scientific Reports, 7*, 44143.
19. Sasaya, T., Sunaguchi, N., Hyodo, K., Zeniya, T., & Yuasa, T. (2017). Multi-pinhole fluorescent X-ray computed tomography for molecular imaging. *Scientific Reports, 7*, 5742.
20. Li, L., Zhang, S., Li, R., & Chen, Z. (2017). Full-field fan-beam X-ray fluorescence computed tomography with a conventional X-ray tube and photon-counting detectors for fast nanoparticle bioimaging. *Optical Engineering, 56*, 043106.
21. Jones, B. L., Manohar, N., Reynoso, F., Karellas, A., & Cho, S. H. (2012). Experimental demonstration of benchtop X-ray fluorescence computed tomography (XFCT) of gold nanoparticle-loaded objects using lead-and tin-filtered polychromatic cone-beams. *Physics in Medicine & Biology, 57*, N457.
22. Agostinelli, S., Allison, J., Amako, K. A., Apostolakis, J., Araujo, H., Arce, P., et al. (2003). GEANT4–A simulation toolkit. *Nuclear Instruments and Methods in Physics Research Section A: Accelerators, Spectrometers, Detectors and Associated Equipment, 506*, 250–303.
23. Chi, Z., Du, Y., Huang, W., & Tang, C. (2018). Focal spot characteristics of Thomson scattering X-ray sources. *Journal of Applied Physics, 124*, 124901.
24. Hartemann, F., Brown, W., Gibson, D., Anderson, S., Tremaine, A., Springer, P., et al. (2005). High-energy scaling of Compton scattering light sources. *Physical Review Special Topics-Accelerators and Beams, 8*, 100702.
25. Chi, Z., Du, Y., Huang, W., & Tang, C. (2017). Energy-angle correlation correction algorithm for monochromatic computed tomography based on Thomson scattering X-ray source. *Journal of Applied Physics, 122*, 234903.
26. Taguchi, K., & Iwanczyk, J. S. (2013). Vision 20/20: Single photon counting X-ray detectors in medical imaging. *Medical Physics, 40*, 100901.
27. Bearden, J. A. (1967). X-ray wavelengths. *Reviews of Modern Physics, 39*, 78.
28. Shepp, L. A., & Vardi, Y. (1982). Maximum likelihood reconstruction for emission tomography. *IEEE Transactions on Medical Imaging, 1*, 113–122.
29. Rose, A. (2013). *Vision: Human and electronic*. Heidelberg: Springer Science & Business Media.
30. Currie, L. A. (1968). Limits for qualitative detection and quantitative determination. Application to radiochemistry. *Analytical Chemistry, 40*, 586–593.
31. Chi, Z., Du, Y., Huang, W., & Tang, C. (2020). Linearly polarized X-ray fluorescence computed tomography based on a Thomson scattering light source: A Monte Carlo study. *Journal of Synchrotron Radiation, 27*, 737–745.
32. Chi, Z., Yan, L., Du, Y., Zhang, Z., Huang, W., Chen, H., et al. (2017). Recent progress of phase-contrast imaging at Tsinghua Thomson-scattering X-ray source. *Nuclear Instruments and Methods in Physics Research Section B: Beam Interactions with Materials and Atoms, 402*, 364–369.
33. Chi, Z., Du, Y., Yan, L., Wang, D., Zhang, H., Huang, W., et al. (2018). Experimental feasibility of dual-energy computed tomography based on the Thomson scattering X-ray source. *Journal of Synchrotron Radiation, 25*, 1797–1802.
34. Jacquet, M. (2014). High intensity compact Compton X-ray sources: Challenges and potential of applications. *Nuclear Instruments and Methods in Physics Research Section B: Beam Interactions with Materials and Atoms, 331*, 1–5.
35. Deitrick, K., Krafft, G., Terzić, B., & Delayen, J. (2018). High-brilliance, high-flux compact inverse Compton light source. *Physical Review Accelerators and Beams, 21*, 080703.
36. See http://www.esrf.eu/Instrumentation/software/dataanalysis/xop2.4 for the basic information (last accessed May 27, 2020)

37. Yuasa, T., Hashimoto, E., Maksimenko, A., Sugiyama, H., Arai, Y., Shimao, D., et al. (2008). Highly sensitive detection of the soft tissues based on refraction contrast by in-plane diffraction-enhanced imaging CT. *Nuclear Instruments and Methods in Physics Research Section A: Accelerators, Spectrometers, Detectors and Associated Equipment, 591*, 546–557.
38. See https://www.nist.gov/pml/x-ray-mass-attenuation-coefficients for the basic information (last accessed June 20, 2020).
39. Hall, E. J., Giaccia, A. J., et al. (2006). *Radiobiology for the radiologist* (Vol. 6). Philadelphia: Lippincott Williams & Wilkins.

3D XRF and Compton Imaging with CdTe and CdZnTe Arrays

Wonho Lee, Younghak Kim, and Changyeon Yoon

Room temperature detectors based on CdTe and CdZnTe has been widely used in radiation imaging owing to their portability, high energy resolution, large atomic number, and density. These detectors require minimal or no cooling which enables them to be applied in realistic fields such as nuclear power plants, nuclear reservoirs, astrophysics, and forensics. The development of multi-pixels and arrays based on CdTe and CdZnTe enables to construct compact 2D and 3D radiation imagers with high detection efficiency and angular resolution. Recent research for X-ray fluorescence (XRF) and Compton imaging with CdTe and CdZnTe arrays are described in this chapter.

Since CdTe arrays with a 10 square centimeters area and 1 mm thickness can be manufactured and operated in room temperature with minimal cooling, they can be applied as radiation detectors of a portable X-ray imaging system to detect characteristic X-rays emitted from target materials and visualize the 2D and 3D distribution of the atomic elements in the materials. The energies of characteristic X-rays emitted from most natural elements are less than 100 keV, and thus, 1 mm thickness CdTe is sufficient to detect the characteristic X-rays. 3D computed tomography of three different metals in a plastic phantom and a PVC pipe based on the detection of both transmitted and characteristic X-rays are presented. A CdTe detector array operates in a pulse mode in which both energy and position information of characteristic X-rays can be obtained, whereas a CsI-coupled CMOS array operates in a current mode in which the amount and position of transmitted X-rays can be measured. Atomic number and structural information of the target

W. Lee (✉) · Y. Kim
Korea University, Seoul, Korea
e-mail: wonhol@korea.ac.kr; dudgkrcjswo@korea.ac.kr

C. Yoon
Korea Hydro & Nuclear Power Co., Seoul, Korea
e-mail: yudoldosa@hanmail.net

© Springer Nature Switzerland AG 2022
K. Iniewski (ed.), *Advanced X-ray Detector Technologies*,
https://doi.org/10.1007/978-3-030-64279-2_10

material is acquired by the CdTe array and a CsI-coupled CMOS array, respectively. Unfiltered backprojection, filtered backprojection, and maximum likelihood expectation and maximization methods are used for image reconstruction. Both structural information and atomic number are represented in fused images and well matched with real distribution. The maximum thickness of the different types of pipes to which 3D XRF is applicable is investigated. 3D rendered images are reconstructed by using analytic and iterative image processing methods, and maximum likelihood method performs better than analytic reconstruction methods.

In the case of CdZnTe, the commercial crystals can be grown up to several centimeters in thickness, and thus they are applicable to detect γ-rays from a tens keV to several MeV γ-rays. Since the major interaction of γ-rays from several hundreds keV to several MeV is Compton scattering, Compton cameras using the Compton kinematics are effective to locate various radiation isotopes emitting γ-rays in that energy range, and hence, many radiation isotopes in the environmental and industrial fields can be investigated by Compton cameras. There are two major methods to construct Compton cameras comprising CdZnTe crystals – a large monolithic crystal and a mosaic array of parallelepiped ones. The monolithic crystal is composed of fine pixel electrodes which enable it to have high spatial resolution, but the achievable size of a single crystal with high energy resolution is limited to approximately 6 cm³. A mosaic array comprising small-size crystals is efficient to build a large-volume detector providing high detection efficiency at a comparatively low production cost compared to monolithic detectors. In addition, degraded crystals in the mosaic array can selectively be replaced without any modification of other crystals. The performance of Compton cameras based on each detector structure is evaluated by measuring various isotopes. The angular resolution of the reconstructed images of Compton cameras comprising a monolithic crystal is generally higher than that of a mosaic array, whereas the absolute detection efficiency of a mosaic array can be higher than that of a monolithic crystal.

1 3D X-Ray Fluorescence Tomography with a CdTe Array

A CdTe array with Schottky contact is widely utilized for X-ray imaging because a large-area CdTe detector comprising numerous micrometer pixels can be manufactured and each pixel shows high energy resolution with minimum cooling system. The thickness of the commercial CdTe array can be up to ~1 mm which is sufficient to detect soft X-rays (< 150 keV) in general purposes. Thus, CdTe array can be applied to reconstruct 2D and 3D fluorescent (i.e., characteristic) X-ray imaging for the analysis of target materials. Principle of characteristic X-ray emission is described in Fig. 1. When an incident radiation interacts with an atom, if the energy of the radiation is higher than the binding energy of the electron and fully transferred to the electron, the electron is ejected and another electron in a higher orbit falls into the vacancy of the ejected electron. The energy difference of the orbit is emitted in the form of an electromagnetic wave called

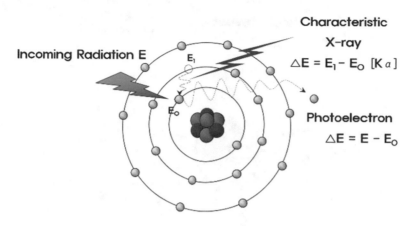

Fig. 1 Process of characteristic X-ray emission after a radiation interaction with an atom [8]

characteristic X-ray. Since the energy difference between the orbits is dependent on the atomic number of the target material regardless of the type and energy of the incident radiation, the atomic number can be investigated by analyzing the energy information of characteristic X-ray without any modification of the target material. If the characteristic X-rays are collimated by a mechanical collimator with high atomic number and density, and then, the collimated X-ray are detected by position-sensitive or multi-pixel detector, the distribution of the target material emitting characteristic X-rays can be reconstructed. The schematic design of 3D XRF imaging system is demonstrated in Fig. 2. When a phantom was exposed to the incident radiations, the transmitted radiations are measured by a position-sensitive detector positioned in the opposite side of the radiation source. The characteristic X-rays are measured by other detectors typically placed perpendicular to the direction of the incident radiation from the radiation source to phantom and transmitted X-ray detector. Based on this geometry, the characteristic X-ray detector is not exposed to the incident radiations from the radiation source. When the phantom is rotated more than 180° during radiation exposure, 3D image of the target material can be reconstructed by using the position and energy information obtained by detectors. The distribution of the materials with different atomic numbers is reconstructed based on the information from the characteristic X-ray detector, whereas transmitted X-ray detector discriminates the attenuation of the incident radiations which is strongly related to the density as well as atomic number. The reconstructed image based on transmitted radiation can precisely investigate densities of the materials, but it is not sufficient to analyze materials such as metals whose attenuation coefficients are high and densities are similar to each other. However, 3D XRF imaging system can combine the distribution of the atomic numbers and densities of target materials which enable to distinguish even those materials with similar densities and attenuation coefficients. In previous XRF research, high-purity germanium (HPGe) detector with cryogenic cooling or small-

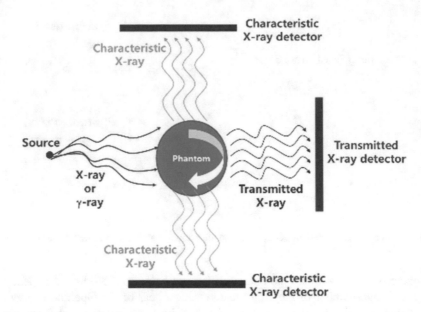

Fig. 2 3D XRF tomography [6]

size semiconductor detectors such as silicon drift detector (SDD) and CdTe detector with minimal cooling were used [1–5]. Recently, with the development of CdTe array, higher efficiency than a point CdTe detector was achieved and measurement time could be reduced [6, 7].

Figure 3 shows the photograph of a 3D XRF system. The X-ray generator was a L8121-03 manufactured by Hamamatsu (Iwata, Japan) whose tube voltage and current are 40–150 kV and 0–500 μA, respectively. The X-ray focal spot size was 50 μm which was sufficiently small to minimize blurring in the reconstructed image attributed to the finite focal spot size. The characteristic X-ray detector was a PID 350 (CdTe) from AJAT (Espoo, Finland) whose active area, pixel size, and thickness are 44.8×44.8 mm^2, 0.35×0.35 mm^2, and 1 mm, respectively. This CdTe detector array operated in a pulse mode in which both energy and position information of characteristic X-rays could be measured. The energy resolution of the CdTe array was 7 keV FWHM at 122 keV. The transmitted X-ray detector was a Xmaru 1215CF+ (CsI-coupled CMOS photodiode) from VATEC (Hwaseong, Korea). The active area, pixel size, and thickness of the photodiode array were 141.96×116.2 mm^2, 0.14×0.14 mm^2, and 0.28 μm, respectively. This CsI-coupled CMOS array operated in a current mode in which the amount and position of transmitted X-rays could be measured. A 2D lead collimator whose height, hole, and entire size were 25 mm, 1×1 mm^2, and 50×50 mm^2, respectively, was located in front of the characteristic X-ray detector to select the incident characteristic X-rays from the phantom. The collimator moved in arbitrary directions during X-ray detection to minimize the shadow of grids remaining in the acquired images.

Fig. 3 Photo of the assembled 3D XRF system [6]

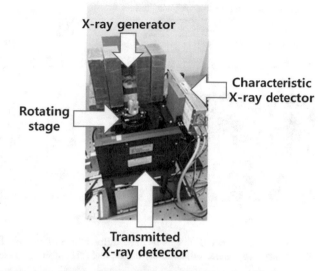

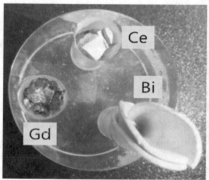
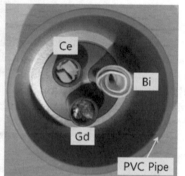

Fig. 4 Photos of phantoms with inner metals [6]

Figure 4 represents photos of a plastic and a PVC phantom with three inner metals (Ce, Gd, and Bi) inside the phantoms. In the case of the plastic phantom (Fig. 4a), both the diameter and the height of the plastic was 40 mm. The phantom simulated a water substance with target metals inside assuming that the phantom was similar to a biological tissue or the absorption in the vessel itself was minimal such as a thin opaque glass. Figure 4b displays a phantom comprising metals in a PVC pipe. The wall thickness of the PVC pipe was 2.5 mm. The diameters of three holes inside the plastic phantom and the PVC pipe were both 10 mm. Table 1 represents the densities and shapes of the target materials.

The distance from the center of the cylindrical phantom to the X-ray source was 150 mm. Each detector was located at a distance of 50 mm from the center of the phantom. The voltage and current of the X-ray tube was 150 kV and 125 μA, respectively. For three-dimensional measurement of transmitted X-rays,

Table 1 Specifications of target materials

Material	Plastic phantom	PVC pipe	Ce	Gd	Bi
Density (g/cm$^{3)}$)	0.941	1.467	6.77	7.90	9.78
Shape	Cylinder	Pipe	Granules	Granules	Flexible board

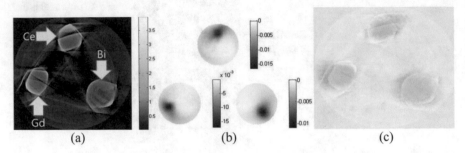

(a)   (b)   (c)

Fig. 5 3D reconstructed images of three different metals in a plastic phantom. (**a**) Transmitted X-ray image, (**b**) characteristic X-ray image, and (**c**) combined image [6]

the phantom was rotated 360° in 1° step for 5 s exposure per each rotational step. In the case of three-dimensional measurement of characteristic X-rays, the phantom was rotated 360° at 2.5° increments for 10 min exposure at each rotational step. For 3D image reconstruction, unfiltered backprojection (UBP) and filtered backprojection (FBP) with Ram-Lak filter were used as the analytic methods and maximum likelihood expectation and maximization (MLEM) methods were utilized as the iterative method. The MLEM method applies expectation maximization to calculate the maximum probability [9–14]. The reconstruction time of analytic methods is faster than that of iterative method and do not need a response function (c_{ij}) which is essential to apply the MLEM method. However, the quality of the reconstructed images processed by the MLEM method is normally higher than that of those reconstructed by analytic methods, especially when the statistical uncertainty of the measured counts is significant owing to the limited amount of detected radiations.

Figures 5 and 6 display 3D reconstructed image of three different metals in a plastic phantom and a PVC pipe. For the reconstruction of transmitted X-ray images (cf. Figs. 5a and 6a), UBP was applied since the amount of transmitted X-rays was sufficiently large to use analytic reconstruction methods, whereas MLEM was utilized to reconstruct characteristic X-rays whose amount was relatively small and statistical uncertainty should be considered for image processing (cf. Figs. 5b and 6b). The iteration number of MLEM was set to 3 to enhance positional resolution without significantly increasing noise in the reconstructed images. Figures 5c and 6c show combined images in which transmitted and characteristic X-ray images were resized based on the edges of the inner metals and overlaid with each other to represent both X-ray images. In the transmitted X-ray images, the position and shape of metals were reconstructed precisely, but atomic numbers of three metals

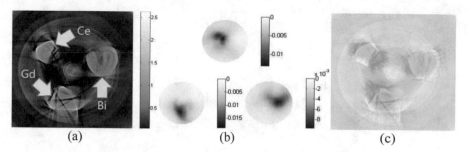

Fig. 6 3D reconstructed images of three different metals in a PVC pipe. (**a**) Transmitted X-ray image, (**b**) characteristic X-ray image, and (**c**) combined image [6]

Table 2 3D XRF applicable thickness for various pipe materials

Pipe material	Thickness [mm]			
	0.7	1.5	2.25	3
PVC	∘	∘	∘	∘
Al	∘	∘	∘	∘
Stainless steel	∘	∘	×	×
Cu	∘	×	×	×

could be distinguished from each other. In contrast, the atomic number of inner metals was clearly discriminated in the characteristic X-ray images, whereas the positional resolution was not high compared to transmitted X-ray images. Both structural information and atomic number were represented in the fused images and well matched with real distribution. Owing to the attenuation and scattering of X-ray in the metals and phantoms, many artifacts and beam hardening effect were presented in the transmitted X-ray images in Figs. 5a and 6a. The amount of artifacts in the PVC pipe was larger than those in the plastic phantom because the metals were located near the center of the PVC pipe and the atomic number and density of the PVC were higher than those of the plastic (plastic, 11.40 and 0.941 g/cm^3; PVC pipe, 29.10 and 1.467 g/cm^3).

Table 2 represents the maximum thickness of various pipe materials in which 3D XRF can be applicable. As the thickness and atomic number increased, the amount of characteristic X-rays penetrating the pipe and detected by CdTe array decreased, and hence, analysis by 3D XRF was limited depending on the thickness and type of the pipe.

Figure 7 demonstrates 3D rendered images constructed by stacking up the 3D XRF reconstructed images. The reconstructed images using UBP were relatively blurred compared to the other images because simple summation of backprojection enhanced low-frequency component. The reconstructed images using FBP showed significantly scattered artifacts because high-frequency component was emphasized by the high-pass filter. As shown in Fig. 4, the Bi layer extended beyond the plastic phantom, and thus, the reconstructed Bi image was extended beyond the FOV. Similar to the 3D cross-sectional images, three different metals were clearly distinguished in the 3D rendered images.

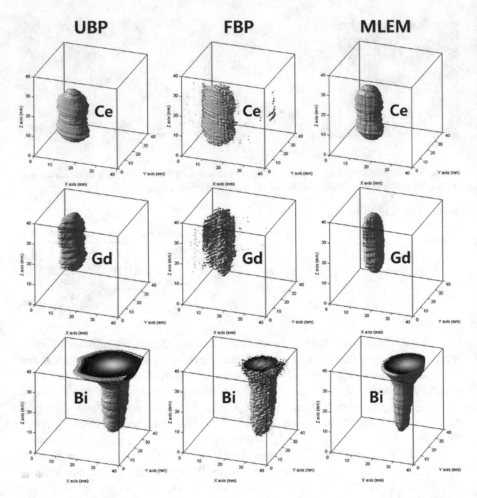

Fig. 7 3D rendered images based on the XRF measurement [6]

2 Multi-isotropic Compton Imaging with CdZnTe Crystals

There are two main collimation methods to reconstruct the original source distribution – mechanical and electronic collimation. In the case of mechanical collimation, the incident radiations from the source are significantly filtered out by the collimator, and thus, a large amount of source radiation cannot reach the detector and the intrinsic detection efficiency is low. In addition, the amount of radiation penetrates the mechanical collimator increased with the energy of the radiation, and the noise of reconstructed image increases. Therefore, a thicker collimator is required to minimize penetration for higher-energy radiation, which results in the decrease of detection efficiency and field of view of the radiation camera. The electronic collimator (i.e., Compton camera) utilizes the kinematics of Compton scattering instead of a mechanical collimator to reconstruct the direction of the incident radiations.

The basic principle of Compton scattering is to analyze a collision between a photon and an electron as a billiard ball collision. Figure 8 demonstrates how such a collision might conceptually occur. When a photon strikes an electron, it is scattered away from its original incident direction while the electron receives the impact and begins to move. If the energy of the incident radiation is E_0, that of the scattered radiation is E_s, and that of the recoil electron energy is E_r, the angle of photon scattering θ can be calculated by energy and momentum conservation using

$$\cos\theta = 1 - \frac{m_e c^2 E_r}{E_0 E_s} \tag{1}$$

The term $m_e c^2$ is the rest mass energy of an electron which equals to 511 keV and the electron is assumed to initially be at rest and unbounded.

Once the energy and the position information of the radiation interactions with detectors are recorded, the source position can be determined as shown in Fig. 8b. The line connecting the positions of the first and the second interaction is the axis of a cone whose opening angle θ is determined from Eq. (1). Based on the information of each sequential interaction in both detectors, a conical line can be drawn in the source sphere or plane. The original source distribution can be reconstructed by sum of many conical lines. A simple backprojection, i.e., the sum of all the cones, is the fastest reconstruction method. However, there are techniques to improve the image quality. Various image processing techniques have been developed to reconstruct a better image with trade-offs between processing time and image quality [15–17]. The detection efficiency of the Compton camera can be one or two orders of magnitude higher than that of mechanical collimation owing to the absence of the physical collimator. There is no loss of the image contrast and quantitative accuracy of the measured activity distribution caused by radiation scattered in the mechanical collimator. There is no artifact of the image due to radiation penetrating through the

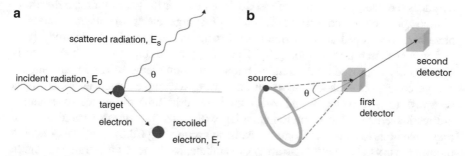

Fig. 8 Schematic diagram of (**a**) Compton scattering and (**b**) Compton camera imaging

collimator, either. Moreover, the angular resolution of the acquired image improves with higher-energy radiation because the energy uncertainty, which is directly related to angular uncertainty (cf. Eq. 1), gradually decreases as the energy of the interaction increases. The probability of the Compton scattering dramatically increases from a few keV to several hundred keV and only slightly decreases up to a few MeV. Therefore, a Compton camera is effective for detecting the higher-energy radiation (800 keV to a few MeV), while mechanical collimation is better at lower energy (<several hundred keV). In contrast to the simple design of the mechanical collimator, Compton cameras require additional hardware for detecting the time coincidence of interaction positions, as well as software for deblurring the overlapping cones.

The first Compton imaging technique was proposed by Pinkau [18] in 1966 and White [19] in 1968 to acquire the image of solar neutrons. Singh and Doria built the first prototype Compton camera in 1983 [20]. The scatter detector was a planar high-purity germanium (HPGe) detector and the second detector consisted NaI(Tl) scintillators without collimation. In 1996, a Compton telescope composed of two HPGe detectors as the scatter and absorption detectors was developed by Phlips et al. [21]. In 2001, Schmid et al. proposed a single coaxial germanium detector for Compton imaging [22]. In 2001, Du et al. [23] developed a prototype Compton imager consisting of two CdZnTe detectors with three-dimensional position sensitivity. In 2004, Lehner [24] proposed 4π Compton imaging using a single 3D position-sensitive CdZnTe detector with multiple Compton scatters.

In the case of CdZnTe, the commercial crystals can be grown up to several centimeters in thickness, and thus they are applicable to detect γ-rays from a tens keV to several MeV γ-rays [25]. CdZnTe material also has a relatively wide band-gap and high resistivity, which enable it to operate in room temperature with long-term stability [26]. However, the charge transport property of CdZnTe semiconductor limits its application. The charge transport property of the semiconductor is one of main factors to evaluate the semiconductor because the semiconductor with high mobility and lifetime provides high energy resolution. In CdZnTe crystals, the mobility of a hole ($120 \text{ cm}^2\text{V}^{-1} \text{ s}^{-1}$) is significantly lower than that of an electron ($1350 \text{ cm}^2\text{V}^{-1} \text{ s}^{-1}$) and the lifetime of the hole is also short, and thus, the hole cannot be collected by electrodes but trapped in the CdZnTe crystal, which severely degrades the energy resolution of the CdZnTe detector [27]. To minimize the effect of incomplete charge collection, several single-charge (electron) collection methods have been developed such as small pixel structure [28], coplanar electrode [29], and virtual Frisch-grid method [30]. The energy resolution of CdZnTe detector is significantly improved by the single-charge collection methods. Among the single-charge collection methods, 3D position sensing technique which uses 2D pixels or arrays to investigate x- and y-direction and drift time difference to calculate z-direction enables the Compton imaging with CdZnTe crystals. There are two major methods to construct Compton cameras comprising CdZnTe crystals – a large monolithic crystal and a mosaic array of parallelepiped ones as illustrated in Fig. 9.

The monolithic crystal is composed of fine pixel electrodes which enable it to have high spatial resolution, but the achievable size of a single crystal with high

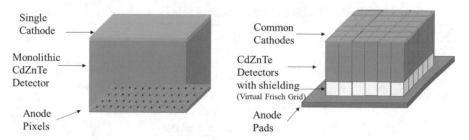

Fig. 9 Schematic diagrams of CdZnTe Compton camera base on (**a**) monolithic crystal [24] and (**b**) mosaic array [31]

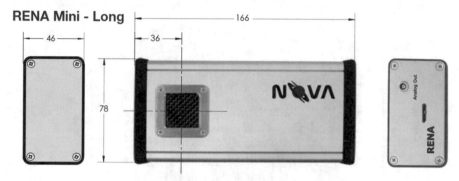

Fig. 10 Photograph of a monolithic CdZnTe system (RENA) [33]

energy resolution is limited to approximately 6 cm^3 [31]. A mosaic array comprising small-size crystals is efficient to build a large-volume detector providing high detection efficiency at a comparatively low production cost compared to monolithic detectors. When side pads are attached to the virtual Frisch grids of the mosaic array and the signals from the side pads are analyzed, less than 100 micron position resolution is achievable [32]. In addition, degraded crystals in the mosaic array can selectively be replaced without any modification of other crystals.

1. Monolithic CdZnTe Detector

Figure 10 illustrates the RENA mini™-long made by Kromek (Durham, UK) including a monolithic CdZnTe crystal. The dimensions of the system and CdZnTe crystal were 166 mm × 78 mm × 46 mm and 20 mm × 20 mm × 5 mm, respectively. The entrance window was a carbon plate for the light-tight enclosure. The CdZnTe crystal was positioned at the backside of the carbon plate. Table 3 describes the specifications of the detector system. The structures of electrodes were an 8 × 8 pixelated pattern on anode and a single plate on cathode. The amplitude and timing information of the interaction event were measured simultaneously but processed by two different circuits with different shaping times at each signal channel. The signal information was transmitted to a computer as a list mode.

Table 3 Specification of a monolithic CdZnTe system (RENA)

Parameter	Description/value
Crystal type	Cadmium zinc telluride
Crystal size	20 mm × 20 mm × 5 mm (8 × 8 pixel pattern)
Energy range	20 keV–3 MeV
Channel	Anode: 64 channels
	Cathode: 1 channel
	Total: 72 channels available
Timing resolution	≤10 ns
Maximum count rate	10^5 count/s
Power consumption	≤6 mW per channel

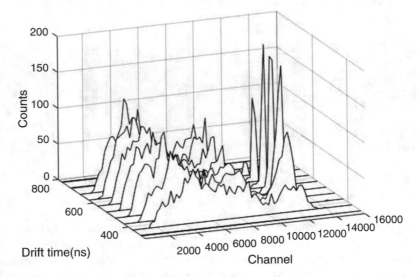

Fig. 11 ^{137}Cs pulse height spectra vs. the drift time in a pixel [33]

The interaction depth in the z-direction was estimated by measuring the drift time which was defined as the time difference between the induced signals owing to the drift of the electron cloud at the cathode and the anode [28]. The spatial resolution in the depth direction limited by the timing resolution was approximately 0.25 mm when the timing bin is set to 50 ns. The pixel size was 2.5 × 2.5 mm which determined the spatial resolution in the x- and y-direction. Since the CdZnTe detector response was dependent on the position, the response was calibrated with the three-dimensional spatial information. The dependency was attributed to nonlinear weighting potential, inclusions and dislocations in the material, electron trapping, and nonuniform response of electronics. Figure 11 represents that ^{137}Cs spectra were dependent on the drift time in a pixel. The dependency was minimized by the drift time correction which was calculated based on the alignment of the photopeak channel of each spectrum to that of the total spectrum [33].

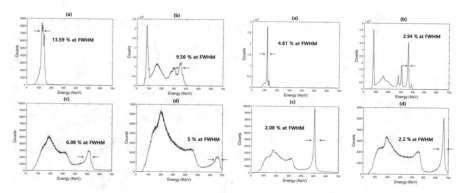

Fig. 12 Energy spectra before (left) and after (right) correction [33]. (**a**)[57]Co, (**b**) [133]Ba, (**c**) [22]Na, and (**d**) [137]Cs

Figure 12 presents measured energy spectra for four kinds of sources – [57]Co, [133]Ba, [22]Na, and [137]Cs – before and after calibration. Single events of Compton scattering or photoelectric absorption were selected and calibrated based on the photopeaks of each source. After calibration, the photopeak of each source became higher and narrower, the energy tailing on the left of photopeak was minimized, and as a result, energy resolution became 2–3 times improved. For example, the photopeaks of [133]Ba from 276 keV to 383 keV were clearly distinguished after calibration. As the energy of the incident radiation increased, the energy resolution of the measured spectrum improved as expected.

In the case of the monolithic CdZnTe detector, real Compton event followed by photoelectric effect and the charge sharing events between directly adjacent pixels were hard to be distinguished from each other, and hence, the signals measured simultaneously in two adjacent pixels were neglected.

2. Mosaic CdZnTe Array

Figure 13 presents the schematic and photo of the mosaic array module [34]. The size of the parallelepiped CdZnTe crystals made by Redlen (Saanichton, Canada) was $5 \times 5 \times 12$ mm^3. The specification of the CdZnTe was listed in Table 4. The sides of the CdZnTe crystals were surrounded by a shrinking tube with 25 μm thickness made of polyethylene terephthalate (PET). The dielectric strength of PET was higher than 150 V per micrometer. The cross-sectional size of the original PET tube was slightly larger than that of CdZnTe crystals. When a CdZnTe crystal was positioned inside of a PET tube, the tube was heated by boiling water (approximately 100 °C). The tube was shrunk by the heat, and hence, the crystal was tightly bound by the tube. The PET wrapping a CdZnTe crystal was covered by a 5-mm-thick Al tape near the anode of the CdZnTe crystal. The Al tape was then a non-contacting electrode to the CdZnTe crystal called a virtual Frisch grid. The virtual Frisch grid was grounded by a single line electrode positioned between the frame and alignment grid. The 6×6 CdZnTe array was held by an

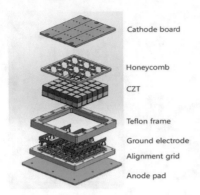
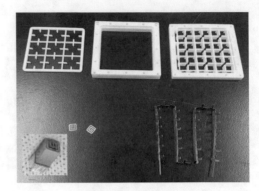

Fig. 13 The schematic (left) and photo (right) of the mosaic array module [34]

Table 4 Specification of Redlen CZT

Parameter	Value
Capacitance	0.19 pF
Resistivity	$\approx 5 \times 10^{10}$ $\Omega \cdot$cm
Size	$5 \times 5 \times 12$ mm^3
Resistance	2.4×10^{11} Ω
Leakage at 1200 V	5 nA

alignment grid, frame, and honeycomb made of Teflon which was an insulating material. The thicknesses of the holders were 5, 7, and 0.5 mm, respectively. The pixel pitch considering the size of CdZnTe crystals and a fabricable septal thickness of the holder was determined to be 7.1 mm. Spring type electrodes were attached to the cathode and anode side of CdZnTe crystals by conductive epoxy for secure contacts to prevent mechanical vibration and electric shock (cf. Fig. 13 (right)). Both single line and spring electrodes were made of beryllium copper (CuBe) plates with 0.1 mm thickness.

The top and bottom of the crystal with spring type electrodes were connected to the cathode and anode board. Each crystal could be simply replaced by a new one without any modification of other crystals and other components in the module.

Figure 14 presents the Rena mini-short made by Kromek. The dimensions of the device were 46 mm × 80 mm × 98 mm. The device had 2 built-in ASIC chips and each chip processes signals from 36 independent channels. Each channel was directed to two signal paths in an ASIC chip. One path with a fast shaping filter extracted timing information, and the other path with a slow shaping filter measured energy information. Both timing and energy information of each signal were simultaneously processed, digitized, and sent to a personal computer. The detector board connecting the CdZnTe mosaic array module to the ASIC device was customized by Nova R&D, a subsidiary of Kromek considering the resistance of the CdZnTe, minimizing electric noise, and maximizing applied voltage (cf. Fig. 15). Each anode of CdZnTe was connected to independent electrodes on the anode board

Fig. 14 Front (right) and side (left) view of Rena mini-short [34]

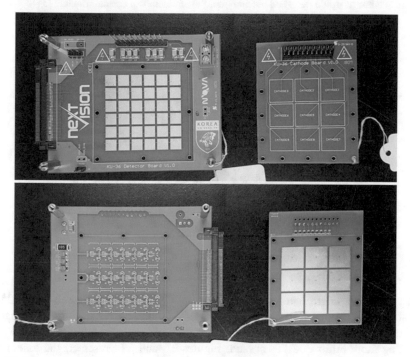

Fig. 15 Front and backside of customized anode (up) and cathode (down) boards [34]

while a common cathode used for 2×2 CdZnTe array to reduce the number of the channels to be processed by ASIC chips.

As the case of the monolithic CdZnTe crystal, the depth of the interaction in the mosaic CdZnTe array was calculated by the time difference of the induced signals between cathode and anodes related with the drift time of the electron cloud. Since the mosaic crystals were placed on 2D direction and the depth information in each crystal could be measured by the timing information of the cathode and anode in the detector module, the 3D positional sensing of each radiation interaction with CdZnTe array could be achieved. In the current design, the spatial resolution in x-

Fig. 16 Photos of inside (left) and outside (middle) of assembled detector module [34]

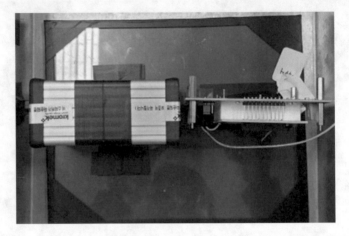

Fig. 17 Photo of Compton camera consisting of Rena mini-short and detector module [34]

and y-direction determined by the finite size of CdZnTe crystals was 5 mm and that in z-direction limited by timing bins was 0.75 mm. The 3D position information was used to minimize the positional dependency of the signal amplitude as the case of using monolithic CdZnTe crystal. Figures 16 and 17 presented the assembled detector module and the entire Compton camera consisting of Rena mini-short and the 2 × 2 CdZnTe detector module.

Figure 18 (left) represents raw energy spectra of the triggered events on anode. Different CZT detector demonstrated different performances but represented a typical shape of non-calibrated spectrum. The energy resolutions of spectra measured by detector 1–4 before calibration were 5.8, 8.6, 12.8, and 6.5%, respectively. To calibrate the energy spectra, we selected the measured events only when the cathode and anode were triggered simultaneously and investigated the 3D positional information of the events. Based on the 3D positional information, the measured signal amplitude was calibrated by a correction factor depending on the measured position. After the calibration, the overall energy spectrum improved as shown in Fig. 18 (right). The photopeaks became higher and narrower and the peak-

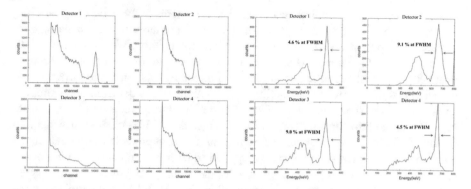

Fig. 18 Energy spectra measured by four CdZnTe detectors before (left) and after (right) correction for ^{137}Cs [34]

to-Compton ratio defined by the height of the photopeak divided by that of the Compton continuum increased. The energy resolutions of corrected spectra measured by detectors 1–4 are 4.6, 9.1, 9.0, and 4.5%. These numbers improved by 20% on average. However, the low-energy region of each spectrum was suppressed because threshold voltage of cathode should be set sufficiently high enough to cut off 15 kHz ripple noises in the cathode output. This noise may be attributed to low current and incomplete rectification of the power supplier the Rena mini-short, and hence, an external high voltage supplier may solve this problem.

3. Compton Imaging

To reconstruct the original source distribution precisely, the MLEM method (cf. Eq. 2) is generally applied for Compton imaging because the amount of γ-rays in most environmental conditions is not sufficiently large enough to use analytic reconstruction methods such as UBP and FBP. MLEM method can minimize the artifact attribute to the uncertainty which is inversely proportional to the square root of the amount of the measured counts.

$$\lambda_j{}^{n+1} = \frac{\lambda_j{}^n}{\sum_i^N c_{ij}} \sum_i^N \frac{Y_i c_{ij}}{\sum_k^M c_{ik}\lambda_k{}^n} \tag{2}$$

$\lambda^n{}_j$ and $\lambda^{n+1}{}_j$ are the pixel values of the source plane before and after n^{th} iteration, respectively. The size and direction of the source plane can be arbitrarily defined by users considering experimental space and calculation time. The initial pixel value $\lambda^0{}_j$ should be positive to apply MLEM algorithm correctly. N is the total number of used events in the reconstruction. c_{ij} is a system matrix composed of probability that a photon radiated from source pixel j is projected to the ith event. The event i is determined by the position and energy information of sequential γ-ray interactions in the CdZnTe detector. Y_i is the total number of counts recorded for event i [13]. In the case of Compton image reconstruction, the amount of possible combinations including the spatial and energy information of scattering and absorption in the CdZnTe detector is excessively large to be calculated by a commercial computer and, thus, list-mode MLEM method, which determines all measured events are unique (i.e., $Y_i = 1$) and the system matrix of each measured event can be calculated based on the angle of incident radiation to the surface of the detector, distance between interactions, attenuations in the detector, and Compton scattering angle (i.e., the Klein-Nishina formula) during reconstruction process [15, 16]. In addition to this conventional list-mode MLEM method, a weighting factor considering the angular uncertainty of the backprojected cone calculated based on each sequential event can be applied to improve angular resolution of the reconstructed image [24]. The total angular uncertainty $\Delta\theta_i$ of the backprojected cone calculated for each event is represented by Eq. (3) and Y_i is determined to the inverse of the total angular uncertainty as Eq. (4).

$$\Delta\theta = \sqrt{\Delta\theta_e^2 + \Delta\theta_g^2 + \Delta\theta_{Doppler}^2} \tag{3}$$

$$Y_i = \frac{1}{\Delta\theta_i} \tag{4}$$

where $\Delta\theta_e$, $\Delta\theta_g$, and $\Delta\theta_{Doppler}$ are the angular uncertainties attributed to energy uncertainty, finite position resolution, and Doppler broadening, respectively. The advanced list-mode MLEM applied to reconstruct Compton imaging is expressed by Eq. (5). A normalization factor $N/\sum_i^N Y_i$ is added to maintain the entire intensity distribution.

$$\lambda_j^{n+1} = \frac{N\lambda_j^n}{\sum_i^N Y_i \sum_i^N c_{ij}} \sum_i^N \frac{Y_i c_{ij}}{\sum_k^M c_{ik}\lambda_k^n} \tag{5}$$

Figures 19 and 20 demonstrate the reconstructed Compton imaging based on the measured data by using a monolithic CdZnTe detector. First, two ^{137}Cs sources (318.7 and 334 kBq) were positioned with various angles from each other to evaluate the angular resolution of the system. Second, ^{133}Ba (241.1 kBq), ^{22}Na (65.7 kBq), and ^{137}Cs (318.7 kBq) were measured simultaneously to estimate the selectivity for different isotopes. The distance between each source and the center of CdZnTe crystal was 22.3 cm, and all isotopes were measured for 8 h. In the case of two ^{137}Cs sources, the angle differences between sources were 20°, 40°, 60°, and 80° and the two sources were separately reconstructed for all angles. At 60° separation angle, the artifacts appeared at the middle point between the two sources because backprojected cones of each source were overlapped at the specific angle. The difference in positions between the reconstructed sources and the actual ones were 20% in average and 28% at maximum. In the case of the multiple sources measured simultaneously, each source was clearly reconstructed and separated from other sources based on the measured energy (cf. Fig. 20b–d). Figure 20a was the summed image of every source normalized by the maximum pixel value for each image.

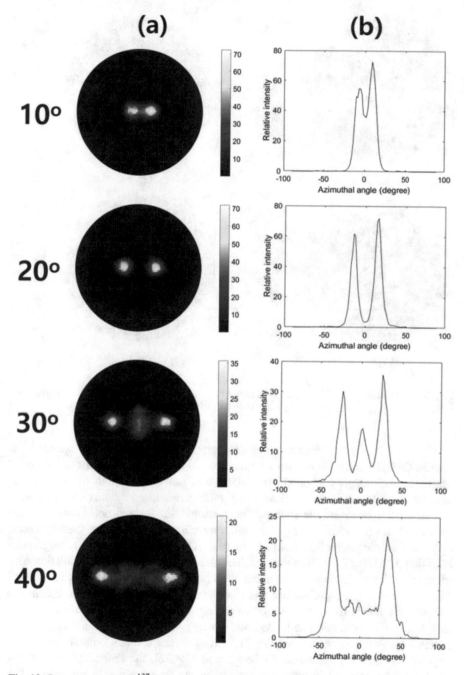

Fig. 19 Reconstructed two ^{137}Cs sources images and their cross sections on azimuthal angle in experiment (20th iteration) [34]

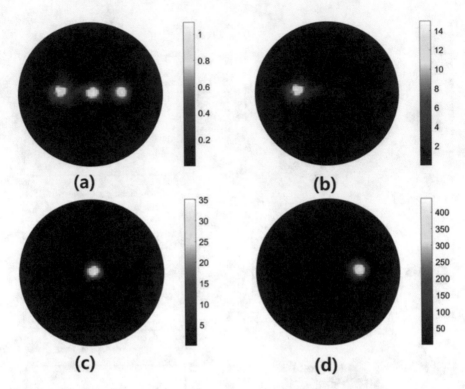

Fig. 20 Reconstructed multiple sources in experiment (20th iteration) [34]. (**a**) Three sources (**b**) [22]Na (511 keV), (**c**) [137]Cs (662 keV), and (**d**) [133]Ba (356 keV)

Figures 21 and 22 present simulation and experimental results based on a mosaic CdZnTe array. Since the distance between interaction positions of Compton scattering followed by photoelectric effect was not long enough to analyze the interaction sequence based on timing information measured by the current system, the sequential order of the multiple interactions was instead determined by the amount of the deposited energy for each interaction. To investigate the sequential order of multiple interaction, a simulation tool – Geant4 application for tomographic emission (GATE) v7.0. – was utilized. The detector structure of the simulation was identical to that of the experiment and a ^{137}Cs point source is positioned with 0°–90° offset. As a result, the probability that Compton scattering deposited more than the followed photoelectric effect was approximately 76% for all positions of the sources and, hence, the interaction deposited more energy was determined as the Compton scattering. As demonstrated in Fig. 21, if it is assumed that Compton scattering deposits more than photoelectric effect (i.e., E1 > E2), the reconstructed sources are located as planned in the simulation, whereas severe artifacts are present for reversed sequence (i.e., E1 < E2). Since spatial resolution in the x- and y-directions (5 mm) was significantly larger than that in the z-direction (0.75 mm), blurring in the reconstructed image was proportional to the offset angle. Figure 22 represents

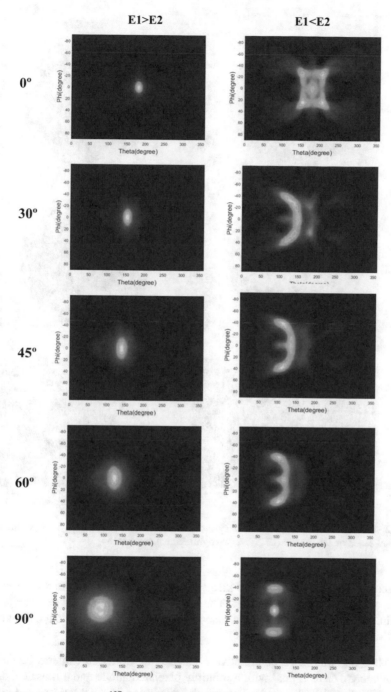

Fig. 21 Compton images of ^{137}Cs with offset angles in simulation (5th iteration) based on the correct (left)/reverse (right) sequential order [34]

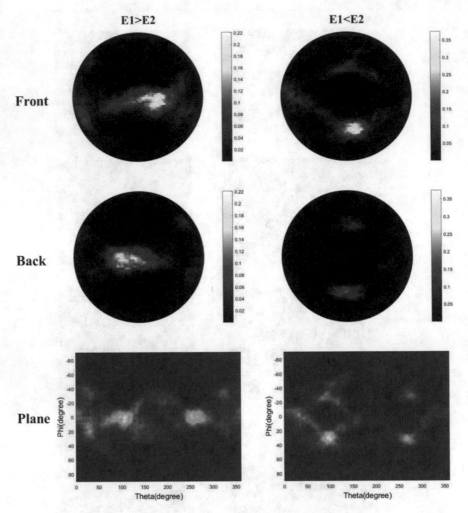

Fig. 22 The 4π and planar Compton images of ^{137}Cs with 90° offset angle in experiment (5th iteration) based on the correct (left)/reverse (right) sequence order [34]

the reconstructed images of a 650-kBq ^{137}Cs source located at 5 cm away from the center of the array with 90° offset. The measurement time of this experiment was 10 min. The artifacts in the reconstructed image were large owing to the limit of the effective events.

Table 5 summarizes a quantitative evaluation of Compton cameras based on a monolithic CdZnTe crystal with a multiply pixelated anode and a mosaic CdZnTe array with virtual Frisch grids. The angular resolution of the reconstructed images and intrinsic detection efficiency of Compton cameras based on a monolithic crystal are generally higher than those of a mosaic array, whereas the absolute detection

Table 5 Comparison of Compton cameras for a 662 keV radiation

System	Electrode	CZT size (pixel size)	Intrinsic efficiency for imaging	Angular resolution (iteration)
Polaris-H100 [35, 36] (H3D)	Pixelated anodes	$20 \times 20 \times 15$ mm^3	$\sim 2.00 \times 10^{-2}$	$\sim 20°$ (—)
Rena mini-long [33] (Kromek)	Pixelated anodes	$20 \times 20 \times 5$ mm^3 ($2.5 \times 2.5 \times 5$ mm^3)	6.70×10^{-4}	$10.18°$ (10)
Virtual Frisch-grid detector array [31] (BNL)	Virtual Frisch grids	$36 \times 36 \times 15$ mm^3 ($6 \times 6 \times 15$ mm^3)	9.40×10^{-3}	$26 \pm 1.30°$ (10)
Virtual Frisch-grid detector array [34] (Korea University)	Virtual Frisch grids	$10 \times 10 \times 12$ mm^3 ($5 \times 5 \times 12$ mm^3)	2.00×10^{-4}	$18.85°$ (5)

efficiency of a mosaic array comprising a large number of crystals can be higher than that of a monolithic crystal.

References

1. Carini, G. A., Chen, W., Geornimo, G. D., Gaskin, J. A., Keister, J. W., Li, Z., Ramsey, B. D., Rehak, P., & Siddons, D. P. (2009, Oct). Performance of a thin-window silicon drift detector X-ray fluorescence spectrometer. *IEEE Transactions on Nuclear Science, 45*(5), 2843–2849.
2. Pausch, G., Plettner, C., Herbach, C. M., Stein, J., Moszynski, M., Nassalski, A., Swiderski, L., Szczesniak, T., Niculae, A., & Soltau, H. (2009, June). Demonstration of a dual-range photon detector with SDD and LaBr3(Ce3+) scintillator. *IEEE Transactions on Nuclear Science, 56*(3), 1229–1237.
3. Marisaldi, M., Labanti, C., Soltau, H., Fiorini, C., Longoni, A., & Perotti, F. (2005, Oct). X- and gamma-ray detection with a silicon drift detector coupled to a CsI(Tl) scintillator operated with pulse shape discrimination technique. *IEEE Transactions on Nuclear Science, 52*(5), 1842–1848.
4. Oh, G., & Lee, W. (2010, Aug). Material investigation and analysis using characteristic X-ray. *Nuclear Engineering and Technology, 42*(4), 426–433.
5. Seller, P., Bell, S., Cernik, R. J., Christodoulou, C., Egan, C. K., Gaskin, J. A., Jacques, S., Pani, S., Ramsey, B. D., Reid, C., Sellin, P. J., Scuffham, J. W., Speller, R. D., Wilson, M. D., & Veale, M. C. (2011). Pixellated Cd(Zn)Te high-energy X-ray instrument. *Journal of Instrumentation, 6*(12), C12009.
6. Yoon, C., Kim, Y., & Lee, W. (2016, June). 3D non-destructive fluorescent X-ray computed tomography with a CdTe array. *IEEE Transactions on Nuclear Science, 63*(3), 1844–1853.

7. Jo, A., & Lee, W. (2018, May). Feasibility of fluorescent X-ray computed tomography to verify the homogeneity of Mo/Gd contained in nuclear fuels: A Monte Carlo simulation. *Nuclear Instruments and Methods in Physics Research A, 902,* 25–32.

8. Yoon, C., & Lee, W. (2014, Jan). Fluorescence X-ray computed tomography (FXCT) using a position-sensitive CdTe detector. *Journal of the Korean Physical Society, 64*(1), 122–128.

9. Dempster, A. P., Laird, N. M., & Rubin, D. B. (1977). Maximum likelihood from incomplete data via the EM algorithm. *Journal of the Royal Statistical Society. Series B: Methodological, 39*(1), 1–38.

10. Parra, L., & Barett, H. H. (1998, Apr.). List-mode likelihood: EM algorithm and image quality estimation demonstrated on 2-D PET. *IEEE Transactions on Medical Imaging, 17*(2), 228–235.

11. Wilderman, S. J., Fessler, J. A., Clinthorne, N. H., LeBlanc, J. W., & Rogers, W. L. (2001, Feb). Improved modeling of system response in list mode EM reconstruction of Compton scatter camera images. *IEEE Transactions on Nuclear Science, 48*(1), 111–116.

12. Yoon, C., & Lee, W. (2012, Aug). Advanced PET using both Compton and photoelectric events. *Journal of the Korean Physical Society, 61*(4), 626–629.

13. Lange, K., & Carson, R. (1984). EM reconstruction algorithm for emission and transmission tomography. *Journal of Computer Assisted Tomography, 8*(2), 306–316.

14. Shepp, L. A., & Vardi, Y. (1982, Oct). Maximum likelihood reconstruction for emission tomography. *IEEE Transactions on Medical Imaging, 1*(2), 113–122.

15. Parra, L., & Barrett, H. H. (1998). List-mode likelihood: EM algorithm and image quality estimation demonstrated on 2-D PET. *IEEE Transactions on Nuclear Science, 17,* 228–235.

16. Wilderman, S. J., Clinthorne, N. H., Fessler, J. A., & Rogers, W. L. (1999). *List-mode likelihood reconstruction of Compton scatter camera images in nuclear medicine.* In IEEE nuclear science symposium conference record, pp. 1716–1720.

17. Parra, L. C. (2000). Reconstruction of cone-beam projections from Compton scattered data. *IEEE Transactions on Nuclear Science, 47,* 1543–1550.

18. Pinkau, K. (1966). Die messung solarer und atmospharischer neutronen. *Zeitschr Fur Naturforschung Part A., 21,* 2100–2106.

19. White, R. S. (1968). An experiment to measure neutrons from the sun. *Bulletin of the American Physical Society Series II, 13,* 714.

20. Singh, M., & Doria, D. (1983). An electronically collimated gamma camera for single photon emission computed tomography. Part II: Image reconstruction and preliminary experimental measurements. *Medical Physics, 10,* 428–435.

21. Phlips, B. F., Inderhees, S. E., Kroeger, R. A., Johnson, W. N., Kinzer, R. L., Kurfess, J. D., Graham, B. L., & Gehrels, N. (1996). Performance of a Compton telescope using position sensitive germanium detectors. *IEEE Transactions on Nuclear Science, 43,* 1472–1475.

22. Schmid, G. J., Beckedahl, D. A., Kammeraad, J. E., Blair, J. J., Vetter, K., & Kuhn, A. (2001). Gamma-ray Compton camera imaging with a segmented HPGe. *Nuclear Instruments and Methods in Physics Research A, 459,* 565–576.

23. Du, Y., He, Z., Knoll, G. F., Wehe, D. K., & Li, W. (2001). Evaluation of a Compton scattering camera using 3-D position sensitive CdZnTe detectors. *Nuclear Instruments and Methods in Physics Research A, 457,* 203–211.

24. Lehner, C. E. (2004). *4π Compton imaging using a single 3-D position sensitive CdZnTe detector.* Ph. D. dissertation, Department of Nuclear Engineering and Radiological Sciences, University of Michigan, Ann Arbor, MI.

25. Bolotnikov, A. E., Camarda, G. S., Chen, E., Cheng, S., Cui, Y., Gul, R., Gallagher, R., Dedic, V., De Geronimo, G., Ocampo Giraldo, L., Fried, J., Hossain, A., MacKenzie, J. M., Sellin, P., Taherion, S., Vernon, E., Yang, G., El-hanany, U., & James, R. B. (2016). CdZnTe position-sensitive drift detectors with thicknesses up to 5 cm. *Applied Physics Letters, 108,* 093504.

26. Amman, M., Lee, J. S., Luke, P. N., Chen, H., Awadalla, S. A., Redden, R., & Bindley, G. (2009). Evaluation of THM-grown CdZnTe material for large-volume gamma-ray detector applications. *IEEE Transactions on Nuclear Science, 56*(3), 795–799.

27. Yücel, H., Birgül, O., Uyar, E., & Çubukçu, S. (2019). A novel approach in voltage transient technique for the measurement of electron mobility and mobility-lifetime product in CdZnTe detectors. *Nuclear Engineering and Technology, 51*, 731–737.
28. He, Z., Li, W., Knoll, G. F., Wehe, D. K., Berry, J., & Stahle, C. M. (1999). 3-D position sensitive CdZnTe gamma-ray spectrometers. *Nuclear Instruments and Methods in Physics Research A, 422*, 173–178.
29. Luke, P. N. (1995). Unipolar charge sensing with coplanar electrodes – Application to semiconductor detectors. *IEEE Transactions on Nuclear Science, 42*, 207–213.
30. McGregor, D. S., He, Z., Seifert, H. A., Wehe, D. K., & Rojeski, R. A. (1998). Single charge carrier type sensing with a parallel strip pseudo-Frisch-grid CdZnTe semiconductor radiation detector. *Applied Physics Letters, 72*, 792–794.
31. Lee, W., Bolotnikov, A. E., Lee, T., Camarda, G. S., Cui, T., Gul, R., Hossain, A., Roy, Y., Yang, G., & James, R. B. (2016). Mini Compton camera based on an array of Frisch-grid CdZnTe detectors. *IEEE Transactions on Nuclear Science, 63*(1), 259–265.
32. Bolotnikov, A. E., Camarda, G. S., Geronimo, G. D., Hossain, A., Giraldo, L. A. O., Yang, G., & James, R.B. (2019). *Position-sensitive Frisch-grid CdZnTe detectors for gamma-ray spectroscopy and imaging*. In Proceeding of SPIE, vol. 11114.
33. Kim, Y., Lee, T., & Lee, W. (2019). Radiation measurement and imaging using 3D position sensitive pixelated CZT detector. *Nuclear Engineering and Technology, 51*, 1417–1427.
34. Kim, Y., & Lee, W. (2020). Development of a virtual Frisch-grid CZT detector based on the Array structure. *Journal of Radiation Protection and Research, 45*(1), 35–44.
35. Wahl, C., Kaye, W. R., Wang, W., Zhang, F., Jaworski, J. M., King, A., Boucher, A. Y., & He, Z. (2015). The Polaris-H imaging spectrometer. *Nuclear Instruments and Methods in Physics Research A, 784*, 377–381.
36. H3D. H100 Gamma-Ray Imaging Spectrometer, [Online] Available: https://h3dgamma.com/H100Specs.pdf, 2019.

Coded Aperture Technique with CdTe Pixelated Detectors for the Identification of the 3D Coordinates of Radioactive Hot-Spots

Ioannis Kaissas

1 Introduction

Physical phenomena should be observed in detail, in order to reveal their causes. Knowledge of their causes leads to more sophisticated practices of observation and hence revealing more details of the phenomena. Modern pixelated detectors facilitate such observations with their fine structure, even for γ-rays, exploiting materials with high atomic number, such as CdTe and CZT. Monte Carlo simulations analyze the virtual experimental setup into case studies, giving the opportunity to finely tune each of its parameters, thus revealing details which would remain unobservable with the actual experimental setup, mostly due to the convolution of the parameters' fluctuations. In the following text, the evolution of a self-feed chain of experiments and simulations is described, in order to reveal technical details, but also the sequence and the coherence of the experiments and the simulations.

Visual spectrum cameras and γ-cameras share a common ancestor, the pinhole camera. As visual light can be focused with lenses and γ-rays do not, the pinhole camera remains at the disposal of γ-camera configurations even nowadays. In the late 1960s, for the sake of evolution, two astrophysicists [1, 7] captured the idea of arranging in a unique way many holes, instead of just one, in order to get more photons on the film of the γ-camera or telescope. Several types of coded aperture emerged through the years of research starting with Dicke's random array and continuing with mathematically sophisticated arrays, like the nonredundant array [12], the uniformly redundant array [9], the modified uniformly redundant array (MURA), [13] and many more. Four examples of coded apertures are presented in Fig. 1. Compared with collimators and pinholes, coded apertures boast about their

I. Kaissas (✉)
Greek Atomic Energy Commission, Patriarchou Grigoriou & Neapoleos, Agia Paraskevi, Attiki, Greece
e-mail: ioannis.kaissas@eeae.gr

© Springer Nature Switzerland AG 2022
K. Iniewski (ed.), *Advanced X-ray Detector Technologies*,
https://doi.org/10.1007/978-3-030-64279-2_11

Fig. 1 Four milestones in the history of coded apertures. From left to right: random array, nonredundant array, uniformly redundant array, and modified uniformly redundant array. The black areas represent the opaque material (e.g., lead), and the white areas represent the transparent material (e.g., air, PMMA)

wider field of view (FOV) combined with great efficiency, great spatial resolution, and accuracy of localizing radioactive sources. Imaging applications of ionizing radiation like X-rays, γ-rays, and neutrons have been developed in astrophysics, nuclear security [11], and nuclear medicine [2]. Also, coded apertures have been used in photography for deblurring out-of-focus regions [16].

1.1 Basic Principles of Coded Aperture Imaging

As mentioned before, coded aperture shall be considered the evolution of the pinhole aperture, as it comprises several pinholes arranged in a special pattern. The specialty lies in the fact that the arrangement of the holes is such that every different subarea of the coded aperture is unique at least by one hole. For instance, in Fig. 2 every single area of the coded aperture, encompassing 19×19 elements, differs from each of the other same-size areas at least by one element. That is, the case for the MURA coded aperture comprises 37×37 elements and with basic (unique) pattern 19×19 elements. The term elements replaces the term holes, because instead of drilling holes on a lead plate, one can place lead elements on a PMMA plate. The benefit of such replacement is that the transparent surface of the coded aperture increases and consequently the photons' flux reaching the pixelated detector increases, allowing shorter photons collection time.

The geometrical combination of a coded aperture with a pixelated detector, particularly the distance between them and their sizes, defines the size of the field of view (FOV). When a radioactive source is placed in the FOV and irradiates the coded aperture-detector pair, a fraction of the coded aperture is projected onto the detector surface forming a 2D image known as shadowgram. Subsequently, the shadowgram is cross correlated with a predefined decoding function, known as G matrix, which has the same configuration of elements with the actual coded aperture [8, 10]. This process is generally referred to as deconvolution. As a result, the direction of the source is revealed as a peak on the correlation matrix. The process is depicted in Fig. 2 for one and two radioactive sources. Taking into account that only

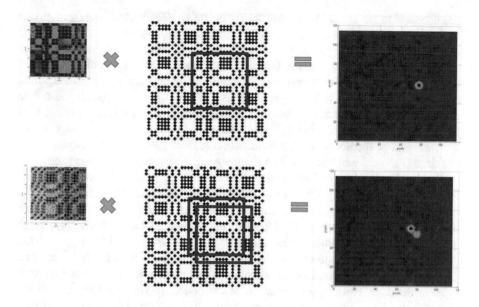

Fig. 2 Top: The projection or shadowgram (left) of a point source on CdTe pixelated detector, through the coded aperture MURA 37 × 37 (center), matches only with the specific surface framed by the red square. The correlation of the shadowgram with the whole coded aperture pattern results in a correlation matrix (right) with a distinct peak known as single point spread function. This peak points out the direction of the source into the fully coded field of view (FCFOV) of the γ-camera. **Bottom:** The same situation for two point sources. The shadowgram is a blurred superposition of the projection of the two sources; nevertheless, through the correlation, two distinct peaks appear on the correlation matrix

the basic pattern (i.e., 19 × 19 elements for the 37 × 37 MURA) has to be projected onto the pixelated detector, the physical dimensions of the detector can be about a quarter of the size of the coded aperture. This is the case of in-focus imaging. When the sources under investigation placed far away from the γ-camera and hence no magnification take place (far-field imaging), the number of elements of the basic pattern determines exactly how much of the quarter. For the 37 × 37 MURA coded aperture, its surface has to be $(37/19)^2 = 3.8$ times the surface of the detector.

The magnification of the shadowgram is another crucial factor affected by the distance between the coded aperture and the pixelated detector, when the near-field imaging comes to the fore (i.e., the ratio: coded aperture-detector distance/source-coded aperture distance is not negligible). Considering the invariable size of the pixelated detector, as the source approaches the γ-camera, the magnification of each element of the coded aperture on the detector increases, and thus the number of the projections of elements consisting the shadowgram decreases, resulting in shadowgram with less elements than the basic pattern has. The correlation of such a shadowgram yields a correlation matrix with undesired side lobes beside the main peak on the correlation matrix. The first step in the development of a γ-camera is the decision to be taken about how far away the sources under investigation will be and then regulating the geometry of the coded aperture-detector pair, in order to obtain clear correlation matrices with no side lobes. This is the optimum distance of the source, and we can name the imaging in-focus. One can think some tricks to extend the range of this optimum distance of the sources under investigation, for instance, choosing quite larger detector for sort optimum distance and then cropping the shadowgram in the case of far placement of the source. Another trick is to increase the element pitch (i.e., the interval between neighbor elements) as the source is placed further away. A key limitation to these games is the matching of the digitized G matrix with the digitized shadowgram during the correlation process.

If each element of a coded aperture does not occupy the entire surface available for the element, but only a part of it, the coded aperture is characterized as NTHT (not two holes touching) in the case of holes for elements or NTOT (not two obscures touching) in the case of opaque elements, i.e., there is always a blank area between two neighbor elements. The MURA example in Fig. 2 is a NTOT coded aperture. In a close-up look, no element is in touch with the neighbor one. This special configuration is used when the segmentation of the pixelated detector is more delicate than the segmentation of the coded aperture and the size of the elements can be reduced to the point their projection on the detector covers at least one pixel. The imaging capacities of these coded apertures remain ideal (as long as they are ideal for the original coded aperture), or even improved, as the G matrix is modified accordingly in a matrix of three levels [10]. A specific advantage is the improvement of the spatial resolution, insomuch the full width at half maximum of the point spread function shrinks while the element size is decreased and also with the use of the deconvolution technique δ-decoding [10].

1.2   Figures of Merit in Coded Aperture Imaging

In order to evaluate the imaging capacities of a γ-camera with coded aperture, some figures of merit are needed. They derived mostly from the correlation matrix or are in fact a special correlation matrix. In bibliography one can find some variations and also some more specific figures of merit that serve better their imaging setup.

The autocorrelation function (ACF) is the correlation matrix produced by the correlation of the basic pattern with the coded aperture. For ideal coded apertures, the result will be a δ-function on a correlation matrix without background noise. This figure of merit evaluates the coded aperture design and the correlation process. For no ideal apertures, there are some modifications that can be applied on the G matrix in order to diminish the background noise [9].

The point spread function (PSF) or single point spread function (SPSF) is the correlation matrix produced when a single point source irradiates the γ-camera. In the case of ideal experimental setup (i.e., coded aperture, pixelated detector, and point source), the result will be also a δ-function. This figure of merit evaluates the coded aperture in combination with the pixelated detector.

The point-source localization accuracy (PSLA or SLA) is the difference between the actual and the detected coordinates of a point source in the FOV of the γ-camera. The PSLA is related directly with the accuracy of calculating the exact coordinates of the peak of the PSF on the correlation matrix. The calculation is usually conducted with fitting methods on the PSF, in order to find its mathematical (e.g., 3D Gaussian) function.

The spatial resolution (SR) or angular resolution (AR) is the minimum distance (or angle) between two sources for which they appear as two distinctive peaks on the correlation matrix. As they are approaching each other, it is just before their two peaks joint into one common peak. Each peak on the correlation matrix is characterized by its full width at half maximum (FWHM). If two peaks with the same height and FWHM approach each other closer than their FWHM, they become indistinguishable. According to Caroli et al. the AR (i.e., the FWHM) is the projection of an element of the coded aperture from the detector to the infinity [6]. In practice, the SR is better when the elements of the coded aperture are smaller, because the surface of their projection on the shadowgram is smaller. Also, the SR is better when the coded aperture-detector distance is increased, because the FOV is decreased and the details of the imaging depicted larger on the correlation matrix, the analogous of zoom-in with an optical camera.

The sensitivity of a γ-camera for a certain photon-energy window W can be determined experimentally accumulating for time T the counts C_W on the pixelated detector from a source with well-known activity A and the counts Bg_W accumulated on the detector for the same time T, with the absence of the source (i.e., only the background):

$$\text{Sensitivity (W)} = \frac{C_W - Bg_W}{T \bullet A} \qquad (1)$$

The greater the ratio of transparent to opaque ratio of the coded aperture and the wider the FOV, the better the sensitivity of the γ-camera.

The signal-to-noise ratio (SNR) is the counts of a source peak on the correlation matrix (e.g., the height of a PSF minus the height of the background noise) to the standard deviation of the background noise of the correlation matrix.

$$\text{SNR} = \frac{\text{Peak Height} - \text{Mean Value of Noise}}{\text{Standard Deviation of Noise}} \qquad (2)$$

Also, other definitions of the SNR can be found in the bibliography, as the case-specific definition might suit the characteristics of the imaging. For instance, in near-field applications, Accorsi et al. include the effect of transparency of coded aperture [2]. SNR is sensitivity dependent and determines the minimum A for which a source is detectable by the γ-camera. This value is usually SNR $\cong$ 10 [14], but not strictly, as it is related to the specific imaging application or experiment. The larger the accumulating time, the better the SNR of the imaging. Increasing the source – γ-camera distance, in favor of spatial resolution and SLA, reduces the photons' flux

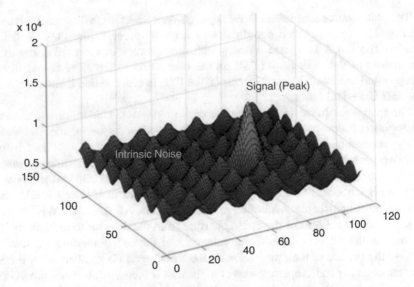

Fig. 3 A correlation matrix with the characteristic peaks and lows of the intrinsic, element-wise noise. The tall peak is the PSF, and its diameter along with the diameter of the peaks and valleys of the intrinsic noise are proportional to the diameter of the coded aperture elements

on the detector, but the implied fluctuations in the peak's height and in the standard deviation of noise are the main factors which will determine whether the SNR will go up or down (Figs. 3 and 4).

The noise or the fluctuation of the background is due to the following:

- The radiation background or other sources outside the FOV
- Other sources inside the FOV
- The imperfections in design and construction of the coded aperture, the pixelated detector, and their geometrical combination
- The deconvolution procedure
- The Poisson noise of the photons reach the detector
- The side lobes in the correlation matrix, which in turn are due to:

  - The extension and the position of the radioactive source
  - The case out-of-focus imaging
  - The deconvolution procedure

The correlation matrix (and the ACF) for a NTOT MURA 37×37 contains a systematic, element-wise noise, i.e., intrinsic noise that lays on the surrounding background of the PSF. Figure 3 shows the details of its shape, which reminds an egg cartel. Its small peaks and valleys have diameter proportional to the size of the elements of the coded aperture. Section 3.3 focuses on this intrinsic noise appeared on the correlation matrix, for these particular coded apertures. The rest of the background noise is defined as random noise.

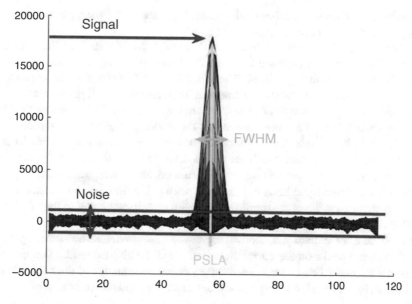

Fig. 4 The basic figures of merit depicted on the face of a correlation matrix with a peak corresponding to a source. SNR is the signal-to-noise ratio

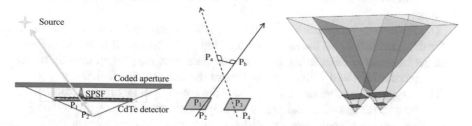

Fig. 5 Left: The correlation matrix with the SPSF projected on the CdTe detector. The P_1 is the center of the peak and P_2 is the vertex of the upside-down virtual pyramid which confines the coded aperture and the detector. The line P_1P_2 points to the source that irradiates the γ-camera. **Middle:** Each γ-camera finds independently the direction of the source. The lines P_1P_2 and the P_3P_4 from the γ-camera pair converge to the location of the source. The middle of the P_aP_b is the estimation of the location of the source. **Right:** The common field of view of the two γ-cameras system (FOV_S) is the common place of the FOV of each γ-camera

1.3 The Parallax Phenomenon

Let me assume that most living creatures are granted two eyes, in order for them to distinguish whether an object is close or far away. The phenomenon is called parallax and a simple method that translates it in mathematical terms called triangulation. The combination of two γ-cameras with coded apertures and the triangulation technique localizes with quite high accuracy radioactive sources or hot-spots in the 3D space.

When two straight lines converge in 3D space, there is a straight-line segment, which is the minimum distance between the two converging lines and is perpendicular to these two straight lines. Therefore finding this straight-line segment of the minimum distance between the lines can be reduced to finding the straight line perpendicular to these two lines. A convenient way is described by Bourke Paul [4].

In the case of more than one source into the FOVs, a PSF for each source arises on the correlation matrix. Consequently, a pair of convergent lines points to each source. In order not to pair two lines which do not belong to the same source, the minimum distances between all lines are calculated, and consequently the long ones are rejected with the hypothesis that they are not a pair pointing to a real source. The rejection terminates when the number of remaining pairs are equal to the number of the PSFs on the correlation matrix.

In the case of extend radioactive sources, i.e., radioactive hot-spots spatially distributed, the peaks on the correlation matrix will be also extended. Despite that, the points P_2 and P_4 can be considered the central coordinates of the peaks; hence the lines P_1P_2 and P_3P_4 converge to the center of the spatial distribution of the hot-spot.

Two of the aforementioned in Section 1.2 figures of merit can be defined also in 3D localization. In this case, the coordinates are the basis of the mathematical definitions.

The 3D source localization accuracy ($|R|$) is the absolute value of the vector defined by the actual and the estimated location of the source. It is directly related with the accuracy of calculating the exact coordinates of the peak of the PSF on the correlation matrices and with the deviations of the triangulation technique.

The 3D spatial or voxel resolution (VR) is the absolute value of the vector defined by the 3D coordinates of a source and its neighbor, insomuch they are so close to each other that the triangulation can distinguish them as two separate sources.

2 Materials and Methods

In the following two sections, an evaluation paradigm of a system of two γ-cameras is presented. The scope of this evaluation was the evolution of the system, in a manner that described in the Introduction's first paragraph. The system flirts with intraoperative nuclear medicine applications, nuclear security identifications, and imaging in decommissioning of nuclear facilities. In previous works, we developed a system of two γ-cameras (Fig. 6) that exploits the parallax phenomenon and combines the two derived directions with a triangulation method, in order to find the 3D coordinates of the radioactive sources [18]. Particular interest emerged for the in-focus and out-of-focus imaging capacities of the system [15] and for a "strange" reduction of the noise for extended hot-spots that lead to the maximization of the

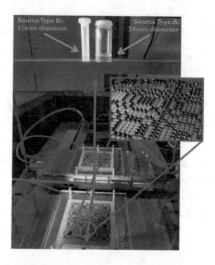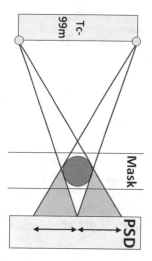

Fig. 6 Left: The two CdTe γ-cameras and the two types of ^{99m}Tc hot-spots with different diameter. **In blue frame**: A close-up detail of NTOT MURA mask. Elements are Pb spheres arranged on a transparent acrylic plate. **Right**: The main shade of the spherical element tends to be eliminated and the penumbra becomes larger insomuch the dimensions of the source increase. For the effect on the shadowgram, see Fig. 9 shadowgrams

Table 1 Specifications of the MURA coded apertures which supply optimum imaging for source distance from the detector 16 cm and 31 cm

Coded aperture type	Number of elements	Surface (mm^2)	Element pitch (μm)	Optimum source distance (cm)
19R-1821	37 × 37	67.4 × 67.4	1821	16
19R-1958	37 × 37	72.5 × 72.5	1958	31

SNR. Both experiments and simulations refine the dependences of the 3D SLA, the spatial resolution and the SNR.

Each γ-camera has an active area of 44 × 44 mm^2 consisting of eight hybrids. The hybrid pixelated detectors are CMOS ASICs with 1-mm-thick CdTe crystals bump bonded. The interval between two neighbor pixels, i.e., the pixel pitch, is 350 μm and the recording rate is 27 frames/s. For these CdTe detectors, the energy resolution of each pixel is 3–4 keV FWHM for the ^{99m}Tc photopeak of 141 keV.

A NTOT MURA coded aperture is placed parallel to the CdTe surface at 2 cm distance from it, on top of each γ-camera. Table 1 presents the specifications of two different types of MURA coded apertures which were constructed with the purpose to localize hot-spots at short and medium distances from the γ-camera. Their basic pattern consists of 19 × 19 elements, and they differ basically in the element pitch, i.e., the interval between two neighbor elements, which determines the optimum source distance (from the detector), for which a point source produces a PSF on the correlation matrix without side lobes. For each of the two coded apertures, the corresponding optimum source distance was used.

Table 2 The characteristics of the sources which are used in the actual experiments as extended hot-spots

Source type	Type A	Type B	Type C
Isotope	Tc-99m	Tc-99m	Co-57
Activity (MBq)	~1.5	~1.5	1.5
Photopeak (keV)	141	141	122
Diameter of cylinder (mm)	24	11	30
Height of cylinder (mm)	9	9	32
Volume (cm^3)	4.072	0.855	22.619

For the experiments with the actual system, two cylindrical glass containers with ^{99m}Tc radioactive uniform solution and a cylindrical uniform gel solution of ^{57}Co, 1.5 MBq activity each, acting as radio-traced sentinel node dummies, were arranged separately in several positions (X, Y, Z) within the FOV$_S$. Table 2 collects the specifications of these sources.

2.1 Simulation

Simulation code was developed in MATLAB to investigate several geometric setups and factors affecting the performance of the actual system. In order to be fast, the code does not consider scattered photons. The comparison with a detailed simulation which considers scattered photons confirms that this intentional omission does not affect the resulting shadowgrams significantly especially in the cases where photoelectric phenomena prevails Compton scattering.

The simulation procedure is comprised by a main body and three subroutines. The main body takes into account the geometric characteristics of the experimental setup (e.g., coded aperture-detector distance, distance between the two detectors), digitizes the selected coded aperture and the G matrix, and calls the subroutines. The first subroutine generates the shadowgram of each γ-camera. Photons are emitted from point sources or hot-spots with random direction limited by the cones which encompass the two coded apertures, and their vertices have the coordinates of the emission points. Photons reaching a transparent region of the coded aperture generate a hit to the corresponding detector pixel. Otherwise, attenuation due to the presence of the opaque elements of the coded apertures is taken into account and a transmission probability is assigned to the photon. In this case, if each photon passes through, it is recorded to the corresponding pixel detector. In this way, the accumulation of the photon counts in every pixel generates the shadowgram. In the second subroutine, the shadowgram undergoes a cross correlation with the G matrix, fitting the correlation matrices with Gaussian 3D distributions, in order to yield the lines in 3D space which point to the direction of the photon emission. In the third subroutine, the estimated source directions are input to the triangulation algorithm, which calculates the 3D coordinates of the "intersections" of the directions. Finally, the main routine collects the results in a datasheet and presents the estimated and the virtual coordinates of the sources for comparison.

Successive simulation experiments can be performed, in order to have the figures of merit's statistics, like the standard deviation, the skewness and the kurtosis of the SLA or the degradation of the SNR as the source move away from the γ-cameras. The results are used to recognize "strange behaviors" like the maximization of the SNR that is analyzed in Section 3.3.

Specifically, the simulation code is used in order to produce shadowgrams of simulated hot-spots with various spatial distributions as presented in Fig. 9, SNR plot. The first subroutine generates a hot-spot of uniform or normal spatial distribution with dimensions from 0 mm to 50 mm or 85 mm, respectively, and accumulates the shadowgram. The simulation experiment was repeated five times, in order to obtain enough statistics, for every incremental step of the hot-spot dimension.

The second part of the code can be fed also with shadowgrams captured by the actual experimental setup. In this manner of operation, the code works as autonomous imaging code.

3 Results and Discussion

An anthology of results is presented in this section, derived from simulations of virtual experiments and real experiments with the actual γ-cameras. Three main studies are unfolded: the 3D SLA, the spatial resolution, and the SNR maximization phenomenon along with a proposition for kernel filtering, which reduces the intrinsic noise on the correlation matrix.

3.1 The 3D Source Localization Accuracy (3D SLA)

The two γ-cameras system was simulated estimating the 3D coordinates of a source placed at several distances from the detectors plane, in order to study the deterioration of the 3D SLA. Figure 7, left, shows that the |R| becomes minimum (maximum accuracy) for the optimum source to detector distance, i.e., in-focus imaging. In accordance with the experimental data, |R| deteriorates for source distance higher than 160 mm, when coded aperture type 19R-1821 is used, because then the shadowgram comprises more than 19×19 elements.

The |R| dependence on the standard deviation of the hot-spot extension is shown in Fig. 7, right. For coded aperture type 19R-1821, |R| deteriorates when the source dimension becomes larger than 25 mm. This happens because the peak on the correlation matrix becomes too wide and the fitting procedure cannot find accurately its center.

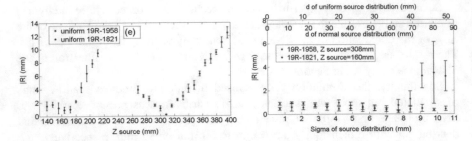

Fig. 7 **Left**: The 3D SLA (|R|) dependence on the source distance from the detector (Z_{source}) for the 19R-1821 and 19R-1958 coded aperture type. In-focus imaging stands for 160 mm and 308 mm where the best SLA is achieved, i.e., the minimum of the curves, for the 19R-1821 and the 19R-1958, respectively. For lower or higher Z_{source}, the imaging turns into out-of-focus imaging and |R| is degraded. **Right:** The |R| dependence on the sigma of the source with uniformly distributed activity for coded aperture type 19R-1821 and coded aperture type 19R-1958

3.2   Spatial Resolution of a Two γ-Cameras System for Extended Sources

With the purpose to study the spatial resolution of the two γ-cameras system, the shadowgrams produced by placing the same source at two neighbor positions are summed into one which undergoes the correlation procedure. The experimental correlation matrices for sources of type A and C are presented, because they are more extended and the effects due to this fact are more pronounced.

For a given coded aperture, as it is mentioned in Section 1.1, when the sources are not at the optimum distance from the detectors plane, the resolution degrades and side lobes appear beside the peaks. In addition, the presence of the side lobes of one source may coincide with the main peak of the other and in this way with its FWHM, resulting in the degradation of the spatial resolution. This can be seen in Fig. 8b and c, as the volume and the lateral distance of the sources differ in such a way that their apparent angle remains constant although the height Z of the sources is different.

With the two sources placed at 308 mm height and the use of the 19R-1958 coded apertures, the FWHM of the peaks on the correlation matrix decreases, so they become sharper, due to the smaller magnification of the elements. The sharpening of the peaks is obvious in the comparison of Fig. 8a with Fig. 8b.

Implementing the triangulation as the final step of the 3D localization procedure for the case of Fig. 8b results in the graphical representation of the 3D coordinates of the two hot-spot centers. In this manner the 3D spatial resolution can be derived.

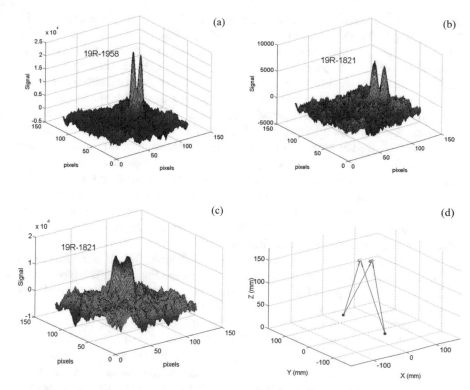

Fig. 8 The correlation matrix for two cylindrical type A sources positioned at the center of FCFOVs, (**a**) at Z = 308 mm and with distance between their centers 42 mm, (**b**) at Z = 160 mm and with distance between their centers 28 mm. (**c**) The correlation matrix for two cylindrical type C sources positioned at Z = 231 mm and with distance between their centers 43 mm. (**d**) The experimental 3D localization of the centers of two type A sources for the case described in (b) with accuracy <5%, after the triangulation. In each graph, the employed coded aperture is indicated

3.3 The Intrinsic Noise

In the current subsection, we focus on the systematic, element-wise noise appearing on the correlation matrix that we call intrinsic noise of the coded aperture pattern (Fig. 3). The rest of the background noise is defined as random noise. The explanation of the origin of the intrinsic noise is attempted in the following text. Also, the optimum combination of element size and source-detector distance, which minimize the intrinsic noise for extended hot-spots with certain dimensions, is presented. Finally, an image filtering method that reduces the intrinsic noise is emerged for the imaging of point sources.

The diameter of the intrinsic noise structure (peaks and valleys in Fig. 3) is proportional to the diameter of the projection of the elements on the pixelated CdTe detector, and consequently it is also proportional to the FWHM of the PSF of the correlation matrix. As the element pitch is increased, the wavelength of the periodic structures of the shadowgram also increases. The SNR of the correlation matrix becomes maximum (Fig. 9, SNR plot) for a certain size of extended hot-spot, for which the penumbra blurs the shade of each single coded aperture element. As far as the extension of the hot-spot eliminates, via the penumbra, the separate projection of each element on the shadowgram (it can be seen in the comparison between the left and the middle shadowgram of Fig. 9), but it does not deteriorate the basic pattern of the coded aperture (this deterioration can be seen in the comparison between the middle and the right shadowgram of Fig. 9), the SNR of the correlation matrix is improved. A further extension of the hot-spot dimension deteriorates the basic pattern on the shadowgram, by hiding the details in the middle range of the spatial frequencies, which are essential for a good correlation with the G matrix.

The 2D FFT of the shadowgram shows clearly the elimination of the high spatial frequencies of each element's projection. Further extension of the hot-spot eliminates also the medium spatial frequencies of the basic pattern of the shadowgram and consequently deteriorates the SNR.

A detailed comparison of experimental data with simulation is difficult, due to the deviations in the activity and the positioning of the hot-spots. The type A and type B sources were placed in the FOV for 184 s each. The derived correlation matrices are presented in Fig. 10. The deviations in the activity and the positioning of the hot-spots slip into the SNR, preventing the verification of the simulation results in the case of small changes of the SNR. For these cases, an alternative and more simplified parameter is used, the standard deviation of the background of the correlation matrix. However, the quite large difference in the SNR of the shadowgrams recorded from two different types of sources with the coded aperture 19R-1821 is in agreement at least qualitatively with the conclusions drawn from the analysis of the simulation results. The type B source has a radius of 5.5 mm and consequently lays on the region of the peak SNR. The SNR for this source is 32. The type A source has a radius of 12 mm and lays on the region of descending SNR right to the peak. The SNR for this source is 20.

The standard deviation of the background of the correlation matrix, which includes the variation of the intrinsic noise, is slightly smaller for the type A source than for the type B source, when the coded aperture 19R-1958 is used. For this combination of optimum source distance, spatial dimensions of hot-spots and element-pitch, the type A source lays on the region of the peak of the SNR, and the type B source lays on the region of its left descending (Fig. 9, SNR plots). The standard deviation of the background of the correlation matrices presented in Fig. 10 is 333 for the type A source and 337 for the type B source, which is not a significant difference. However, the peaks and valleys of the intrinsic noise, like the ones appearing in Fig. 3, are evident on the background of Fig. 10a, while they are eliminated on the background of Fig. 10b.

Following the existence of the two components in the background, the fast Fourier transform of the correlation matrices also has the two components: (1) the intrinsic and (2) the random noise signal for each spatial frequency. The elimination of the peaks and valleys should appear as a reduction of the signal of their specific spatial frequencies, even though the random noise of these frequencies still exists.

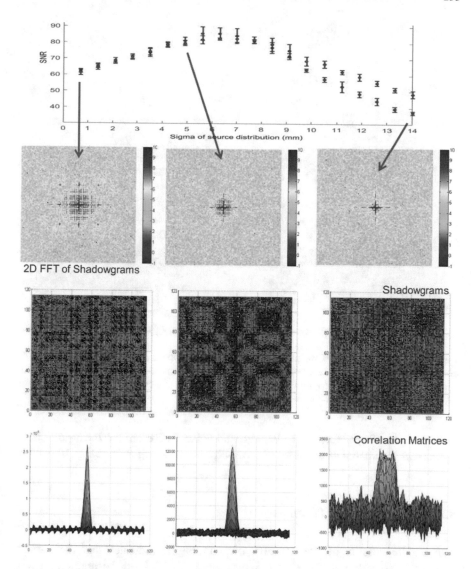

Fig. 9 SNR plot: The signal-to-noise ratio of the correlation matrix as function of the extent of the hot-spot, i.e., the standard deviation (sigma) of the source distribution. Blue: uniform distribution. Red: normal distribution. All the simulations performed with the coded aperture 19R-1821 and the hot-spot placed at 160 mm distance from γ-camera. The error bars stand for the standard deviation of the results of five-time repeated simulation experiments. **2D FFT plots**: In every column (top to bottom): the XY projection of the fast Fourier transform (FFT) of the shadowgram, the XY projection of the actual shadowgram, the YZ projection of the correlation matrix. In every row (left to right): Extended Tc-99m hot-spot with 0.7 mm, 5 mm and 14 mm standard deviation

Indeed, the side lobes from Fig. 10c and d are reduced and specifically a reduction of 5% (from 11.93 to 11.36) is evident for the signal with a spatial frequency correlated

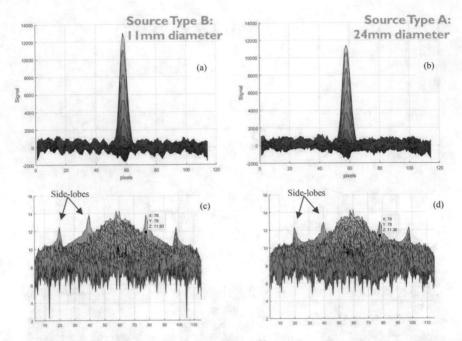

Fig. 10 Top: Experimental correlation matrix for the ^{99m}Tc type B and type A hot-spots, with the coded aperture 19R-1958 and the hot-spots placed at 308 mm distance from the γ-camera. **Bottom**: Their fast Fourier transforms with the indication of a spatial frequency which corresponds to the "wave" structure and the side lobes

with the "wave" structure of the background noise. Therefore the intrinsic noise is reduced as the spatial dimensions of the source become larger from Fig. 10a and b.

3.4   The Optimum Geometry

The middle column of the 2D FFT plots in Fig. 9 indicates that there exists a geometrical setup that yields the maximum SNR of the correlation matrix. Specifically, for source dimension around 2 cm and with the coded aperture 19R-1958, the source has to be placed around a distance of 30 cm from the γ-camera. If the coded aperture 19R-1821 is used, the source must be placed around a distance of 16 cm from the γ-camera and its dimension has to be around 1 cm.

In nuclear medicine and intraoperative imaging, the objects under investigation have certain spatial dimensions [8]. For example, the lymph nodes can be considered as spheres with diameter around 1–2 cm [9, 10]. Therefore, different geometrical configurations of the coded aperture γ-camera can be used for the imaging of the sentinel lymphs, or the thyroid residue, in order to achieve the best SNR possible.

3.5 The Kernel Filtering

The reduction of the intrinsic noise resulting from the increase of the dimensions of the hot-spot can be exploited to reduce this noise, also, in the correlation matrix of point-like sources, where this noise is the highest. Due to the periodicity of the peaks and valleys on the correlation matrix, scanning the whole surface of the correlation matrix with an appropriate kernel filter should reduce the intrinsic noise. Since the dimension of the peaks and valleys are proportional to the FWHM of the PSF of the correlation matrix, this filter (Fig. 11) is chosen to be the cropped and normalized peak of the autocorrelation function (ACF) of each coded aperture. For the 19×19 basic pattern NTOT MURA, pixel size, it is a matrix of 6 by 6 pixels. The result is the slight expansion of FWHM of the peak, as the peak undergoes a convolution with the normalized, cropped ACF and the significant reduction of the intrinsic noise on the correlation matrix, as one can see in Fig. 11.

In many cases of astronomy or security applications of coded apertures, the sources under investigation are considered point-like. The aforementioned filtering

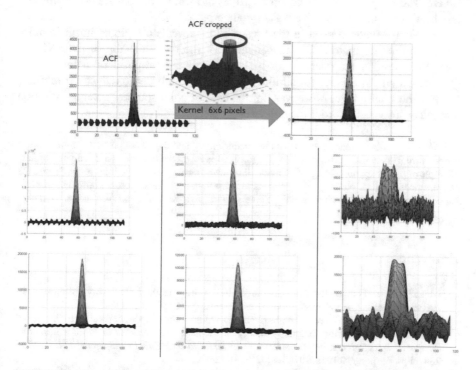

Fig. 11 The kernel filter consists of the cropped normalized peak of the ACF. Its application on the correlation matrix reduces the intrinsic noise: Top: kernel filter is applied on the ACF. In each column from left to right: kernel filter is applied on the correlation matrix of extended Tc-99m hot-spot with 0.7 mm, 5 mm, and 14 mm standard deviation

can be useful for smoothing the background of the correlation matrix, which includes one or more peaks, related to point-like sources.

4 Summary

Coded aperture γ-cameras reveal characteristics unbeatable by the collimators and pinholes ones, such as the wide FOV, the great sensitivity, and the fine spatial resolution. Exploiting the parallax phenomenon with two γ-cameras finds the 3D coordinates of radioactive hot-spots located into the FOV.

An important adjustment for coded aperture γ-cameras is the geometric configuration of the aperture and the pixelated detector, in order for the imaging to be in-focus, i.e., without side lobes in the correlation matrix. The spatial resolution is strongly affected by the distance of the source from the γ-camera in the near-field applications, due to this effect.

NTOT coded apertures are more luminous in terms of γ- or X-rays than the NTHT. The proposed NTOT coded apertures are simple in construction terms and less heavy than their respective NTHT counterparts.

The MURA pattern used in the present work reveals a periodic intrinsic noise on the correlation matrix. The investigation of the maximization of the SNR for spatially extended radioactive hot-spots leads to the proposal of using certain geometrical characteristics for similar sized sources, in order to reduce the intrinsic noise in the final image. Moreover, for the case of point sources, a kernel filtration method is proposed in order to achieve better SNR.

Acknowledgments A team's effort for the understanding and the development of an innovative coded aperture system commenced 9 years ago. This team was the environment for my research in the topic of coded aperture imaging. I thank all of them for the patient, the joint research, and the conversations we had. Namely, many thanks to Haris Lambropoulos, Costas Potiriadis, Christos Papadimitropoulos, Costas Karafasoulis, Yiannis Kazas, Aris Kyriakis, and Alexandros Clouvas.

References

1. Ables. (1968). Fourier transform photography: A new method for X-ray astronomy. In *Proceedings of the Astronomical Society of Australia* (pp 172–173).
2. Accorsi, R., Gasparini, F., & Lanza, R. C. (2001). Optimal coded aperture patterns for improved SNR in nuclear medicine imaging. *Nuclear Instruments and Methods in Physics Research, Section A: Accelerators, Spectrometers, Detectors and Associated Equipment, 474*(3), 273–284. https://doi.org/10.1016/S0168-9002(01)01326-2.
3. Accorsi, R., & Lanza, R. C. (2001). Near-field artifact reduction in planar coded aperture imaging. *Applied Optics, 40*(26), 4697–4705. https://doi.org/10.1364/AO.40.004697.
4. Bourke, P. (1998). *The shortest line between two lines in 3D*. Retrieved June 17, 2017, from http://paulbourke.net/geometry/pointlineplane/

5. Cannon, T. M., & Fenimore, E. E. (1979). Tomographical imaging using uniformly redundant arrays. *Applied Optics, 18*(7), 1052–1057. https://doi.org/10.1364/AO.18.001052.
6. Caroli, E., Stephen, J. B., Di Cocco, G., Natalucci, L., & Spizzichino, A. (1987). Coded aperture imaging in X- and gamma-ray astronomy. *Space Science Reviews, 45*(3–4), 349–403. https://doi.org/10.1007/BF00171998.
7. Dicke, R. H. (1968). Scatter-hole cameras for X-rays and gamma rays. *The Astrophysical Journal, 153*(2), 101–106. https://doi.org/10.1007/s13398-014-0173-7.2.
8. Fenimore, E. E. (1978). Coded aperture imaging: Predicted performance of uniformly redundant arrays. *Applied Optics, 17*(22), 3562–3570.
9. Fenimore, E. E., & Cannon, T. M. (1978). Coded aperture imaging with uniformly redundant arrays. *Applied Optics, 17*(3), 337–347.
10. Fenimore, E. E., & Cannon, T. M. (1981). Uniformly redundant arrays: Digital reconstruction methods. *Applied Optics, 20*(10), 1858–1864. https://doi.org/10.1364/AO.20.001858.
11. Gmar, M., Agelou, M., Carrel, F., & Schoepff, V. (2011). GAMPIX: A new generation of gamma camera. *Nuclear Instruments and Methods in Physics Research, Section A: Accelerators, Spectrometers, Detectors and Associated Equipment, 652*(1), 638–640. https://doi.org/10.1016/j.nima.2010.09.003.
12. Golay, M. (1971). Point arrays having compact, nonredundant autocorrelations. *Journal of the Optical Society of America, 61*, 272–273.
13. Gottesman, S. R., & Fenimore, E. E. (1989). New family of binary arrays for coded aperture imaging. *Applied Optics, 28*(20), 4344–4352. https://doi.org/10.1364/AO.28.004344.
14. Gros, A., Goldwurm, A., Soldi, S., Gotz, D., Caballero, I., Mattana, F., & Heras, J. A. Z. (2013). The IBIS / ISGRI source location accuracy. In *Proceedings of science* (pp. 1–6). Retrieved from http://arxiv.org/abs/1302.6915
15. Kaissas, I., Papadimitropoulos, C., Potiriadis, C., Karafasoulis, K., Loukas, D., & Lambropoulos, C. P. (2017). Imaging of spatially extended hot spots with coded apertures for intra-operative nuclear medicine applications. *Journal of Instrumentation, 12*(1). https://doi.org/10.1088/1748-0221/12/01/C01059.
16. Levin, A., & Freeman, W. T. (2007). Image and depth from a conventional camera with a coded. *Aperture, 26*(3). https://doi.org/10.1145/1239451.1239521.
17. Mu, Z., & Liu, Y. H. (2006). Aperture collimation correction and maximum-likelihood image reconstruction for near-field coded aperture imaging of single photon emission computerized tomography. *IEEE Transactions on Medical Imaging, 25*(6), 701–711. https://doi.org/10.1109/TMI.2006.873298.
18. Papadimitropoulos, C., Kaissas, I., Potiriadis, C., Karafasoulis, K., Loukas, D., & Lambropoulos, C. P. (2015). Radioactive source localization by a two detector system. *Journal of Instrumentation, 10*(12). https://doi.org/10.1088/1748-0221/10/12/C12022.

Positron Emission Tomography (PET) Imaging Based on Sub-millimeter Pixelated CdZnTe Detectors

Yongzhi Yin and Sergey Komarov

1 Positron Emission Tomography

The positron emission tomography (PET) has been widely used in both clinical diagnosis and animal research [1–6]. The emergence of PET/CT and PET/MR has taken the application of PET to new heights [7, 8]. With the in-depth use of PET in clinical research, higher requirements have been put forward for PET image resolution, system sensitivity, counting rate, and other metrics.

PET is a functional imaging device that reconstructs the spatial distribution of radiopharmaceuticals (e.g., Fluorodeoxyglucose F-18 Injection, FDG) in humans or animals and then converts it to medical images in three-dimensional images by a consistent measurement of a pair of outgoing (180°) 511 keV gamma ray photons emitted by positron annihilation. FDG is an F-18-labeled glucose molecule that is absorbed by the lesion via the human metabolism after injection into the human body. Because the tumor is abnormally active in glucose metabolism, its enrichment in tumor cells is higher than that in common tissue cells. Therefore, PET reflects the metabolism of organ lesions. PET instruments detect the two gamma ray photons from positron annihilations. Most of the positrons emitted by nuclides (e.g., F-18) are annihilated into two gamma ray photons; less than 2% of positrons are annihilated into three gamma ray photons. An illustration of the PET instrument is shown in Fig. 1.

The first PET used a NaI(Tl) scintillator as a gamma ray detector material, which was later replaced by BGO, LSO, and LYSO scintillator materials. Regardless of

Y. Yin (✉)
School of Nuclear Science and Technology, Lanzhou University, Lanzhou, China
e-mail: yinyzh@lzu.edu.cn

S. Komarov
Department of Radiology, Washington University, St. Louis, MO, USA

© Springer Nature Switzerland AG 2022
K. Iniewski (ed.), *Advanced X-ray Detector Technologies*,
https://doi.org/10.1007/978-3-030-64279-2_12

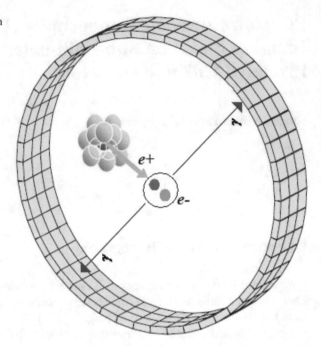

the type of detector, the core is a pair of gamma ray photons generated by positron
annihilation. Two opposite detectors that conform to the measurements give the
probability of positron annihilation interaction. The connection between this pair of
detectors is called the line of response (LOR). If the point emits multiple positrons,
the location of the positron annihilation reaction can be calculated in addition to the
LOR passing through the point in question. More counts make the statistical error
of a spatial distribution of positron drugs smaller.

The three main aspects that affect the performance of the PET detector are
positron range, accolinearity effect, and intrinsic spatial resolution of the gamma
ray detector. (1) Positron range is accompanied by a certain energy emission in
the human body, so that the positron will travel a distance before annihilation
occurs. This distance is not measurable in PET data acquisition, which limits the
spatial resolution of the PET device to some extent. The positron energy emitted
by individual radioactive sources is different. The maximum energy of the positron
emitted by the C-11 is 0.96 MeV (i.e., decay energy), while the positron range in
water is 2.1 mm. For F-18, the maximum energy of the positron emitted is 0.64 MeV,
and the positron range in water is 1.2 mm. Positrons have partial decay energy
for most positron emissions, and positron ranges contribute less to PET position
resolution, for example, about 0.2 mm, as a result of the changing direction of the
positron in the collision process. (2) It is generally considered that two gamma ray
photons are emitted in opposite directions at 180°. In practice, the positron may
have a certain momentum at annihilation, and the conservation of momentum will
cause the two photon emission angles to deviate from 180°; the uncertainty of this

angle is about 0.23° [9, 10]. PET scanners with a larger radius suffer from a larger contribution of the nonlinear effect terms. The noncollinear term contribution for smaller diameter PET devices is smaller. (3) The intrinsic spatial resolution of the detector and the geometric factor of the relative position of the detector are also contributed. The resolution of a pair of detectors is determined by the point spread function (PSF) of the detector pair [11].

The resolution of the PET system can be represented by the following formula [12]:

$$R_{sys} \approx \sqrt{R_{src}^2 + R_{180}^2 + R_{det}^2} \tag{1}$$

where R_{src} is the positron range, R_{180} is the noncollinear effect, and R_{det} is the intrinsic resolution of the detector pair. Usually the intrinsic resolution of the detector is determined by the size of the detector pixel. The spatial resolution of the detector is close to the size of the detector pixel when the annihilation position is close to one detector. The spatial resolution of the detector in the middle of the two detectors is half the size of the detector pixel.

Another factor limiting the resolution of PET images is parallel error [11], i.e., the uncertainty of the depth of the interaction (DOI) of gamma ray photons in the detector. For a pair of coincidence detectors in the opposite directions, gamma ray photons incident from and react within the detector. This makes a small parallel error. However, for two coincidence detectors close to the edge of the PET device, gamma ray photons may incident in the edge of the detector and deposit energy within several crystal arrays. For the coincidence event measurement of PET, it is not clear whether the gamma ray photons interact with the crystal arrays near the surface or in the inner layer. This makes the difference between the photon incident crystal pixel and the pixel that actually interacts with the photons. The parallel error results to a LOR uncertainty, which in turn results in decreased PET image resolution. (Fig. 2)

The timing window is set during PET data acquisition to exclude the random coincidence events. The narrower the width of the timing window, the less chance there is of random coincidence events being recorded. The minimum timing window is determined by the timing resolution of the PET detector, $\Delta T = 2\tau$ [12, 13]. However, the true coincidence events suffer loss if the timing window is set too narrow in PET data acquisition.

Not all of the gamma ray photons arriving in the coincidence timing window are emitted from the prompt positron annihilation. There are mainly three categories of coincidence events [12]. (1) Scatter coincidence events correspond to the pair of coincidence events from one positron annihilation, but one or two gamma ray photons are scattered in the human body. (2) In random coincidence events, two photons are detected, but these two photons are from different annihilation events. (3) In true coincidence events, two photons from the same annihilation are detected and none of them are scattered. The aim of PET imaging is to use true coincidence events to reconstruct the spatial distribution of radioisotopes in the human body. It

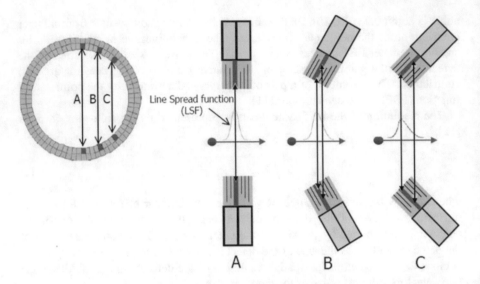

Fig. 2 Parallax error in PET measurement, depth of interaction (DOI) of gamma ray photons in the positions of A, B, and C. The uncertainty of DOI increases when the diameter of PET scanner decreases

is necessary to reject the scatter events and random events as much as possible via a process named scatter correction and random correction [14, 15]. In PET emission scanning, one needs to minimize the timing window and use the appropriate amount of radioactive drugs.

To reduce the timing window, fast time-response gamma ray detectors and short pulse-shaping times are used to reduce signal pileup, for example, in TOF-PET [16], the timing resolution reaches hundreds of picoseconds. Meanwhile, in order to obtain high count rates of true coincident events, the commonly used methods in PET scanning also include (1) increasing the dose of radiopharmaceuticals, especially in small animal PET imaging experiments; (2) using the gamma ray detector with high detection efficiency, such as thicker BGO/LSO crystals; (3) using a wider energy window to include part of the Compton scattering counts in addition to the photoelectric events, but at the same time increasing the uncertainty of the location of the photons detected in PET detector; and (4) increasing the solid angle of gamma ray detection, making the detector closer to the radioactive source, and also increasing the DOI errors.

2 Pixelated CdZnTe Detectors in PET

Cadmium zinc tellurium (CdZnTe), as a room temperature gamma ray detector, has been widely used in particle physics, nuclear physics, astrophysics, radiation safety, medical imaging, and other fields [17–23]. The choice of the CdZnTe detector as a

Fig. 3 Pixelated CdZnTe detectors

potential candidate for high-resolution PET imaging applications is due to its room temperature operation, high energy resolution, relatively high gamma ray detection efficiency, and high spatial resolution [24–28]. Advances in consumer electronics and achievements of CdZnTe fabrication processes in the recent years have made the semiconductor widely available, and highly pixelated CdZnTe detectors are becoming more economically feasible [29–31]. Several pixelated CdZnTe detectors are shown in Fig. 3.

CdZnTe detectors have many advantages in gamma ray detection and applications: (1) the energy resolution is extremely high. Compared with scintillator detectors, the photoelectric peak of the energy spectrum of CdZnTe is sharp. In PET data acquisition, only photopeak events are recorded. The sharp photoelectric peak of CdZnTe makes it easier to exclude Compton scattering events in PET applications. It is also easier to identify various radionuclides in radiation safety applications [24], (2) because there is no need for sensors to convert the gamma ray photon to an electron. The conventional PET/SPECT module consists mainly of scintillators and PMTs, whereas semiconductor detectors directly convert the incident gamma photons into charges and generate electric signals on metal electrodes evaporated on the CdZnTe surface [32]. Therefore, the package size of CdZnTe is smaller than the size of the scintillator detector, which makes the CdZnTe detector easier for arranging the geometry when setting up the whole PET device. In the specific imaging of the region of interests (ROIs) and small animal applications, a CdZnTe detector is a potential candidate for the development of a prototype PET device. (3) When used at room temperature, different from HPGe detectors will need low temperature cooling. (4) The average atomic number of a CdZnTe detector is approximately 49, which is a relatively high gamma ray detection efficiency [32]. (5) The position resolution of the CdZnTe detector depends on the size of the anode pixel [33], and the pixel size can be as small as possible under the conditions permitted by readout electronics, e.g., 0.35 mm or less. The scintillator detector is very difficult to segment to such a small size.

There are many kinds of gamma-ray semiconductor detectors that are used at room temperature, including but not limited to HgI_2, CdTe, and CdZnTe. With the relatively low charge mobility-lifetime product and charge polarization phenomenon of CdTe detectors, its application is limited to the detector with thin thickness [34]. In the 1990s, the doping of Zn in CdTe crystals significantly increased the bandgap energy and the impedance of the bulk crystal, so the mobility-lifetime product and polarization effect were significantly improved [35, 36]. Though the charge trapping is still the main problem in the $Cd_{1-x}Zn_xTe$ detector, improving the impedance of the CdZnTe detector and increasing voltage at both ends of the detector have significantly enhanced the charge collection of CdZnTe [37].

The low-energy tail phenomenon of CdZnTe detectors has been significantly improved by the appearance of small pixel effect and single-polarity charge sensing [38–40]. The small pixel effect was found by H. H. Barret et al. [38]. If the size of the anode pixel is much smaller than the detector thickness, then the electrode collection will be sensitive only to the electrons. In this way, the trapping of holes will be greatly reduced, so that the energy resolution of the CdZnTe detector will be greatly improved. A monopolar charge-sensing structure similar to Frisch grid called coplanar structures was originally used in CdZnTe detectors by Luke et al. [39]. By using the relative gain method, the energy resolution of CdZnTe detectors is greatly improved by adjusting the multiples of the collection and noncollection electrodes. Concurrently, He et al. developed a depth sensing method for CdZnTe detectors [41]. By adjusting the depth information of the gamma photons at the peak position in the CdZnTe detector, the energy resolution of the CdZnTe detector was further improved by correcting the DOI of gamma rays.

CdZnTe detectors also have some disadvantages, such as (1) relatively poor timing resolution. The timing resolution of scintillator detectors is usually 1 ns FWHM, and the timing resolution of zinc telluride detectors is usually more than 10 ns after offline correction. This makes a wider coincidence timing window in PET applications of CdZnTe detectors, which causes a huge amount of random coincidence·events to be recorded in PET imaging experiments [27]. (2) The detection efficiency of gamma rays in CdZnTe detectors is not superior to that of BGO/LSO crystals currently used in large quantities in PET applications. (3) The growth of uniform and large volume crystals is difficult. Although the growth of large volume CdZnTe crystals has improved greatly in recent years, it is still expensive [42, 43]. (4) The direct electrode readout of electrical signals is an advantage for small pixel anode CdZnTe detectors, but thousands of channels of high-density readout electronics remain a challenge in data acquisition [44]. (5) The electrical signal is much smaller than the PMT/scintillator output signal and requires complex low-noise amplifying electronics equipment, such as ASIC [45–47].

Although CdZnTe detectors have been greatly improved in crystal growth, the application of CdZnTe detector is still in the very early stages. There are three primary difficulties at this time. (1) The signal readout of CdZnTe detectors is very complex. The control of the detector's dark current and surface leakage current is sensitive. When the voltage of the CdZnTe detector increases, the crosstalk between anode pixels is more severe [48]. (2) The response time is poor. In recent years, fast scintillator detectors have developed rapidly and are widely used in TOF-PET. The conformity of measurement by CdZnTe detectors is still one of the chief obstacles in PET imaging experiments [42, 49]. (3) The position resolution of the CdZnTe detector when its pixel anode is very small needs improvement. Particularly, the processing of charge-sharing events between the anodes will be another challenging problem in the PET imaging experiments and the application of the 3D position sensitive CdZnTe detector [50–54].

Before the invention of small pixel effect, the energy resolution of CdZnTe detectors was relatively poor due to the appearance of the low-energy tail. In recent years, the energy resolution of CdZnTe detector combined with the depth-sensing correction of gamma ray interaction has been improved to ~1% at 662 keV gamma rays. At the same time, the 2D anode of the pixelated detector provides better spatial resolution. In order to reduce the charge loss in the pixel gap, steering grids are often added to the pixel gap. Additionally, a compensation voltage is added to the steering grid to change the electric field distribution of the detector, thereby improving the charge collection efficiency and reducing the pixel size [55].

The full-energy peak of the γ energy spectrum of the CdZnTe detector often has a tail at the low-energy end, resulting in a reduction in energy resolution, mainly due to incomplete charge collection. On the other hand, from the analysis of the Hecht equation, it is known that the signal of CdZnTe is related to the position of the incident photon in the detector, i.e., to the DOI. If the gamma ray reaction takes place very close to the cathode, the probability of the hole collection will increase greatly. In contrast, if the gamma ray reaction takes place close to the anode, the probability of the hole collection will decrease sharply. In both cases the electrons can be fully collected. For photons deposited with the same energy at different depths, it is necessary to reduce the low-energy tail caused by incomplete charge collection in the experiment.

The geometrical structure of this CdZnTe detector, which is only sensitive to electron collection and insensitive to hole collection, is called a single-polarity charge-sensing device [24]. Two methods are often used to achieve unipolar charge induction. The first type is a strip-shaped anode structure, which reduces the weighting potential of adjacent electrodes, so that the detector is sensitive to electrons at most detection depths. The second type is the pixel anode structure. This structure has a small pixel effect. When the charge drifts close to the anode, the weighting potential changes sharply, and the weighting potential is almost constant in most of the detection depth.

As shown in Fig. 4, the weighting potential of the coplanar cathode of the pixelated CdZnTe detector remains linear. The weighting potential of the small pixel anode slowly increases over a long distance when the charge drifts to the anode. Only when the electron nears the anode, such as within one pixel, does the weighting potential increase rapidly from 0 to 1. Therefore, the pixel detector is unipolar sensitive. In order to compensate for the trapping of electrons, it is ensured that the signal does not change with depth in the direction of the detectors. He et al. developed the depth-sensing method based on using electronic trapping and DOI to maintain linearity. The depth-sensing methods have two categories: cathode to anode ratio and electron drift time. (Fig. 4)

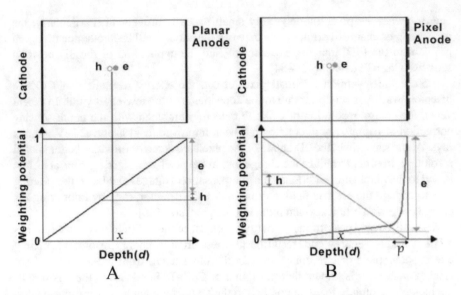

Fig. 4 Weighting potential of the CdZnTe detector: (**a**) planar anode and (**b**) pixel anode

I. Ratio of Cathode to Anode

The weighting potential of the planar cathode is linearly related to the thickness of the detector. After subtracting the weighting potential of two adjacent anodes, the signal is not correlated with the thickness of the CdZnTe detector. According to the Shockley-Ramon theory [56, 57] and ignoring the contribution of holes, the signal amplitudes of the cathode and anode can be expressed by the following formula:

$$A_c = E_d (d - x), \quad A_a = E_d \tag{2}$$

Then, the ratio of the cathode to the anode reflects the DOI information:

$$R_{C/A} = \frac{A_c}{A_a} = d - x \tag{3}$$

In practice, the ratio of the cathode to anode is not an absolute DOI due to factors such as the existence of electron trapping, crystal nonuniformity, and electrode nonuniformity. However, it provides a relatively accurate method to estimate the DOI and compensate for electronic trapping. This method is also applicable to pixel anode detectors. During data processing, the overall thickness of the detector is usually subdivided into several small segments according to the ratio of the cathode and anode. By measuring the position of the full-energy peak in the energy spectrum of each small segment and setting the gain on each segment to make the full-energy peak normalized, the deposited energy in the CdZnTe detector is scaled to the gamma ray energy, and the energy resolution shows a significant improvement.

The cathode to anode ratio method is only applicable to single-anode events. For multi-anode events, such as charge-sharing events, the energy of the incident photon deposition will be detected by multiple anodes, but the cathode signal is still the sum of the deposited energy. Thus, the method of cathode-anode ratio will no longer be applicable. The electron drift time method described below is applicable to both multi-anode events and single-anode events, but the estimation of the DOI is usually not as accurate as the cathode-anode ratio method.

II. Electron Drift Time

The signal of the semiconductor detector is caused by the charge drifting in the electric field and inducing a charge on the electrode. Therefore, the response time of the signal is related to charge drift and electric field in the crystal. For a CdZnTe detector with a thickness of 5 mm, the maximum drift time of electrons is 500 ns at an electron mobility of $1 \times 10^3 cm^2/V \cdot s$ under a high voltage of -1000 V.

For a pixelated CdZnTe detector, the electron drift time at the same photon interaction depth will be a fixed value, whether it is a single-anode event or a charge-sharing event. As a result, by measuring the electron drift time, it is possible to estimate the DOI of the event. Due to small pixel effect, when the charge begins to move, the cathode will sense the charge, and the cathode signal will start to rise. However, the anode can only sense the charge when the electrons drift close, so the anode signal will be delayed a lot. Experimentally, the electron drift time is obtained by measuring the time difference between the cathode signal and the anode signal.

The position of the full-energy peak at different electron drift times will vary with the DOI. Using the same method as the cathode-anode ratio method, the overall thickness of the detector is divided into several small sections, and the depth resolution correction can improve the energy resolution of the CdZnTe detector. On the other hand, by measuring the maximum electron drift time, the electron mobility value of the CdZnTe detector can be estimated.

3 CdZnTe PET Detector Characterizations

The CdZnTe crystals used in our PET experiments were purchased from the former Orbotech Medical Solutions, from a modified high-pressure Bridgman grown 2 cm $\times$ 2 cm wafer. The initial wafer electrode was an indium 8 $\times$ 8 pixel-anode. We fabricated it in a class in 100 clean rooms of the Washington University Physics Department. The basic process is as follows: after polishing the anode surface of the detector, the CdZnTe wafer is etched in a 5%–95% Br-methanol solution. This process can improve the electrical performance of the wafer and the bonding ability of the metal electrode. We then applied the standard photolithographic process and the electron beam evaporator process to make the anode [58, 59]. The finished pixelated CdZnTe detector has a coplanar cathode gold surface with a thickness of 125 nm and a 100 nm titanium metal pixel anode. Gold is a high-work-function metal, which has the advantages of low leakage current and high energy resolution in the experimental measurements. Titanium is a relatively low work function metal [60]. Although the pixel anodes are separated from each other in electrical conductivity, they maintain a similar potential. When a high voltage is applied to the cathodes, the parallel plate electric field is maintained between the cathodes and anodes.

3.1 Charge Sharing

Energy calibration and measurement were performed for CdZnTe detectors with 0.35 mm and 0.6 mm pixels. As illustrated by gamma ray sources of Am-241 (59.5 keV), Co-57 (122 keV and 136 keV), and Na-22 (511 keV), the noncollimated gamma ray source tests showed the following [51, 61]:

1. Both 0.35 mm and 0.6 mm pixelated CdZnTe detectors showed small pixel effect. The energy resolution of the 0.6 mm pitch detector was better than the energy resolution of the 0.35 mm pitch detector. This seems to contradict the theory of small pixel effect. However, many experiments and theoretical simulations have shown that the energy resolution of the pixelated CdZnTe detector is getting worse. When the pixel size is reduced to less than 1 mm, such as 0.6 mm and 0.35 mm in this experiment, charge sharing between adjacent pixels becomes the main factor affecting the energy resolution, even exceeding the small pixel effect.
2. Both 0.35 mm and 0.6 mm pitch detectors showed a large amount of charge sharing, but the ratio of charge sharing events to the total reaction events increased sharply as the pixel size decreased. This agreed with the theoretical calculation.
3. For a specific pixelated CdZnTe detector, when the gamma ray energy increased, the charge sharing between the anodes also increased. For example, in the 59.5 keV test, there was almost no charge sharing. While in the 511 keV measurement, charge sharing was prevalent.
4. When the energy was held constant, such as at 122 keV, the charge-sharing event between adjacent anodes was proportional to the distance between pixels. For example, the charge sharing between the middle pixel and the neighboring pixel was significantly more than the charge sharing between the middle pixel and the corner pixel.

Table 1 shows the ratio of a single-pixel event to charge-sharing event, R_{12}, for 0.35 mm and 0.6 mm CdZnTe detectors at an irradiation of 122 keV and 59.5 keV sources. For the low-energy gamma ray photons at 59.5 keV, the charge sharing of 0.35 mm and 0.6 mm pitch detectors remained unchanged at 4.65:1 and 4.67:1, respectively. For high-energy gamma ray photons at 122 keV, the proportion of the charge-sharing events of 0.35 mm and 0.6 mm pitch detectors increased significantly to 2.85:1 and 1.38:1, respectively.

The two-dimensional charge-sharing spectra of the 0.6 mm and 0.35 mm detectors when exposed to a collimated beam at different locations are shown in

Table 1 The ratio of single-pixel event to charge-sharing event (R_{12}) for 0.35 mm and 0.6 mm CdZnTe detectors with irradiation by 59.5 keV and 122 keV gamma ray sources

Pixel size	R_{12}	
	122 keV	59.5 keV
0.6 mm	1.38:1	4.67:1
0.35 mm	2.85:1	4.65:1

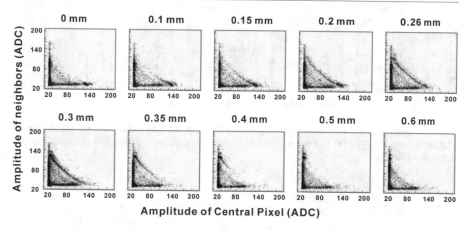

Fig. 5 Charge sharing of the 0.6 mm pixelate CdZnTe detector when the collimated beam scanned across two pixels. The measurements above each plot are the locations of the collimated beam in millimeters

Figs. 5 and 6, respectively. In the measurement, the beam scanned across two neighboring pixels. The two-dimensional spectrum of the signal amplitudes of two adjacent pixels clearly showed the changes of charge sharing at different collimated beam locations.

In the measurement of the 0.6 mm detector, we used a rectangular beam with an effective size of 0.16 mm × 0.16 mm. In the measurement of the 0.35 mm detector, in order to obtain a narrower scanning beam, we used a 0.8 mm × 0.625 mm rectangular beam. The anode gap of the two CdZnTe detectors was 0.1 mm, so a narrower direction (parallel to the gap between the two pixels) was used in the experimental measurement.

The 0.6 mm and 0.35 mm anode detectors showed the same change in charge sharing. When the collimated gamma ray beam hits the middle of the pixel anode, there were a few charge-sharing events. As the gamma beam moved toward the gap, the charge-sharing event became significant; when the gamma beam hits the gap, the charge-sharing event reached its peak. The charge sharing of the 0.35 mm pixel anode detector is more than that of the 0.6 mm detector. For the collimated beam of Co-57, the 122 keV and the shared full-energy peak at 136 keV were clearer than the shared full-energy peak in the two-dimensional spectrum of 0.35 mm.

Figure 7 shows a comparison of the charge-sharing curves of a 0.6 mm and a 0.35 mm pixelated CdZnTe detector. The 122 keV collimated gamma beam scanned

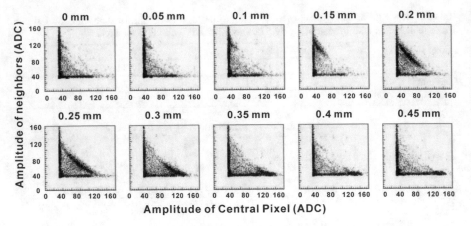

Fig. 6 Charge sharing of the 0.35 mm pixelate CdZnTe detector when the collimated beam scanned across two pixels. The measurements above each plot are the locations of the collimated beam in millimeters

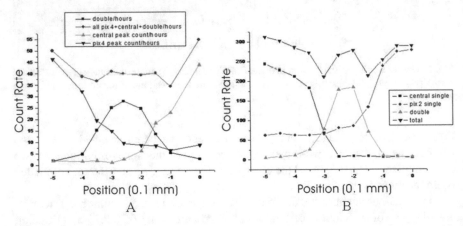

Fig. 7 (a) Charge-sharing range of 0.6 mm pixelated CdZnTe detector. (b) Charge-sharing range of 0.35 mm pixelated CdZnTe detector

the gap between the two adjacent pixels. Figure 7a shows the single-anode full-energy peak event count rate of the middle pixel anode, the single-anode full-energy peak event count rate of the neighboring pixel anode, the charge-sharing full-energy peak event count rate curve and the count rate of all triggered full-energy peak events. None of count rate curves for the trigger events between two adjacent pixels remain constant. This phenomenon indicates that the charge loss was caused by the anode gap of 100 μm. The charge-sharing curves of the two sizes of detectors show that the relationship between the count rate and position when the gamma beam scans across the gap and the proportion of charge-sharing full-energy peak events in all full-energy peak events is close to 70%. It is assumed that there is a relationship between the range of charge sharing P_{share}, the actual measurement range $P_{measured}$, and the width of the gamma beam $P_{beam\ width}$:

$$P_{share}^2 = P_{measured}^2 - P_{beam\ width}^2 \qquad (4)$$

After subtracting the beam width, the charge-sharing range of the 0.35 mm and 0.6 mm pixel anode detectors were almost equal. This shows that for low-energy gamma ray photons, such as 122 keV, the range of charge sharing depends on the size of the gap rather than the size of the pixel size. Yet for high-energy gamma ray photons, such as 511 keV, the influencing factors of the charge-sharing range are more complicated, because of the larger charge cloud and more Compton scattering.

3.2 Spatial Resolution

In order to accurately measure the three-dimensional spatial resolution of the 0.35 mm pixel anode CdZnTe detector for 511 keV positron annihilation photons, we used a 3.5-cm-thick tungsten collimator to collimate the gamma ray beam. The gamma ray collimator is composed of four pieces of tungsten metal [25]. The surfaces of the four tungsten blocks are polished very smoothly with an error of less than 1 μm. In the middle of the collimator, the tungsten blocks form a square hole of 0.17×0.17 mm^2.

The measurement results of the 511 keV collimated gamma beam and 122 keV collimated gamma beam are shown in Fig. 8. In the data acquisition, the threshold of 511 keV measurement was 55 keV, and the threshold of 122 keV measurement was 22.5 keV. Figure 8 shows the single-pixel full-energy event curve (gray) of the middle 0.35 mm pixel and the double-pixel charge-sharing full-energy event curve (black) of the central pixel and adjacent pixel in the 122 keV (a) and 511 keV (b) gamma beam tests. The following conclusions can be made from these experiments:

1. The measurement results for 511 keV and 122 keV showed that the peak of the single-event resolution range appeared in the center of the anode of the middle pixel, and the peak of the charge-sharing event resolution range appeared in the center of the anode gap and between the two peaks. The distance is about 0.175 mm, which is half of 0.35 mm. The experimental values agreed well with theoretical expectations.
2. The charge-sharing range of 511 keV gamma ray photons was greater than the charge-sharing range of 122 keV gamma ray photons in the measurement for the 0.35 mm pixelated CdZnTe detector.

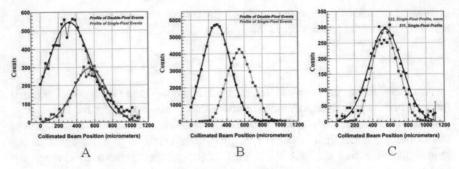

Fig. 8 Measurements of the charge-sharing range of the 0.35 mm pixelated CdZnTe detector for 122 keV (**a**) and 511 keV (**b**) collimated gamma ray beam, and the comparison of the single-pixel profiles for two collimated gamma ray beams (**c**)

Table 2 Spatial resolution measurement of single events and charge-sharing events of 0.35 mm CdZnTe detector

Gamma ray beam	Position resolution of single-anode events (FWHM)		Position resolution of charge-sharing events (FWHM)	
	$P_{measured}$[a]	P_{share}[a]	$P_{measured}$	P_{share}
122 keV	0.34 mm	0.207 mm	0.36 mm	0.235 mm
511 keV	0.41 mm	0.308 mm	0.52 mm	0.444 mm

[a]$P_{measured}$ is the measured value of gamma ray beam; P_{share} is the position resolution subtracted from the measured value; $P_{beam\ width}$ is the width of gamma-ray beam, 0.27 mm × 0.27 mm.

3. The peak value of the charge-sharing event distribution in the 511 keV measurement was about twice the peak value of the single-event distribution, and the 511 keV measurement showed that the proportion of the charge-sharing events in all triggered events was greater than that of the 122 keV measurement.
4. The detection efficiency of 511 keV gamma photons was lower than that of 122 keV gamma photons. Therefore, there were better statistics and smaller errors in the 122 keV measurement.

Table 2 shows the FWHM of the spatial resolution of the 0.35 mm pixel anode detector. For 122 keV gamma photons, the spatial resolution range measured by single events was 0.34 mm, and the spatial resolution range measured by charge-sharing events was 0.36 mm. For 511 keV gamma ray photons, the spatial resolution range of single events was 0.41 mm, and the charge-sharing range was 0.52 mm. If the width of the gamma beam (0.27 mm × 0.27 mm) was subtracted from the measured value, it was the same as the width of charge sharing.

3.3 Depth of Interaction

For gamma ray photons deposited at the same energy with different depths, the depth of interaction (DOI) can be obtained from the signal amplitude cathode to anode ratio. If the CdZnTe detector is divided into several layers in depth direction with the pixels of detector, the three-dimensional position information of photons interaction can be obtained [62].

$$R_{C/A} = \frac{A_c}{A_a} = d - x \tag{5}$$

Due to the different distances of charge drift deposited at different depths, the events with a longer charge drift length, that is, events with deeper DOI, have a greater chance of charge loss, so the position of this event in the energy spectrum will likely be closer to the lower energy end. Figure 9 shows the relationship between the signal amplitude of the full-energy peak and the cathode-anode ratio when the 122 keV and 511 keV collimated beams hit the middle of 0.35 mm pixel. Events with smaller ratios occurred on the side near the anode, while events with larger ratios occurred on the side near the cathode. We can clearly observe the following:

(1) The signal of the cathode had better DOI characteristics, and the position of its full-energy peak increased linearly with the increase of the ratio $R_{C/A}$. At different DOIs, the position of the full-energy peak of the anode signal remained unchanged, and the amplitude of the signal represented the deposition energy.

(2) The amplitude of the full-energy peak of the anode signal decreased with the DOI. A large amount of charge trapping occurs when the electron drift time increases, in the measurement of both 122 keV and 511 keV. This trapping was related to the drift path length, external electric field, and crystal.

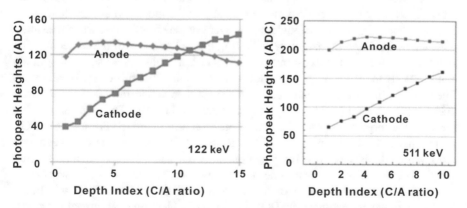

Fig. 9 DOI analysis of 0.35 mm pixelated CdZnTe detector: (**a**) 122 keV collimated beam and (**b**) 511 keV collimated beam

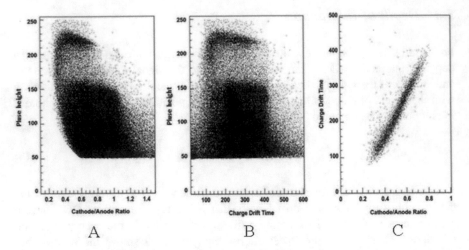

Fig. 10 The measurement results of the single-pixel event of the 0.35 mm pixelated CdZnTe detector with a 511 keV collimated gamma beam. (**a**) Pulse height vs cathode/anode ratio. (**b**) Pulse height vs electron drift time. (**c**) The linear relationship between the cathode-anode ratio and electron drift time for full-energy peak events

(3) On the large ratio side, the amplitude of the 122 kcV full-energy peak signal decreased more rapidly than that of the 511 keV full-energy peak signal, indicating that charge tapping of 122 keV photons were more severe than that of 511 keV photons, which was the same as theoretical expectations. Because most of the 122 keV photon reaction occurs on the side closer to the cathode, that is, the upper layer of the photon incidence direction, the electron drift time was longer; while the 511 keV photon reaction occurs deeper, closer to the anode side, the electron drift time was relatively shorter.

Next, we examined the relationship of electron drift time and cathode-anode ratio. Figure 10 shows the two-dimensional ratio of the anode signal amplitude to the cathode-anode signal for the single-anode event of the middle pixel when measured with a 511 keV gamma beam on a 0.35 mm pixel anode CdZnTe detector. The spectrum, as well as the two-dimensional spectrum of the anode signal amplitude and the electron drift time, clearly shows the location of the full-energy peak event. There was a linear relationship between the cathode-anode ratio and the electron drift time for the full-energy peak event, as shown in Fig. 10c. The graph shows that for the full-energy peak event, the time of electron drift increased linearly with DOI.

The linear relationship between the electron drift time and the cathode-anode ratio was suitable not only for single-anode events but also for charge-sharing events, because electron drift time at different depths was the same whether it was a full-energy peak event or a Compton scattering event. Using this linear relationship, we can estimate the DOI and derive the electron drift time of the event, by subtracting the electron drift time from the detection time of the anode signal of the photon in the CdZnTe detector. Thus, the time resolution of the CdZnTe detector from the anode signal can be significantly improved. The interaction time of the photon in the CdZnTe detector can be given by the time difference in the coincidence measurement between the CdZnTe detector and the fast scintillator detector.

The linear relationship between the cathode anode ratio and the electron drift time is very useful in the time performance measurement of the CdZnTe detector and in the PET imaging experiment. In the PET imaging experiment, the PET image was reconstructed using a single-pixel full-energy peak event and a charge-sharing full-energy peak event that conformed to the measurements for the CdZnTe detector and LSO scintillator detector.

3.4   Timing Resolution

The time characteristics of the CdZnTe detector are an issue for PET imaging applications. For the cathode signal, the drift of the starting position of the signal is small. Many groups have developed methods based on the fitting of the cathode signal waveform and extrapolating the start time of the signal to measure the time resolution of the CdZnTe detector, but the online time performance measurements are not yet mature [27, 63]. The disadvantage is that the rise of the cathode signal is slow, the leading edge of the signal is directly related to the energy deposition, the amplitude of the signal varies with the position, and the time resolution is hard to improve.

The anode signal has a rapidly rising edge due to the existence of the small pixel effect, and the timing is relatively accurate. The disadvantage in this case is that the signal shape will be very different at different depths with energy deposition. This error due to electron drift time is hundreds of nanoseconds. Experiments confirmed that for the full-energy peak event, the electron drift time was linearly related to the cathode-anode ratio; therefore, subtracting the electron drift time from the measurement time of the anode signal can greatly improve the time resolution of the anode signal. Moreover, this method is applicable to gamma rays at different depths. In view of this, this paper develops a method to improve the time resolution of CdZnTe detector based on the coincidence measurement of the CdZnTe detector and a fast scintillator detector.

For the coincidence events of the CdZnTe detector and fast scintillator detector (such as LSO), the relative time of the anode signal of photon interacting in CdZnTe detector and the scintillator signal can be given by the coincidence measurement. The electron drift time in the CdZnTe detector can be obtained by the signal waveform of the flash ADC system. Using the time correction method, we made a preliminary measurement of the time resolution of the CdZnTe detector. It used the Siemens Inveon LSO PET detector module and the 0.35 mm CdZnTe detector and the time-amplitude converter (TAC) for time-amplitude conversion. Then, we used the ADC to measure the time difference between the two detector signals. The time resolution of the Siemens Inveon LSO scintillator detector was about 1 ns; the electron drift time of the 5-mm-thick CdZnTe detector was about 400 ns.

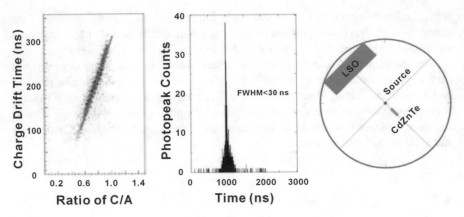

Fig. 11 The linear relationship between the cathode-anode ratio and electron drift time of full-energy peak event. (**a**) The line is a fit. (**b**) The time resolution after correction was better than 30 ns FWHM. (**c**) A coincidence measurement is set between CdZnTe detector and LSO detector. The source is the Na-22

For the full-energy peak event, the linear relationship between the cathode-anode ratio and electron drift time is shown in Fig. 11. The line in Fig. 11a is a fit. The time resolution before correction is about 150 ns FWHM, and the time resolution after correction was better than 30 ns FWHM. This demonstrated that the time correction method is effective, and the time resolution after the correction was significantly improved.

4 Evaluation of CdZnTe PET Imaging Resolution

The conformity measurement of VP-PET with the Siemens Inveon LSO PET detector module was used to discern the image resolution that can be obtained after integrating the 0.35 mm CdZnTe detector as a PET insert device into the Siemens Inveon PET device. The geometrical configuration that meets the measurement is shown in Fig. 11.

A 0.35 mm pixelated CdZnTe detector was used as a simulated insert device, and a Siemens Inveon LSO scintillator detector was used as a simulated PET scanner device. They were placed in two PET simulators. On the rotating arm, the center of rotation corresponded to the axis of the PET scanner [64, 65]. For the convenience of data reconstruction, the axis of the insert device was also on the axis of rotation. A NEMA NU4 Na-22 point source was placed between the two detectors [26] with an intensity of 11 μCi and a diameter of 250 μm, encapsulated in a 10 mm square propylene plastic. The distance from the point source to the CdZnTe insert detector was 23 mm, and the distance from the point source to the LSO Scanner detector was 127 mm. The point source was placed near the rotation axis of the PET simulator.

In the experiment, the imaging device was obtained by fixing the distance between the CdZnTe detector and the rotation center and adjusting the distance between the LSO detector and the rotation axis to keep the two detectors and the point source approximately in a straight line.

The Siemens Inveon LSO PET detector module consists of an array of 20 × 20 LSO crystals; each crystal is 1.6 mm × 1.6 mm × 10 mm. The effective area of the entire LSO detector is 32 mm × 32 mm [66, 67]. The detection solid angle of the Siemens Inveon PET scanner was increased, and the sensitivity of the system was improved. The scintillator and position sensitive photomultiplier tube were coupled through a trapezoidal light guide. The design of the trapezoidal light guide mainly considered that the larger area scintillator was optically coupled to the sensitive area of the smaller area photomultiplier tube, which allowed the center of gravity of the optical signals from different crystals to be separated from each other.

In order to overcome the difficulty of coincidence measurement due to the difference in time response of semiconductor and scintillator detectors, we used a relatively long shaping time. The shaping time of the LSO signal was 1 μs, and the shaping time of CdZnTe signal is 3 μs. In this way, the signal had a wide Gaussian-shaped peak plateau area. Because the activity of the Na-22 point source is very low, the long shaping time did not cause serious accumulation of the LSO signal with a larger solid angle. Experimental measurements showed that the average energy resolution of the 511 keV photon by the LSO detector was better than 20%, which was close to the 18% energy resolution of normal use.

In the PET imaging experiment of the CdZnTe detector and the LSO detector, the selection of the trigger mode was one of the main issues, because the CdZnTe detector had a long electron drift time and the time resolution is poor. In this experiment, we used the logical signals of the CdZnTe detector and the LSO detector to trigger the signal peak sampling of the data acquisition system, but the trigger signals of this logical signal were different so different time walks were introduced. We tested three conforming trigger modes [24, 55]: (1) the trigger logic came from the LSO signal, (2) the trigger logic came from the anode signal of CdZnTe, and (3) the trigger logic came from the CdZnTe cathode signal.

For the logic signal triggered by the LSO, the time resolution of the peak sampling of the LSO energy signal was 2 ns, so the peak sampling error from the LSO was the smallest. There was also a fixed delay for the CdZnTe cathode signal. The time error for the peak sampling of the CdZnTe anode energy signal was the superposition of the electron drift time of the CdZnTe and the time resolution of the LSO, because there was a shaping time of 3 μs. The Gaussian peak area was very wide, and the peak sampling error had little effect on the energy resolution.

Figure 12 shows the fusion PET image of the Na-22 point source measured twice; the point sources were 1.2 mm apart. We can see that the two point sources can be clearly distinguished. The FWHM of the PET image reconstructed by single-pixel events was 670 μs. If there is charge-sharing event, the resolution of the PET image dropped to 730 μs, which was a reduction of about 9%. At the same time, the system's count rate was increased by 2.5 to 3 times. The energy window was 350–650 keV.

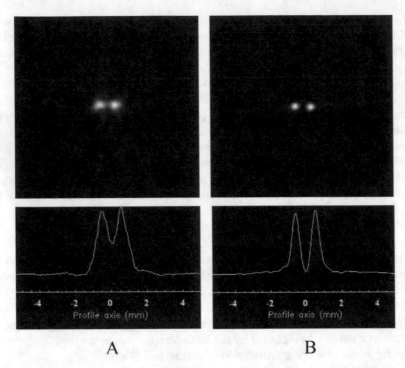

Fig. 12 The FBP reconstructed PET images of two Na-22 point sources (**a**) and simulated data from MC (**b**). Two point sources are 1.2 mm apart

For the 0.35 mm CdZnTe detector, the charge-sharing event is about twice as much as the single-anode event. Therefore, how to use the charge-sharing event became one of the main challenges in the PET imaging experiment for the CdZnTe detector. From the previous analysis, we knew that when the charge-sharing event was included in the reconstruction of the PET image, it was equivalent to increasing the effective width of the anode of the middle pixel. Thus, the resolution of the PET image should be reduced, but at the same time, the count rate of the system should increase by 2.5 to 3 times.

The analysis of the charge-sharing event used the charge sharing event between the intermediate pixel A5 and the adjacent pixel A4, as an example. The pixel structure is shown in Fig. 13. The signal amplitude ratio of the two pixels is divided into three categories: (1) $1 < R_{54} < 3$, region A; (2) $3 < R_{54} < 9$, region B; and (3) $9 < R_{54}$, region C. The sum of the anode signal amplitude of the two pixels of the charge-sharing event was 511 keV. The signal amplitude of the adjacent anode was close to the experimentally set threshold of 50 keV. Here we called the event in region C the single-anode event; region A and region B were a charge-sharing event.

The selection of charge-sharing areas also included the consideration of counting statistics. The proportion of events in the three areas was 35%, 25%, and 40%, as

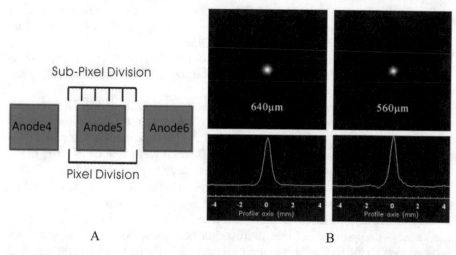

Fig. 13 Interpolation algorithms for charge-sharing events and reconstruction of point source PET images. (**a**) The pixel area (pixel division) was divided into five small areas (sub-pixel division) according to the interpolation algorithm. (**b**) The resolution of the PET image reconstructed using the pixel division was 0.64 mm. The PET image resolution of the sub-pixel division was 0.56 mm. After deducting the point source diameter and positron range, the interpolation algorithm gave a PET image resolution of 0.46 mm

Table 3 The selection of the charge-sharing events of the CdZnTe detector, the corresponding PET image resolution and the count rate in this area

Region	PET image resolution (μm @ FWHM)	Count percentage (%)
A	640	~35%
B	620	~25%
C	590	~40%

Experimental results

shown in Table 3. We divide the shared event partitions, and each partition was calculated according to the size of the pixel anode of a unit during the image reconstruction process. This way, an actual detection of a small area was regarded as a larger area for image reconstruction, which may magnify the point source image. The distribution of the single-pixel full-energy peak event to the gamma incident position may be less than 0.35 mm. We used 0.35 mm for image reconstruction, which may result in the images of 5–10 μm error. (Table 3)

According to the theory of pixel detectors [68], if the charge cloud is detected by multiple pixels, then the charge collected by each pixel is proportional to the signal amplitude of the pixel. A reasonable interpolation algorithm can improve the resolution of the pixel detector. Through interpolation, the size of the pixel will be further subdivided according to the interpolated area, which appears that the relative size of the pixel is reduced, thereby improving the position resolution of the detector and the image resolution.

For the 0.35 mm pixelated CdZnTe detector used in the experiment, when a charge-sharing event occurs, according to the Shockley-Ramo theory, the signal size of the adjacent pixel anode is proportional to the collected charge [69]. Therefore, we tried using interpolation to further partition the charge-sharing events of the 0.35 mm CdZnTe detector, and we used the 2D FBP algorithm for image reconstruction.

A 0.35 mm pixel can be divided into five areas, A, B, C, D and E. The selection of areas A, B, and C was based on the ratio of the signal amplitude between the intermediate pixel A5 and the adjacent pixel A4, which was the same as in the previous section. The selection of the regions D and E was based on the ratio of the signal amplitude between the anode of the intermediate pixel A5 and the adjacent pixel A6, and the principle was the same as the selection of the regions B and A. $R_{56} =$ Central/Pixel$_6$, area D is $3 \ R_{56} < 9$, and area E is $1 \ < R_{56} < 3$.

Experimentally, we tried to make the count statistics of the selected regions the same. We approximated that each region had the same geometric width for interpolation. The size of each region (sub-pixel division) was approximately $350/5 = 70 \ \mu$m, as shown in Fig. 13. After 2D FBP reconstruction [70], the 0.56 mm FWHM PET image resolution was obtained. This resolution included the size of the Na-22 point source (0.25 mm diameter) and positron range. The actual resolution of the imaging system was: $R = \sqrt{560^2 - 250^2 - 200^2} \approx 460 \ \mu$m, of which 250 was the point source size item, 200 was the positron range item, and the unit was micrometers.

5 Summary

With the advantages of the CdZnTe detector used in X-ray and gamma ray detection, the application of pixelated CdZnTe detectors for high-resolution PET imaging is increasingly investigated by researchers and industry. There are still many limiting factors for CdZnTe detectors in PET applications, such as charge sharing and timing resolution.

When the pixel size of the CdZnTe detector is made very small (e.g., less than 1 mm), the charge-sharing events will degrade the spatial resolution of the pixelated CdZnTe detector. Thus, if one includes coincidence events from CdZnTe charge-sharing events, the PET image resolution will degrade, but the count rate increases dramatically. We developed an interpolating algorithm for the charge-sharing events between two neighbor pixels. We divided the pixel pitch of the CdZnTe detector into several regions based on the ratio of the signal amplitude of the central pixel to the neighbor pixel, as well as the count rate of each region. Then, we reconstructed the PET image using the events from different regions individually. By interpolating the location of the charge-sharing events, the reconstructed PET image was improved. More importantly, the count rate capability of the CdZnTe

PET scanner was increased by including the charge-sharing events in the PET image reconstruction.

The long drift time of electrons and holes in thick CdZnTe detectors limits their timing resolution. For PET imaging applications that require fast timing for coincidence detection and the rejection of random events, the timing performance of CdZnTe is still suboptimal. As a result, the use of highly pixelated CdZnTe detectors for high-resolution PET imaging remains a challenging task. A proposed timing correction method was introduced in this chapter. By measuring the time difference between CdZnTe anode signal and fast coincidence detector signal based on the coincidence setup, combined with measuring the electron drift time and the cathode-anode ratio from a DAQ system, the timing resolution of CdZnTe detector was improved significantly. Moreover, if the DOI of gamma ray photons in the CdZnTe detector was calculated, the uncertainty of timing response of photopeak events in a specific DOI region was reduced, and the timing resolution of the CdZnTe detector was improved further.

Advances in CdZnTe crystal growth, consumer electronics, and detector fabrications in the recent years have made the CdZnTe detector a very promising candidate for PET imaging application. Several research groups and companies have investigated CdZnTe for PET imaging and have shown promising results. The CdZnTe PET scanner will begin to see clinical use in a near future.

References

1. Phelps, M. E., Hoffman, E. J., et al. (1976). Tomographic images of blood pool and perfusion in brain and heart. *Journal of Nuclear Medicine, 17*, 603–612.
2. Ter-Pogossian, M. M., Phelps, M. E., et al. (1975). A positron-emission transaxial tomograph for nuclear imaging (PETT). *Radiology, 114*(1), 89–98.
3. Weissleder, R., Ross, B. D., Rehemtulla, A., & Gambhir, S. S. (2010). *Molecular imaging – Principles and practice.* Shelton: People's Medical Publishing House.
4. Muehllehner, G., & Karp, J. S. (2006). Positron emission tomography. *Physics in Medicine and Biology, 51*, R117–R137.
5. Cherry, S. R. (1997). MicroPET: A high resolution PET scanner for imaging small animals. *IEEE Transactions on Nuclear Science, 44*, 1161–1166.
6. Cherry, S. R. (2004). In vivo molecular and genomic imaging: New challenged for imaging physics. *Physics in Medicine and Biology, 49*, 13–48.
7. Beyer, T., Townsend, D. W., et al. (2000). A combined PET/CT scanner for clinical oncology. *Journal of Nuclear Medicine, 41*, 1369–1379.
8. Shao, Y., et al. (1997). Simultaneous PET and MR imaging. *Physics in Medicine and Biology, 42*, 1965–1970.
9. Moses, W. W., & Derenzo, S. E. (1993). Empirical observation of resolution degradation in positron emission tomographs utilizing block detectors [abstract]. *Journal of Nuclear Medicine, 34*(suppl), 101P.
10. Levin, C. S., & Hoffman, E. J. (1999). Calculation of positron range and its effect on the fundamental limit of positron emission tomography system spatial resolution. *Physics in Medicine and Biology, 44*(3), 781–799.
11. Wernick, M. N., & Aarsvold, J. N. (2004). *Emission tomography the fundamentals of PET and SPECT.* Burlington: Elsevier Academic Press.

12. Bailey, D. L., Townsend, D. W., Valk, P. E., & Maisey, M. N. (2005). *Positron emission tomography basic sciences*. London: Springer.
13. Knoll, G. F. (2000). *Radiation detection and measurement* (3rd ed.). New York: John Wiley & Sons.
14. Watson, C. C. (2000). New, faster, image-based scatter correction for 3D PET. *IEEE Transactions on Nuclear Science, 47*(4), 1587–1594.
15. Markiewicz, P. J., Tamal, M., et al. (2007). High accuracy multiple scatter modelling for 3D whole body PET. *Physics in Medicine and Biology, 52*, 829–847.
16. Lewellen, T. K. (1998). Time-of-flight PET. *Seminars in Nuclear Medicine, 28*, 268–275.
17. Gu, Y., Matteson, J. L., et al. (2011). Study of a high-resolution, 3D positioning cadmium zinc telluride detector for PET. *Physics in Medicine and Biology, 56*, 1563–1584.
18. Zhang, F., He, Z., et al. (2005). Feasibility study of using two 3-D position sensitive CZT detectors for small animal PET. *IEEE Nuclear Science Symposium Conference Record*, 1582–1585.
19. Kastis, G. A., Wu, M. C., et al. (2002). Tomographic small-animal imaging using a high resolution semiconductor camera. *IEEE Transactions on Nuclear Science, 49*(1), 172–175.
20. Wagenaar DJ (2004) Chapter 15, CdTe and CdZnTe Semiconductor Detectors for Nuclear Medicine Imaging, EMISSION TOMOGRAPHY The Fundamentals of PET and SPECT, Miles N. Wernick, John N. Aarsvold, Elsevier Inc.
21. Jo, W. J., Jeong, M., et al. (2016). Preliminary research of CZT based PET system development in KAERI. *Journal of Radiation Protection and Research, 41*(2), 81–86.
22. Kim, K. H., Hwang, S., et al. (2016). The effect of low-temperature annealing on a CdZnTe detector. *IEEE Transactions on Nuclear Science, 63*(4), 2278–2282.
23. Bolotnikov, A. E., Ackley, K., et al. (2015). High-efficiency CdZnTe gamma-ray detectors. *IEEE Transactions on Nuclear Science, 62*(6), 3193–3198.
24. He, Z., Knoll, G. F., et al. (1997). Position-sensitive single carrier CdZnTe detectors. *Nuclear Instruments and Methods A, 388*, 180–185.
25. Yin, Y., Chen, X., et al. (2014). Evaluation of PET imaging resolution using 350 μm pixelated CZT as VP-PET insert detector. *IEEE Transactions on Nuclear Science, 61*(1), 154–161.
26. Yin, Y., Chen, X., et al. (2013). 3D spatial resolution of 350um pitch pixelated CdZnTe detectors for imaging applications. *IEEE Transactions on Nuclear Science, 60*(1), 9–15.
27. Cai, L., Lai, X., et al. (2014). MRC-SPECT: A sub-500 μm resolution MR-compatible SPECT system for simultaneous dual-modality study of small animals. *Nuclear Instruments and Methods A, 734*, 147–151.
28. Yang, S., Li, M., et al. (2020). Effect of CZT system characteristics on Compton scatter event recovery. *IEEE Transactions on Radiation and Plasma Medical Sciences, 4*(1), 91–97.
29. Vernekohl, D., Abbaszadeh, S., et al. (2019). Robust detector calibration for a novel PET system based on cross-strip CZT detectors. *IEEE Transactions on Radiation and Plasma Medical Sciences, 3*(6), 626–633.
30. Espagneta, R., Frezzaa, A., et al. (2017). Conception and characterization of a virtual coplanar grid for a 11×11 pixelated CZT detector. *Nuclear Instruments and Methods A, 860*, 62–69.
31. Abbaszadeh, S., & Levin, C. S. (2017). Direct conversion semiconductor detectors for radiation imaging. In *Semiconductor radiation detectors: Technology and applications* (pp. 1–20). Boca Raton: CRC Press.
32. Spieler, H. (2005). *Semiconductor detector systems*. Oxford: Oxford Science Publications.
33. Prokesch, M., Soldner, S. A., et al. (2016). CdZnTe detectors operating at X-ray fluxes of 100 million photons/(mm2.sec). *IEEE Transactions on Nuclear Science, 63*(3), 1854–1859.
34. Schlesinger, T. E., Toney, J. E., et al. (2001). Cadmium zinc telluride and its use as a nuclear radiation detector material. *Materials Science and Engineering, 32*, 103–189.
35. Yoon, H., Goorsky, M. S., et al. (1999). Resistivity variation of semi-insulating Cd1-xZnxTe in relationship to alloy composition. *Journal of Electronic Materials, 28*(6), 838–842.
36. Eisen, Y., & Shor, A. (1998). CdTe and CdZnTe materials for room temperature X-ray and gamma-ray detectors. *Journal of Crystal Growth, 184–185*, 1302–1312.

37. Toney, J. E., Schlesinger, T. E., et al. (1998). Elementary analysis of line shapes and energy resolution in semiconductor radiation detectors. *Materials Research Society Symposium Proceedings, 487*, 193–198.
38. Barrett, H. H., Eskin, J. D., & Barber, H. B. (1995). Charge transport in arrays of semiconductor gamma-ray detectors. *Physical Review Letters, 75*(1), 156.
39. Luke, P. N. (1995). Unipolar charge sensing with coplanar electrodes application to semiconductor detectors. *IEEE Transactions on Nuclear Science, 42*(4), 207–213.
40. Luke, P. N. (1994). Single-polarity charge sensing in ionization detectors using coplanar electrodes. *Applied Physics L, 65*(22), 2884–2886.
41. He, Z., Knoll, G. F., & Wehe, D. K. (1998). Direct measurement of product of the electron mobility and mean free drift time of CdZnTe semiconductors using position sensitive single polarity charge sensing detectors. *Journal of Applied Physics, 84*(10), 5566.
42. Sellin, P. J., Prekas, G., et al. (2010). Performance of CZT wafers grown by vapour phase transport. *IEEE Nuclear Science Symposium Conference Record, R01-6.*
43. Zappettini, A., Marchini, L., et al. (2011). Growth and characterization of CZT crystals by the vertical Bridgman method for X-ray detector applications. *IEEE Transactions on Nuclear Science, 58*(5), 2352–2356.
44. Llopart, X., Campbell, M., et al. (2002). Medipix2, a 64 k pixel readout chip with 55 micron square elements working in single photon counting mode. *IEEE Transactions on Nuclear Science, 49*(5), 2279–2283.
45. Groll, A., & Levin, C. S. (2018). Calibrations of the integrated circuit readout for a high resolution preclinical CZT PET imaging system. *IEEE Nuclear Science Symposium Conference Record.* https://doi.org/10.1109/NSSMIC.2018.8824635.
46. Gao, W., & Liu, H. (2014). Design of a multichannel low-noise front-end readout ASIC dedicated to CZT detectors for PET imaging. *IEEE Transactions on Nuclear Science, 61,* 2532–2539.
47. Cenkeramaddi, L. R., Genov, G., et al. (2012). Low-energy CZT detector array for the ASIM mission. *IEEE Instrumentation and Measurement Technology Conference.* https://doi.org/10.1109/I2MTC.2012.6229184.
48. Garson, A., Li, Q., et al. (2007). Leakage currents and capacitances of thick CZT detectors. *IEEE Nuclear Science Symposium Conference Record, 2258–2261.*
49. Ojha, N., Griesmer, J., et al. (2010). PET performance of the Gemini TF PET-MR: The world's first whole body PET-MRI scanner. *IEEE Nuclear Science Symposium Conference Record, M03-2.*
50. Iniewski, K., Chen, H., et al. (2007). Modeling charge-sharing effects in Pixellated CZT detectors. *IEEE Nuclear Science Symposium Conference Record, 2007,* 4608–4611.
51. Yin, Y., Komarov, S., et al. (2009). Characterization of highly pixelated CdZnTe detectors for sub-millimeter PET imaging. *IEEE Nuclear Science Symposium Conference Record,* 2411–2414.
52. Matteson, J. L., Gu, Y., et al. (2008). Charge collection studies of a high resolution CZT-based detector for PET. *IEEE Nuclear Science Symposium Conference Record,* 503–510.
53. Chen, C. M. H., Boggs, S. E., et al. (2002). Numerical modeling of charge sharing in CdZnTe pixel detectors. *IEEE Transactions on Nuclear Science, 49*(1), 270–276.
54. Kim, J. C., Anderson, S. E., et al. (2011). Charge sharing in common-grid pixelated CdZnTe detectors. *Nuclear Instruments and Methods A, 654,* 233–243.
55. Lee, K., Matteson, J., et al. (2010). Precision measurements of the response of a pixelated CZT detector with an Al2O3 insulated steering grid. *IEEE Nuclear Science Symposium Conference Record, R05-40.*
56. Shockley, W. (1938). Currents to conductors induced by a moving point charge. *Journal of Applied Physics, 9,* 635–636.
57. Ramo S (1939) Currents induced by electron motion, proceedings of the I.R.E., p 584.
58. Li, Q., Beilicke, M., et al. (2011). Study of thick CZT detectors for X-ray and Gamma-ray astronomy. *Astroparticle Physics, 34,* 769–777.

59. Jung, I., Krawczynski, H., et al. (2007). Detailed studies of pixelated CZT detectors grown with the modified horizontal Bridgman method. *Astroparticle Physics, 28*(4–5), 397–408.
60. Jung I, Garson A et al (2006) Test of thick pixelated Orbotech detectors with and without steering grids, arXiv:Astro-ph/0608673v1.
61. Yin, Y., Liu, Q., et al. (2014). Charge sharing effect on 600 um pitch pixelated CZT detector for imaging applications. *Chinese Physics C, 38*(11), 116002.
62. Carrascal, J., Castilla, J., et al. (2014). Energy and DOI calibrations for high spatial resolution CZT detectors. *IEEE Transactions on Nuclear Science, 61*(1), 518–527.
63. Okada, Y., Takahashi, T., et al. (2001). CdTe and CdZnTe detectors for timing measurement. *IEEE Transactions on Nuclear Science, 49*(4), 1986–1992.
64. Tai, Y. C., Wu, H., et al. (2008). Virtual-Pinhole PET. *Journal of Nuclear Medicine, 49*(3), 471–479.
65. Wu, H., Pal, D., et al. (2008). A feasibility study of a prototype PET insert device to convert a general purpose animal PET scanner to higher resolution. *Journal of Nuclear Medicine, 49*(1), 79–87.
66. Constantinescu, C. C., & Mukherjee, J. (2009). Performance evaluation of an Inveon PET preclinical scanner. *Physics in Medicine and Biology, 54*, 2885–2899.
67. Visser, E. P., Disselhorst, J. A., et al. (2009). Spatial resolution and sensitivity of the Inveon small-animal PET scanner. *Journal of Nuclear Medicine, 50*(1), 139–147.
68. Rossi, L., Fischer, P., Rohe, T., & Wermes, N. (2006). *Pixel detectors from fundamentals to applications*. Berlin, Heidelberg: Springer.
69. He, Z. (2001). Review of the Shockley-Ramo theorem and its application in semiconductor gamma-ray detectors. *Nuclear Instruments and Methods A, 463*, 250–267.
70. Pal, D., O'Sullivan, J. A., et al. (2007). 2D linear and iterative reconstruction algorithms for a PET-insert scanner. *Physics in Medicine and Biology, 52*, 4293–4310.

Index

© Springer Nature Switzerland AG 2022
K. Iniewski (ed.), *Advanced X-ray Detector Technologies*,
https://doi.org/10.1007/978-3-030-64279-2

Printed in the United States
by Baker & Taylor Publisher Services